Pitman Research Notes Series

Main Editors
H. Brezis, Université de Paris
R.G. Douglas, State University of New York at Stony Brook
A. Jeffrey, University of Newcastle upon Tyne *(Founding Editor)*

Editorial Board
R. Aris, University of Minnesota
A. Bensoussan, INRIA, France
S. Bloch, University of Chicago
B. Bollobás, University of Cambridge
W. Bürger, Universität Karlsruhe
S. Donaldson, University of Oxford
J. Douglas Jr, Purdue University
R.J. Elliott, University of Alberta
G. Fichera, Università di Roma
R.P. Gilbert, University of Delaware
R. Glowinski, Université de Paris
K.P. Hadeler, Universität Tübingen
K. Kirchgässner, Universität Stuttgart
B. Lawson, State University of New York at Stony Brook
W.F. Lucas, Claremont Graduate School
R.E. Meyer, University of Wisconsin-Madison
S. Mori, Nagoya University
L.E. Payne, Cornell University
G.F. Roach, University of Strathclyde
J.H. Seinfeld, California Institute of Technology
B. Simon, California Institute of Technology
S.J. Taylor, University of Virginia

Submission of proposals for consideration
Suggestions for publication, in the form of outlines and representative samples, are invited by the Editorial Board for assessment. Intending authors should approach one of the main editors or another member of the Editorial Board, citing the relevant AMS subject classifications. Alternatively, outlines may be sent directly to the publisher's offices. Refereeing is by members of the board and other mathematical authorities in the topic concerned, throughout the world.

Preparation of accepted manuscripts
On acceptance of a proposal, the publisher will supply full instructions for the preparation of manuscripts in a form suitable for direct photo-lithographic reproduction. Specially printed grid sheets can be provided and a contribution is offered by the publisher towards the cost of typing. Word processor output, subject to the publisher's approval, is also acceptable.

Illustrations should be prepared by the authors, ready for direct reproduction without further improvement. The use of hand-drawn symbols should be avoided wherever possible, in order to maintain maximum clarity of the text.

The publisher will be pleased to give any guidance necessary during the preparation of a typescript, and will be happy to answer any queries.

Important note
In order to avoid later retyping, intending authors are strongly urged not to begin final preparation of a typescript before receiving the publisher's guidelines. In this way it is hoped to preserve the uniform appearance of the series.

Longman Scientific & Technical
Longman House
Burnt Mill
Harlow, Essex, CM20 2JE
UK
(Telephone (0279) 426721)

Titles in this series. A full list is available on request from the publisher.

51 Subnormal operators
   J B Conway
52 Wave propagation in viscoelastic media
   F Mainardi
53 Nonlinear partial differential equations and their applications: Collège de France Seminar. Volume I
   H Brezis and J L Lions
54 Geometry of Coxeter groups
   H Hiller
55 Cusps of Gauss mappings
   T Banchoff, T Gaffney and C McCrory
56 An approach to algebraic K-theory
   A J Berrick
57 Convex analysis and optimization
   J-P Aubin and R B Vintner
58 Convex analysis with applications in the differentiation of convex functions
   J R Giles
59 Weak and variational methods for moving boundary problems
   C M Elliott and J R Ockendon
60 Nonlinear partial differential equations and their applications: Collège de France Seminar. Volume II
   H Brezis and J L Lions
61 Singular Systems of differential equations II
   S L Campbell
62 Rates of convergence in the central limit theorem
   Peter Hall
63 Solution of differential equations by means of one-parameter groups
   J M Hill
64 Hankel operators on Hilbert Space
   S C Power
65 Schrödinger-type operators with continuous spectra
   M S P Eastham and H Kalf
66 Recent applications of generalized inverses
   S L Campbell
67 Riesz and Fredholm theory in Banach algebra
   B A Barnes, G J Murphy, M R F Smyth and T T West
68 Evolution equations and their applications
   K Kappel and W Schappacher
69 Generalized solutions of Hamilton–Jacobi equations
   P L Lions
70 Nonlinear partial differential equations and their applications: Collège de France Seminar. Volume III
   H Brezis and J L Lions
71 Spectral theory and wave operators for the Schrödinger equation
   A M Berthier
72 Approximation of Hilbert space operators I
   D A Herrero
73 Vector valued Nevanlinna theory
   H J W Ziegler
74 Instability, nonexistence and weighted energy methods in fluid dynamics and related theories
   B Straughan
75 Local bifurcation and symmetry
   A Vanderbauwhede
76 Clifford analysis
   F Brackx, R Delanghe and F Sommen
77 Nonlinear equivalence, reduction of PDEs to ODEs and fast convergent numerical methods
   E E Rosinger
78 Free boundary problems, theory and applications. Volume I
   A Fasano and M Primicerio
79 Free boundary problems, theory and applications. Volume II
   A Fasano and M Primicerio
80 Symplectic geometry
   A Crumeyrolle and J Grifone
81 An algorithmic analysis of a communication model with retransmission of flawed messages
   D M Lucantoni
82 Geometric games and their applications
   W H Ruckle
83 Additive groups of rings
   S Feigelstock
84 Nonlinear partial differential equations and their applications: Collège de France Seminar. Volume IV
   H Brezis and J L Lions
85 Multiplicative functionals on topological algebras
   T Husain
86 Hamilton–Jacobi equations in Hilbert spaces
   V Barbu and G Da Prato
87 Harmonic maps with symmetry, harmonic morphisms and deformations of metric
   P Baird
88 Similarity solutions of nonlinear partial differential equations
   L Dresner
89 Contributions to nonlinear partial differential equations
   C Bardos, A Damlamian, J I Díaz and J Hernández
90 Banach and Hilbert spaces of vector-valued functions
   J Burbea and P Masani
91 Control and observation of neutral systems
   D Salamon
92 Banach bundles, Banach modules and automorphisms of C^*-algebras
   M J Dupré and R M Gillette
93 Nonlinear partial differential equations and their applications: Collège de France Seminar. Volume V
   H Brezis and J L Lions
94 Computer algebra in applied mathematics: an introduction to MACSYMA
   R H Rand
95 Advances in nonlinear waves. Volume I
   L Debnath
96 FC-groups
   M J Tomkinson
97 Topics in relaxation and ellipsoidal methods
   M Akgül
98 Analogue of the group algebra for topological semigroups
   H Dzinotyiweyi
99 Stochastic functional differential equations
   S E A Mohammed

100 Optimal control of variational inequalities
   V Barbu
101 Partial differential equations and dynamical systems
   W E Fitzgibbon III
102 Approximation of Hilbert space operators Volume II
   C Apostol, L A Fialkow, D A Herrero and D Voiculescu
103 Nondiscrete induction and iterative processes
   V Ptak and F-A Potra
104 Analytic functions – growth aspects
   O P Juneja and G P Kapoor
105 Theory of Tikhonov regularization for Fredholm equations of the first kind
   C W Groetsch
106 Nonlinear partial differential equations and free boundaries. Volume I
   J I Díaz
107 Tight and taut immersions of manifolds
   T E Cecil and P J Ryan
108 A layering method for viscous, incompressible L_p flows occupying R^n
   A Douglis and E B Fabes
109 Nonlinear partial differential equations and their applications: Collège de France Seminar. Volume VI
   H Brezis and J L Lions
110 Finite generalized quadrangles
   S E Payne and J A Thas
111 Advances in nonlinear waves. Volume II
   L Debnath
112 Topics in several complex variables
   E Ramírez de Arellano and D Sundararaman
113 Differential equations, flow invariance and applications
   N H Pavel
114 Geometrical combinatorics
   F C Holroyd and R J Wilson
115 Generators of strongly continuous semigroups
   J A van Casteren
116 Growth of algebras and Gelfand–Kirillov dimension
   G R Krause and T H Lenagan
117 Theory of bases and cones
   P K Kamthan and M Gupta
118 Linear groups and permutations
   A R Camina and E A Whelan
119 General Wiener–Hopf factorization methods
   F-O Speck
120 Free boundary problems: applications and theory. Volume III
   A Bossavit, A Damlamian and M Fremond
121 Free boundary problems: applications and theory. Volume IV
   A Bossavit, A Damlamian and M Fremond
122 Nonlinear partial differential equations and their applications: Collège de France Seminar. Volume VII
   H Brezis and J L Lions
123 Geometric methods in operator algebras
   H Araki and E G Effros
124 Infinite dimensional analysis–stochastic processes
   S Albeverio
125 Ennio de Giorgi Colloquium
   P Krée
126 Almost-periodic functions in abstract spaces
   S Zaidman
127 Nonlinear variational problems
   A Marino, L Modica, S Spagnolo and M Degliovanni
128 Second-order systems of partial differential equations in the plane
   L K Hua, W Lin and C-Q Wu
129 Asymptotics of high-order ordinary differential equations
   R B Paris and A D Wood
130 Stochastic differential equations
   R Wu
131 Differential geometry
   L A Cordero
132 Nonlinear differential equations
   J K Hale and P Martinez-Amores
133 Approximation theory and applications
   S P Singh
134 Near-rings and their links with groups
   J D P Meldrum
135 Estimating eigenvalues with *a posteriori/a priori* inequalities
   J R Kuttler and V G Sigillito
136 Regular semigroups as extensions
   F J Pastijn and M Petrich
137 Representations of rank one Lie groups
   D H Collingwood
138 Fractional calculus
   G F Roach and A C McBride
139 Hamilton's principle in continuum mechanics
   A Bedford
140 Numerical analysis
   D F Griffiths and G A Watson
141 Semigroups, theory and applications. Volume I
   H Brezis, M G Crandall and F Kappel
142 Distribution theorems of L-functions
   D Joyner
143 Recent developments in structured continua
   D De Kee and P Kaloni
144 Functional analysis and two-point differential operators
   J Locker
145 Numerical methods for partial differential equations
   S I Hariharan and T H Moulden
146 Completely bounded maps and dilations
   V I Paulsen
147 Harmonic analysis on the Heisenberg nilpotent Lie group
   W Schempp
148 Contributions to modern calculus of variations
   L Cesari
149 Nonlinear parabolic equations: qualitative properties of solutions
   L Boccardo and A Tesei
150 From local times to global geometry, control and physics
   K D Elworthy

151 A stochastic maximum principle for optimal control of diffusions
    U G Haussmann
152 Semigroups, theory and applications. Volume II
    H Brezis, M G Crandall and F Kappel
153 A general theory of integration in function spaces
    P Muldowney
154 Oakland Conference on partial differential equations and applied mathematics
    L R Bragg and J W Dettman
155 Contributions to nonlinear partial differential equations. Volume II
    J I Díaz and P L Lions
156 Semigroups of linear operators: an introduction
    A C McBride
157 Ordinary and partial differential equations
    B D Sleeman and R J Jarvis
158 Hyperbolic equations
    F Colombini and M K V Murthy
159 Linear topologies on a ring: an overview
    J S Golan
160 Dynamical systems and bifurcation theory
    M I Camacho, M J Pacifico and F Takens
161 Branched coverings and algebraic functions
    M Namba
162 Perturbation bounds for matrix eigenvalues
    R Bhatia
163 Defect minimization in operator equations: theory and applications
    R Reemtsen
164 Multidimensional Brownian excursions and potential theory
    K Burdzy
165 Viscosity solutions and optimal control
    R J Elliott
166 Nonlinear partial differential equations and their applications: Collège de France Seminar. Volume VIII
    H Brezis and J L Lions
167 Theory and applications of inverse problems
    H Haario
168 Energy stability and convection
    G P Galdi and B Straughan
169 Additive groups of rings. Volume II
    S Feigelstock
170 Numerical analysis 1987
    D F Griffiths and G A Watson
171 Surveys of some recent results in operator theory. Volume I
    J B Conway and B B Morrel
172 Amenable Banach algebras
    J-P Pier
173 Pseudo-orbits of contact forms
    A Bahri
174 Poisson algebras and Poisson manifolds
    K H Bhaskara and K Viswanath
175 Maximum principles and eigenvalue problems in partial differential equations
    P W Schaefer
176 Mathematical analysis of nonlinear, dynamic processes
    K U Grusa
177 Cordes' two-parameter spectral representation theory
    D F McGhee and R H Picard
178 Equivariant K-theory for proper actions
    N C Phillips
179 Elliptic operators, topology and asymptotic methods
    J Roe
180 Nonlinear evolution equations
    J K Engelbrecht, V E Fridman and E N Pelinovski
181 Nonlinear partial differential equations and their applications: Collège de France Seminar. Volume IX
    H Brezis and J L Lions
182 Critical points at infinity in some variational problems
    A Bahri
183 Recent developments in hyperbolic equations
    L Cattabriga, F Colombini, M K V Murthy and S Spagnolo
184 Optimization and identification of systems governed by evolution equations on Banach space
    N U Ahmed
185 Free boundary problems: theory and applications. Volume I
    K H Hoffmann and J Sprekels
186 Free boundary problems: theory and applications. Volume II
    K H Hoffmann and J Sprekels
187 An introduction to intersection homology theory
    F Kirwan
188 Derivatives, nuclei and dimensions on the frame of torsion theories
    J S Golan and H Simmons
189 Theory of reproducing kernels and its applications
    S Saitoh
190 Volterra integrodifferential equations in Banach spaces and applications
    G Da Prato and M Iannelli
191 Nest algebras
    K R Davidson
192 Surveys of some recent results in operator theory. Volume II
    J B Conway and B B Morrel
193 Nonlinear variational problems. Volume II
    A Marino and M K V Murthy
194 Stochastic processes with multidimensional parameter
    M E Dozzi
195 Prestressed bodies
    D Iesan
196 Hilbert space approach to some classical transforms
    R H Picard
197 Stochastic calculus in application
    J R Norris
198 Radical theory
    B J Gardner
199 The C^*-algebras of a class of solvable Lie groups
    X Wang
200 Stochastic analysis, path integration and dynamics
    K D Elworthy and J C Zambrini

201 Riemannian geometry and holonomy groups
   S Salamon
202 Strong asymptotics for extremal errors and polynomials associated with Erdös type weights
   D S Lubinsky
203 Optimal control of diffusion processes
   V S Borkar
204 Rings, modules and radicals
   B J Gardner
205 Two-parameter eigenvalue problems in ordinary differential equations
   M Faierman
206 Distributions and analytic functions
   R D Carmichael and D Mitrovic
207 Semicontinuity, relaxation and integral representation in the calculus of variations
   G Buttazzo
208 Recent advances in nonlinear elliptic and parabolic problems
   P Bénilan, M Chipot, L Evans and M Pierre
209 Model completions, ring representations and the topology of the Pierce sheaf
   A Carson
210 Retarded dynamical systems
   G Stepan
211 Function spaces, differential operators and nonlinear analysis
   L Paivarinta
212 Analytic function theory of one complex variable
   C C Yang, Y Komatu and K Niino
213 Elements of stability of visco-elastic fluids
   J Dunwoody
214 Jordan decomposition of generalized vector measures
   K D Schmidt
215 A mathematical analysis of bending of plates with transverse shear deformation
   C Constanda
216 Ordinary and partial differential equations. Volume II
   B D Sleeman and R J Jarvis
217 Hilbert modules over function algebras
   R G Douglas and V I Paulsen
218 Graph colourings
   R Wilson and R Nelson
219 Hardy-type inequalities
   A Kufner and B Opic
220 Nonlinear partial differential equations and their applications: Collège de France Seminar. Volume X
   H Brezis and J L Lions
221 Workshop on dynamical systems
   E Shiels and Z Coelho
222 Geometry and analysis in nonlinear dynamics
   H W Broer and F Takens
223 Fluid dynamical aspects of combustion theory
   M Onofri and A Tesei
224 Approximation of Hilbert space operators. Volume I. 2nd edition
   D Herrero
225 Operator theory: proceedings of the 1988 GPOTS-Wabash conference
   J B Conway and B B Morrel
226 Local cohomology and localization
   J L Bueso Montero, B Torrecillas Jover and A Verschoren
227 Nonlinear waves and dissipative effects
   D Fusco and A Jeffrey
228 Numerical analysis 1989
   D F Griffiths and G A Watson
229 Recent developments in structured continua. Volume III
   D De Kee and P Kaloni
230 Boolean methods in interpolation and approximation
   F J Delvos and W Schempp
231 Further advances in twistor theory. Volume I
   L J Mason and L P Hughston
232 Further advances in twistor theory. Volume II
   L J Mason and L P Hughston
233 Geometry in the neighborhood of invariant manifolds of maps and flows and linearization
   U Kirchgraber and K Palmer
234 Quantales and their applications
   K I Rosenthal
235 Integral equations and inverse problems
   V Petkov and R Lazarov
236 Pseudo-differential operators
   S R Simanca
237 A functional analytic approach to statistical experiments
   I M Bomze
238 Quantum mechanics, algebras and distributions
   D Dubin and M Hennings
239 Hamilton flows and evolution semigroups
   J Gzyl
240 Topics in controlled Markov chains
   V S Borkar
241 Invariant manifold theory for hydrodynamic transition
   S Sritharan
242 Lectures on the spectrum of $L^2(\Gamma\backslash G)$
   F L Williams
243 Progress in variational methods in Hamiltonian systems and elliptic equations
   M Girardi, M Matzeu and F Pacella
244 Optimization and nonlinear analysis
   A Ioffe, M Marcus and S Reich
245 Inverse problems and imaging
   G F Roach
246 Semigroup theory with applications to systems and control
   N U Ahmed
247 Periodic-parabolic boundary value problems and positivity
   P Hess
248 Distributions and pseudo-differential operators
   S Zaidman
249 Progress in partial differential equations: the Metz surveys
   M Chipot and J Saint Jean Paulin
250 Differential equations and control theory
   V Barbu

251 Stability of stochastic differential equations with respect to semimartingales
   X Mao
252 Fixed point theory and applications
   J Baillon and M Théra
253 Nonlinear hyperbolic equations and field theory
   M K V Murthy and S Spagnolo
254 Ordinary and partial differential equations. Volume III
   B D Sleeman and R J Jarvis
255 Harmonic maps into homogeneous spaces
   M Black
256 Boundary value and initial value problems in complex analysis: studies in complex analysis and its applications to PDEs 1
   R Kühnau and W Tutschke
257 Geometric function theory and applications of complex analysis in mechanics: studies in complex analysis and its applications to PDEs 2
   R Kühnau and W Tutschke
258 The development of statistics: recent contributions from China
   X R Chen, K T Fang and C C Yang
259 Multiplication of distributions and applications to partial differential equations
   M Oberguggenberger
260 Numerical analysis 1991
   D F Griffiths and G A Watson
261 Schur's algorithm and several applications
   M Bakonyi and T Constantinescu

Joseph Wiener
University of Texas-Pan American, USA

and

Jack K Hale
Georgia Institute of Technology, USA

(Editors)

Partial differential equations

Copublished in the United States with
John Wiley & Sons, Inc., New York

Longman Scientific & Technical
Longman Group UK Limited
Longman House, Burnt Mill, Harlow
Essex CM20 2JE, England
and Associated companies throughout the world.

*Copublished in the United States with
John Wiley & Sons Inc., 605 Third Avenue, New York, NY 10158*

© Longman Group UK Limited 1992

All rights reserved; no part of this publication may be reproduced, stored in
a retrieval system, or transmitted in any form or by any means, electronic,
mechanical, photocopying, recording, or otherwise, without the prior written
permission of the Publishers, or a licence permitting restricted copying in
the United Kingdom issued by the Copyright Licensing Agency Ltd,
90 Tottenham Court Road, London, W1P 9HE

First published 1992

AMS Subject Classification: 35Jxx, 35Kxx, 35Lxx, 35Pxx, 34Gxx

ISSN 0269-3674

ISBN 0 582 09114 4

British Library Cataloguing in Publication Data

A catalogue record for this book is
available from the British Library

Library of Congress Cataloging-in-Publication Data

Partial differential equations / J. Wiener (editor) and J. Hale
  (editor).
   p. cm. -- (Pitman research notes in mathematics series, ISSN
0269-3674)
   1. Differential equations, Partial--Congresses. I. Wiener, J.
(Joseph) II. Hale, Jack K. III. Series.
QA370.P36 1992
515'.353--dc20   92-4752
                                                               CIP

Printed and bound in Great Britain
by Biddles Ltd, Guildford and King's Lynn

Contents

Preface

J Anderson
The effects of convection on reaction-diffusion equations 1

G Auchmuty
Minimax results and differential inclusions 6

R C Brown and D B Hinton
Finding good upper bounds for the best constant in a generalized Hardy Littlewood inequality 11

T A Burton
Admissible controls in a PDE of Lurie type 16

C Y Chan and D T Fung
Quenching for coupled parabolic problems 29

C Y Chan and L Ke
Existence of solutions and critical lengths for time-periodic semilinear parabolic problems 34

C Y Chan and K K Nip
Quenching for semilinear Euler–Poisson–Darboux equations 39

R Estrada
Series of Dirac delta functions 44

H Gingold and V Trutzer
The large time pointwise behaviour of linear degenerate hyperbolic systems 49

J-P Gossez
Nonresonance in some semilinear Neumann problems 54

Q Du, M Gunzburger and J Peterson
On the Ginzburg–Landau equations of superconductivity 58

J H Kim and K S Ha
Comparison theorems for solutions of one dimensional stochastic differential and integral equations of Ito's type 63

P W Hammer
Applications of quasilinearization to parameter identification in partial differential equations 68

D W Brewer and T L Herdman
A parameter dependence problem in neutral functional differential equations 74

J H Jaroma, V Lj Kocic and G Ladas
Global asymptotic stability of a second-order difference equation 80

B S Lalli and B G Zhang
Existence of nonoscillatory solutions for odd order neutral equations 85

Y Latushkin
Exact Lyapunov exponents and exponentially separated cocycles 91

U Ledzewicz-Kowalewska
Optimality conditions for abnormal optimal control problems with additional equality constraints 96

X B Lin
Exponential dichotomy and stability of long periodic solutions in predator-prey models with diffusion 101

A Lunardi
Stability of travelling waves in free boundary parabolic problems arising as combustion models 106

A N Lyberopoulos
On the ω-limit set of solutions of scalar balance laws on S^1 111

M Maesumi
Triple-shock elementary waves for two-phase incompressible flow in porous media 116

G N Hile and C Matawa
Existence and uniqueness of solutions to quasilinear elliptic equations 121

A Fonda and J Mawhin
An iterative method for the solvability of semilinear equations in Hilbert spaces and applications 126

P S Milojevic
Nonlinear Fredholm theory and applications   133

V H Moll and A L Fogelson
Activation waves and threshold phenomena in platelet aggregation   153

J Berkovits and V Mustonen
On the asymptotic bifurcation for a class of wave equations with sublinear nonlinearity   159

O A Oleinik
Some mathematical problems of elasticity and Korn's inequalities   163

C D Pagani, M Verri and A Zaretti
On a problem of seakeeping   180

E Avgerinos and N S Papageorgiou
Optimal control and relaxation of nonlinear evolution equations   185

F Rothe
A priori estimates for nonlinear Poisson equations and phase plane analysis of the Emden–Fowler equation   192

B Ruf
Butterflies in banach space and forced secondary bifurcation   201

N G Medhin and M Sambandham
Numerical treatment of random integro-differential equations   206

R Saxton
Finite time boundary blowup for a degenerate, quasilinear Cauchy problem   212

S Shao and H K Cheng
The Maxwell transfer equations: modification of Grad's system, I   216

P X Sheng
Fixed point theorems and topological degrees   221

M W Smiley
On the asymptotic behavior of nonautonomous dissipative abstract differential equations   226

T Svobodny and F Ahmed
A numerical method for control of a distributed parameter equation   231

T J Tabara
On an existence problem for the boundary value problem of a second order linear
differential equation with polynomial coefficients … 236

L Lin, B Temple and J Wang
On the convergence of Glimm's method and Godunov's method when wave speeds
coincide … 241

C Vargas and M Zuluaga
On a nonlinear Dirichlet problem type at resonance and bifurcation … 248

J Vosmansky
Zeros of solutions of linear differential equations as continuous functions of the
parameter k … 253

H Warchall
Induced lacunas in multiple-time initial value problems and unbounded domain
simulation … 258

G Harris, W Hudson and B Zinner
Existence of wavefronts for the discrete Fishers equation … 265

Preface

This is the second volume in the Proceedings of the International Conference on Theory and Applications of Differential Equations which was held at the University of Texas-Pan American, Edinburg, Texas 78539, U.S.A., on May 15-18 1991. The book includes papers that were delivered as invited survey lectures and research contributions in the area of partial differential equations and their applications to important problems of mathematical and theoretical physics. The main purpose of the conference was to provide a forum for the discussion of recent work and modern trends in the rapidly developing field of differential equations and dynamical systems. A large number of experts from twenty-seven countries attended the meetings, and more than one hundred and forty presentations were made. The published communications reflect the contemporary achievements and problems in theory and applications of differential equations and discuss new ideas and directions for future research activity.

The present volume features recent results in the theory and practice of partial differential equations. The articles will be of interest to many mathematicians, scientists, and engineers. Indeed, the talks cover topics such as: mathematical problems in elasticity, asymptotic stability of partial differential equations, reaction-diffusion equations, existence theorems and the number of solutions for elliptic equations, localization of dependence for hyperbolic equations and decay of solutions for hyperbolic systems, existence of solutions for periodic parabolic problems and finite extinction time for nonlinear parabolic equations. Some of the equations represent mathematical models in the life and natural sciences: reaction-diffusion equations in ecology, problems in climate modeling, diffusively perturbed predator-prey models. There are also articles concerned with optimal control, nonlinear evolution equations, variational methods, and numerical techniques. Specialists in mathematical and theoretical physics will certainly enjoy the reports on limit sets of scalar conservation laws, resonant nonlinear systems of conservation laws, incompressible flow in porous media, shock transition zones, travelling fronts in platelet aggregation, and models of superconductivity. The papers employ a wide array of deep mathematical theories and techniques. These include methods from linear and nonlinear functional analysis, a number of topological and topological degree techniques, as well as asymptotic and other classical analysis tools.

On behalf of the participants, we acknowledge with gratitude the financial support provided by the University of Texas System Chancellor's Council, The University of Texas-Pan American, and the U.S. Army Research Office. Our special thanks are due to the faculty and staff of the Mathematics Department, UT-Pan American, for the enormous help in organizing excellent local arrangements. Finally, our appreciation is extended to the staff of Longman, for their cooperation.

Joseph Wiener
Jack K. Hale

J ANDERSON

The effects of convection on reaction-diffusion equations

There have been many studies which show that some solutions of reaction-diffusion equations blow up in finite time. Recently, Chen, Levine, and Sacks have shown that addition of a convective term to a reaction-diffusion equation results in a situation where all solutions are uniformly bounded and asymptotically approach zero. Convection has caused a reversal from blow up to decay. Levine and Sacks have proven results on the blow up of solutions of reaction-diffusion equations which include the case of degenerate diffusion, i.e. degenerate parabolicity. In this paper we show that such a situation can again be reversed by the addition of convection.

1 Introduction.

The focus of this paper will be an analysis of the long-time behavior of nonnegative solutions of

$$\begin{aligned} u_t &= (u^m)_{xx} + \epsilon(u^n)_x + au^p & 0 < x < 1, t > 0 \\ u(0,t) &= u(1,t) = 0 & t > 0 \\ u(x,0) &= u_0(x) \geq 0 & 0 \leq x \leq 1 \end{aligned} \qquad (1.1(\epsilon))$$

where $a, \epsilon \geq 0$ and $m, n, p \geq 1$. In [3], the case of $(1.1(\epsilon))$ with $m = 1$ was investigated. Therein it was shown that if $n \geq p > 1$ and $\epsilon > 0$, then all solutions of $(1.1(\epsilon))$ with $u_0 \in L^\infty((0,1))$ are uniformly bounded for $t > 0$. Moreover, there exists $\epsilon_0 > 0$ such that for $\epsilon \geq \epsilon_0$ all solutions of $(1.1(\epsilon))$ decay to zero as $t \to \infty$. This is very striking when contrasted to the case of $\epsilon = 0$ and $p > 1$ in which all solutions, with u_0 appropriately chosen, blow up in finite time [6]. The goal of this work is to show that such a result is still true for $m \geq 1$. Herein, we will also show that the convective term plays an important role in the long time behavior of solutions in the remaining cases $p > n$. Specifically, our main results are the following.

Theorem 1.1 *Let $m, n, p \geq 1$. For each $M > 0$ there exists $\epsilon_0 = \epsilon_0(M) > 0$ such that any solution of $(1.1(\epsilon))$ with $\|u_0\|_\infty \leq M$ and $\epsilon \geq \epsilon_0$ is uniformly bounded on $[0,1] \times [0,\infty)$.*

Theorem 1.2 *Let $m, n, p \geq 1$. If either $m < p \leq n$ or $n \leq p < m$, and if $\epsilon \geq \epsilon_1$, then any solution of $(1.1(\epsilon))$, $u(x,t)$, with $\|u(\cdot,0)\|_\infty \leq M$, has $\lim_{t \to \infty} u(x,t) = 0$ for all $x \in [0,1]$. Here, $\epsilon_1 \equiv \max\left\{a^{(n-m)/(p-m)}, \epsilon_0(M)\right\}$.*

Theorem 1.3 *Suppose $n \geq p > m \geq 1$. If $u(x,t)$ is a nonnegative solution of $(1.1(\epsilon))$ for some $\epsilon > 0$, then $u(x,t)$ is uniformly bounded on $[0,1] \times [0,\infty)$. Moreover, there exists $\epsilon_0 > 0$ such that if $\epsilon \geq \epsilon_0$, then $\lim_{t \to \infty} u(x,t) = 0$ for all $x \in [0,1]$. (Here, ϵ_0 is independent of $\|u(\cdot,0)\|_\infty$.)*

In [8], it was shown that whenever $p > m$, solutions of (1.1(ϵ)) blow up in finite time if u_0 is sufficiently large. By Theorem 1.3, we see that the addition of convection has completely reversed the situation. Moreover, as discussed in [7, p.p.143-144], Theorem 1.1 reveals a situation in which uniformly bounded solutions and solutions which blow up in finite time co-exist.

These results have interesting implications when (1.1(ϵ)) is interpreted in a physical setting. For example, in studying the propagation of a bacterial culture in a saturated porous column one arrives at the one dimensional partial differential equation $u_t = [D(u)u_x - vu]_x + R(u)$ where u is the population density, $D(u)$ is the local diffusion, $R(u)$ is the local growth rate, and v is the fluid velocity [10]. See also [2] and [12]. If we assume $D(u) = mu^{m-1}$ and $R(u) = au^p$, then this model equation is contained in (1.1(ϵ)). With conditions at the top and bottom of the column which are lethal to bacteria, we have $u(0,t) = u(1,t) = 0$ (after appropriate scaling in x and t). Theorem 1.2 now states that whenever $p < m$ there is a flow velocity which will lead to eventual extinction of the bacteria population. Furthermore, without restriction on the exponents p and m, Theorem 1.1 yields a global existence result for sufficiently large velocities v. Finally, in the development of all of these results, we shall see the population density stabilizes to equilibrium.

In general we do not expect classical solutions of (1.1(ϵ)). Thus, we first define what is meant by a solution of (1.1(ϵ)). In the process, subsolutions and supersolutions of the model problem are also defined. To do this we will use the class of "test functions", $\mathcal{T}$, as introduced in [1]. Briefly these are nonnegative functions satisfying continuity and integrability conditions which are sufficient for the following definition to make sense.

Definition 1.1 *A function $u(x,t)$, defined everywhere on $[0,1] \times (0,T]$ and a.e. (almost everywhere) on $[0,1] \times \{0\}$, is called a subsolution (supersolution) of (1.1(ϵ)) on $(0,1) \times (0,T)$ if the following all hold.*

i.) $u \in L^\infty((0,1) \times (0,T))$.

ii.) $u(x,0) \leq (\geq) u_0(x)$ for $x \in [0,1]$, and $u(0,t), u(1,t) \leq (\geq) 0$ for $t > 0$.

iii.) *For every $\xi \in \mathcal{T}$ and every $t \in [0,T]$,*

$$\int_0^1 [u(x,t)\xi(x,t) - u_0(x)\xi(x,0)] \, dx \leq (\geq) \int_0^t \int_0^1 \{u\xi_s + u^m \xi_{xx} - \epsilon u^n \xi_x + au^p \xi\} \, dx \, ds.$$

A function $u(x,t)$ is called a solution of (1.1(ϵ)) on $(0,1) \times (0,T)$ if it is both a subsolution and a supersolution of (1.1(ϵ)) on $(0,1) \times (0,T)$.

The main result of [1] is that nonnegative solutions of (1.1(ϵ)) can be constructed as monotone limits of the following problem

$$\begin{aligned} u_t &= (u^m)_{xx} + \epsilon(u^n)_x + au^p & 0 < x < 1, t > 0 \\ u(0,t) &= u(1,t) = k & t > 0 \\ u(x,0) &= u_0(x) + k & 0 \leq x \leq 1, \end{aligned} \qquad (1.1(\epsilon)_k)$$

where $k > 0$. If u_k is the solution of (1.1(ϵ)$_k$), then $u_k \geq u_l \geq l > 0$ for $k \geq l$ and $u(x,t) = \lim_{k \to 0^+} u_k(x,t)$ is a solution of (1.1(ϵ)). Although this local existence result is not new in the case of (1.1(ϵ)), see [5,11], its constructive proof yields comparison results which are useful in the sequel.

2 Equilibrium states and long time behavior of solutions.

There are three main facts about solutions of (1.1(ϵ)) which we establish in this section. Ultimately each of these facts play a key role in the proof of Theorems 1.1 - 1.3. Firstly, any solution, $u(x,t)$, of (1.1(ϵ)) such that $u(\cdot,0)$ is a supersolution of (1.1(ϵ)) is monotone decreasing in t. Hence, $w(x) \equiv \lim_{t\to\infty} u(x,t)$ exists and is an equilibrium solution of (1.1(ϵ)). This is not the most general result one could try to prove, i.e., the ω-limit set of any solution consists of equilibria. However, when combined with the comparison theorem from [1], we obtain asymptotic stability of $w(x)$ from above.

Secondly, for any $M > 0$ and sufficiently large ϵ, there exists a time-independent supersolution of (1.1(ϵ)), $u_0(x)$, such that $u_0(x) \geq M$ on $[0,1]$. To establish this we use an argument which is very similar to that used in [3].

Finally, we show that all stationary solutions of (1.1(ϵ)) are classical, and for $m < p \leq n$ or $n \leq p < m$ and $\epsilon > 0$ sufficiently large, the only equilibrium state is the trivial solution. This will complete the proof of Theorems 1.1 - 1.3.

Lemma 2.1 *Assume that $m, p, n \geq 1$ and $a \geq 0$. If $u_0(x)$ is a supersolution of (1.1(ϵ)) such that $u_0 \geq M > 0$ on $[0,1]$, then the solution $u(x,t)$ of (1.1(ϵ)) with $u(\cdot,0) = u_0(\cdot)$ is monotone decreasing in t, i.e. for each $x \in [0,1]$ and $t_1 \geq t_2 \geq 0$, we have $0 \leq u(x,t_1) \leq u(x,t_2)$.*

Proof: To establish this result, we need only observe that u_0 is a supersolution of (1.1(ϵ)$_k$) for each $k \in (0, M)$. Thus, if u_k is the solution of (1.1(ϵ)$_k$) with $u_k(\cdot,0) = u_0(\cdot)$, then $u_k(x,t) \leq u_0(x)$ for each $0 \leq x \leq 1, t > 0$. Also, it is a standard argument to show that $u_k(x,t)$ is monotone decreasing in t. Since $u_0 \geq M > 0$, it follows by the same methods as used in [1] that $u(x,t) \equiv \lim_{k\to 0} u_k(x,t)$ is the solution of (1.1(ϵ)). Hence it follows that $u(x,t) \leq u_0(x)$ for all $t > 0$, and $u(x,t)$ is decreasing in t. $\square$

The result of Lemma 2.1 leads to a situation in which $\lim_{t\to\infty} u(x,t)$ exists. We now show that this limit is, in fact, an equilibrium state for (1.1(ϵ)).

Lemma 2.2 *Assume $p, m, n \geq 1$. If $u(x,t)$ is a solution of (1.1(ϵ)) such that $u \in L^\infty((0,1) \times (0,\infty))$, $w(x) = \lim_{t\to\infty} u(x,t)$ exists for each $x \in [0,1]$, and $w \in L^\infty((0,1))$, then $w(x)$ is an equilibrium solution for (1.1(ϵ)).*

Proof: By our definition of solution, $\int_0^1 [u(x,t) - u_0(x)] \xi(x) dx = \int_0^t \int_0^1 \{u^m \xi'' - \epsilon u^n \xi' + a u^p \xi\} dx ds$, for each test function $\xi(x,t) \equiv \xi(x)$. Define $F(t) \equiv \int_0^t \int_0^1 \{u^m \xi'' - \epsilon u^n \xi' + a u^p \xi\} dx ds$. Then $\lim_{t\to\infty} F(t) = \int_0^1 [w(x) - u_0(x)] \xi(x) dt$. Since u is continuous on $(0,1) \times (0,\infty)$, see [4,11], $F(t)$ is differentiable on $(0,\infty)$. By simple analysis, there must exist a sequence $t_n \to \infty$ as $n \to \infty$ such that $\lim_{n\to\infty} F'(t_n) = 0$. Thus, $\int_0^1 \{w^m \xi'' - \epsilon w^n \xi' + a w^p \xi\} dx = \lim_{n\to\infty} F'(t_n) = 0$, and the result is established. $\square$

3

As a result of the two lemmas above and Theorem 2.2 in [1] it follows that if $u(x,t)$ and $w(x)$ are a solution and supersolution of $(1.1(\epsilon))$, respectively, with $u(x,0) \leq w(x)$, then $\lim_{t\to\infty} u(x,t)$ is a steady state which is stable from above. Such a technique has been used a great deal in the past. For example, see [3,9]. In the direction of building the supersolutions, $w(x)$, we present the following construction which is very much like that of [3, Lemma 3.4].

Lemma 2.3 *Let $M > 0$ be given. There exists $\epsilon_0 = \epsilon_0(M) > 0$ such that for any $\epsilon \geq \epsilon_0$, $(1.1(\epsilon))$ has a time independent supersolution, $w(x)$, for which $w(x) \geq M$ on $[0,1]$.*

Proof: We will construct $w(x)$ which satisfies

$$(w^m)_{xx} + \epsilon(w^n)_x + aw^p \leq 0 \qquad (1)$$

on $(0,1)$, i.e., $w(x)$ is a "classical" supersolution. We seek $w(x)$ in the form $w(x) = (Ae^{-x})^{1/m}$, where $A > 0$. By (1) we see that A must be chosen to satisfy $1 + a(Ae^{-x})^{q-1} \leq \epsilon r(Ae^{-x})^{r-1}$ and $(Ae^{-1})^{1/m} \geq M$, where $r \equiv n/m$ and $q \equiv p/m$. Define $C \equiv \max\{e^{r-q}, A^{q-r}\}$. Then $(Ae^{-x})^{q-1} \leq C(Ae^{-x})^{r-1}$ on $[0,1]$ for $A \geq 1$. Hence (1) will follow if A satisfies

$$1 + [aC - \epsilon r](Ae^{-x})^{(r-1)} \leq 0. \qquad (2)$$

Set $A = \max\{1, M^m e\}$ and choose $\epsilon_0 \equiv r^{-1}(aC + 2D^{-1})$, where $D \equiv A^{(r-1)} \min\{1, e^{-(r-1)}\}$. It is now straightforward to verify that (2) is true for any $\epsilon \geq \epsilon_0$. □

It is worth noting here that the exact procedure used in [3] can be replicated above in the case of $p \leq n$, with $n > m \geq 1$. This yields the stronger result that Lemma 2.3 is true for any $\epsilon > 0$.

The combination of Lemmas 2.1 - 2.3 form the proof of Theorem 1.1. In order to prove Theorems 1.2 and 1.3, we need to show that for sufficiently large $\epsilon > 0$, $(1.1(\epsilon))$ has only the zero steady state. This fact we obtain as a result of the following two lemmas.

Lemma 2.4 *If $w(x)$ is a stationary solution of $(1.1(\epsilon))$, then*

$$\left[(w^m)_x + \epsilon w^n + \int_0^x aw^p(y)\,dy\right]_x = 0$$

on $[0,1]$. In particular, $w(x)$ is a solution in the classical sense.

Proof: By definition, $w(0) = w(1) = 0$, and $\int_0^1 \{w^m \xi_{xx} - \epsilon w^n \xi_x + aw^p \xi\} = 0$ for every $\xi \in \mathcal{T}$. Since $w_x^m \in L^2((0,1))$, [1, Lemma 5.1], we may integrate by parts to obtain $\int_0^1 \{w_x^m(x) + \epsilon w^n(x) + \int_0^x aw^p(y)\,dy\} \xi'(x)\,dx = 0$. Hence, $w_x^m(x) + \epsilon w^n(x) + \int_0^x aw^p(y)\,dy =$ constant. Finally, that w is continuous on $[0,1]$ follows from $w_x^m \in L^2((0,1))$. □

Since all stationary solutions of $(1.1(\epsilon))$ are classical, we may easily establish the final lemma.

Lemma 2.5 *If $m < p \leq n$ or $n \leq p < m$ and if $\epsilon \geq a^{(n-m)\setminus(p-m)}$, then $(1.1(\epsilon))$ has no nontrivial, nonnegative stationary solutions.*

Proof: Suppose $w(x)$ is a nontrivial, nonnegative solution of $(1.1(\epsilon))$. Let $M = w(\xi) \equiv \max_{0 \leq x \leq 1} w(x) > 0$. By Lemma 2.4 it follows that

$$w_x^m(x) + \epsilon w^n(x) = \epsilon M^n + \int_x^\xi a w^p(y)\, dy. \tag{3}$$

Now $w_x^m(1) \leq 0$ and equation (3) yield $\epsilon M^n \leq \int_\xi^1 a w^p(y)\, dy < a M^p$. Thus $M^{n-p} < a/\epsilon$.

We can derive another inequality which M must satisfy by using (3) in a slightly different manner. It is clear that we may find $\rho \in (\xi, 1)$ for which $(1-\xi) w_x^m(\rho) = w^m(1) - w^m(\xi) = -M^m$. Hence equation (3) with $x = \rho$ yields
$M^m = \left[\int_\xi^\rho a w^p(y)\, dy - \epsilon(M^n - w^n(\rho)) \right](1-\xi) \leq a M^p (\rho - \xi)(1-\xi)$. Therefore, $M^{m-p} < a$.

Using the inequalities $M^{n-p} < a/\epsilon$ and $M^{m-p} < a$, the lemma is easily established. □

References

[1] J.R. Anderson, Local existence and uniqueness of solutions of degenerate parabolic equations. *Communications in Partial Differential Equations* **16**, 105-143 (1991).

[2] R.S. Cantrell and C. Cosner, Diffusive logistic equations with indefinite weights: population models in disrupted environments II. *SIAM Journal of Mathematical Analysis*, **22**, no. 4, 1043-1064, 1991.

[3] T.F. Chen, H.A. Levine, and P.E. Sacks, Analysis of a convective reaction diffusion-equation. *Nonlinear Analysis, Theory, Methods, and Applications* **12**, 1349-1370 (1988).

[4] E. DiBenedetto, Continuity of weak solutions to a general porous medium equation. *Indiana University Mathematics Journal* **32**, 83-118 (1983).

[5] A.S. Kalashnikov, Some problems of the qualitative theory of non-linear degenerate second-order parabolic equations. *Russian Math. Surveys*, **42**, no. 2, 169-222 (1987).

[6] H.A. Levine, Some nonexistence and instability theorems for solutions of formally parabolic equations of the form $Pu_t = -Au + F(u)$. *Archives of Rational Mechanics and Analysis* **51**, 371-386 (1973).

[7] H.A. Levine, L.E. Payne, P.E. Sacks, and B. Straughan, Analysis of a convective reaction-diffusion equation II. *SIAM Journal of Mathematical Analysis* **20**, 133-147 (1989).

[8] H.A. Levine and P.E. Sacks, Some existence and nonexistence theorems for solutions of degenerate parabolic equations. *Journal of Differential Equations* **52**, 135-161 (1984).

[9] H. Matano, Asymptotic behavior and stability of solutions of semilinear diffusion equations. *Publ. RIMS, Kyoto Univ* **15**, 401-454 (1979).

[10] A.A. Metry, Predictive tools for contaminant transport in groundwater, in *Permeability and Groundwater Contaminant Transport*, T.F. Zimmie and C.O. Riggs eds. American Society for Testing and Materials, Philadelphia, PA, 1981.

[11] P.E. Sacks, Continuity of solutions of a singular parabolic equation. *Nonlinear Analysis, Theory, Methods, and Applications* **7**, 387-409 (1983).

[12] J.F. Sykes, S. Soyupak, and G.J. Farguhar, Modeling of leachate organic migration and attenuation in groundwaters below sanitary landfills. *Water Resources Research* **18**, 135-145 (1982).

G AUCHMUTY
Minimax results and differential inclusions

This note describes some existence results for differential inclusions involving convex functionals. First it has shown how solutions of such inclusions may be characterized as minimax points of a specific function. Then an extension of Ky Fan's minimax theorem is used to develop existence criteria. These results are used to provide a simple, direct proof of the existence of stationary solutions of the Navier-Stokes' equations.

1 MINIMAX CHARACTERIZATIONS

Let X be a real Hausdorff locally convex topological vector space and X^* be its dual. Our interest is in solving the inclusion

$$F(u) \in \partial\phi(u) \tag{1.1}$$

where

(A1): $\phi : X \to (-\infty, \infty]$ is a lower semicontinuous (l.s.c.) convex function, with nonempty essential domain K_o.

(A2): K is a non-empty closed convex subset of X and $F : K \to X^*$ is a function.

We seek a solution $\hat{u}$ in $K_1 = K \cap K_o$ of (1.1) with $\partial\phi(u)$ being the subdifferential of ϕ at u. Associated with this problem consider the function $L : K_1 \times K_o \to \mathbf{R}$ defined by

$$L(u, v) = \langle F(u), v - u \rangle + \phi(u) - \phi(v) \tag{1.2}$$

A point $(\hat{u}, \hat{v})$ in $K_1 \times K_o$ is said to be a minimax point of L provided

$$L(\hat{u}, \hat{v}) = \inf_{u \in K_1} \sup_{v \in K_o} L(u, v) \tag{1.3}$$

Theorem 1 *Suppose* (A1) − (A2) *hold and* $(\hat{u}, \hat{v})$ *is a point in* $K_1 \times K_o$ *obeying*

$$L(\hat{u}, v) \leq L(\hat{u}, \hat{v}) = 0 \tag{1.4}$$

for all v in K_o. Then $\hat{u}$ will be a solution of (1.1), $(\hat{u}, \hat{v})$ is a minimax point of L and
$$F(\hat{u}) \in \partial\phi(\hat{v}). \tag{1.5}$$

Proof: Define $J : K_1 \to \mathbf{R}$ by $J(u) = \sup_{v \in K_o} L(u, v)$. Then (1.2) and the definition of K_o implies that
$$J(u) = \phi(u) + \phi^*(F(u)) - \langle u, F(u) \rangle$$
where ϕ^* is the conjugate convex functional of ϕ. Thus $J(u) \geq 0$ for all u in K_1, and $J(\hat{u}) = 0$ if and only if (1.1) holds from the generalized Young's inequality (proposition 51.2 in Zeidler (1985)).

When (1.4) holds then $J(\hat{u}) = L(\hat{u}, \hat{v}) = 0$, so $\hat{u}$ will be a solution of (1.1) and $(\hat{u}, \hat{v})$ is a minimax point of L.

Condition (1.5) is the extremality condition obeyed at the maximizers $\hat{v}$ of $L(\hat{u}, .)$ on K_o.

2 EXISTENCE THEOREMS.

Here we shall describe some simple conditions on F and ϕ that guarantee the existence of solutions of (1.1). These theorems are based on results of Ky Fan (1972). First an abstract minimax theorem is required.

Theorem 2 *Let $K_1 \subseteq K_2$ be nonempty closed convex sets in a Hausdorff locally convex topological vector space X. Let $L : K_1 \times K_2 \to \mathbf{R}$ be a function obeying*

(i): *for all u in K_1, $L(u, u) = 0$, $L(u, .)$ is u.s.c. and quasi-concave on K_2,*

(ii): *for all v in K_2, $L(., v)$ is l.s.c. on K_1, and*

(iii): *for some v_o in K_2, the set $E_o = \{u \in K_1 : L(u, v_o) \leq 0\}$ is compact in K_1.*

Then there is a $\hat{u}$ in K_1 such that
$$\sup_{v \in K_2} L(\hat{u}, v) = 0 = \inf_{u \in K_1} \sup_{v \in K_2} L(u, v) \tag{2.1}$$

Proof: For each v in K_2, let $E(v) = \{u \in K_1 : L(u, v) \leq 0\}$. Conditions (i) and (ii) imply that each $E(v)$ is a closed, nonempty set in K_1

From the first part of (i), $\sup_{v \in K_2} L(u,v) \geq L(u,u) = 0$ as $K_1 \subseteq K_2$. Thus (2.1) will hold if and only if there is a $\hat{u}$ in $\cap_{v \in K_2} E(v)$ or, equivalently, that this intersection is nonempty.

This will be done by verifying the conditions of lemma 1 in Ky Fan (1972)- which is a result proven in Ky Fan (1961). It is only necessary to verify that the convex hull of any finite subset $\{v_1, v_2, \ldots, v_n\}$ of K_2 lies in $\cup_{j=1}^n E(v_j) = E_n$. Suppose otherwise, that there is a convex combination $\sum_{j=1}^n c_j v_j$ with $\sum_{j=1}^n c_j v_j$ not in this E_n.

Then $L(\sum_{j=1}^n c_j v_j, v_k) > 0$ for $1 \leq k \leq n$.

This implies $L(z,z) > 0$ with $z = \sum_{j=1}^n c_j v_j$ upon using the quasi-concavity of $L(z,.)$ on K_2. This contradicts the fact that $L(z,z) = 0$ so z is in E_n and thus $\cap_{v \in K_2} E(v)$ is nonempty from Ky Fan's lemma. Each element in this intersection will be a solution of (2.1).

Corollary 1 *Let X be a real reflexive Banach space; K_1, K_2 be nonempty closed convex subsets of X with $K_1 \subseteq K_2$. Assume $L : K_1 \times K_2 \to \mathbf{R}$ is a function obeying (i), (ii)': for each v in K_2, $L(.,v)$ is weakly l.s.c. on K_1, and (iii)': there is a v_o in K_2 such that $E(v_o) = \{u \in K_1 : L(u,v_o) \leq 0\}$ is bounded in K_1, then there is a $\hat{u}$ in K_1 obeying (2.1).*

Proof: Take the topology on X to be the weak topology. Then (ii)' implies (ii) and (iii)' implies (iii) as X is reflexive and $L(.,v)$ is weakly l.s.c. Thus this corollary follows from the theorem.

For the particular function defined by (1.2), an existence result is obtained by requiring that X be a reflexive Banach space and

(A3): Assume that for each v in K_o, the functional $g_v : K_1 \to \mathbf{R}$ defined by $g_v(u) = \langle F(u), v - u \rangle$ is weakly l.s.c. on K_1.

(A4): There is a v_o in K_o such that $\{u \in K_1 : L(u,v_o) \leq 0\}$ is bounded in K_1.

Corollary 2 *Assume X is a real reflexive Banach space, K_1, K_o, L are as defined in section 1 and (A1) - (A4) hold. Then there is a $\hat{u}$ in K_1 obeying (1.1) and (2.1).*

Proof: We must verify that L obeys the conditions of corollary 1, with K_o in place of K_2. Condition (i) holds as ϕ is convex on K_o. (ii)' holds when ϕ is weakly l.s.c. on K_1 and (A3) holds and (A4) implies (iii)'.

Hence from corollary 1, there is a $\hat{u}$ in K_1 obeying (2.1) and thus (1.4). Then theorem 1 implies the result.

3 STATIONARY NAVIER-STOKES EQUATIONS.

To illustrate the use of these results, we shall apply this theory to obtain existence theorems for the stationary Navier-Stokes equations.

Let Ω be a bounded, connected open set in $\mathbf{R}^3$ with a uniformly Lipshitz boundary $\partial\Omega$ and let V_1 be the subspace of divergence-free vector fields in $H_o^1(\Omega, \mathbf{R}^3)$ defined as in Temam (1977), Chapter 1, Section 1.4. V_1 is a closed subspace and define $\Phi : V_1 \to \mathbf{R}$ by

$$\Phi(u) = \int_\Omega \sum_{j=1}^3 \left[\frac{\nu}{2} \sum_{k=1}^3 \left(\frac{\partial u_j}{\partial x_k}\right)^2 - f_j u_j \right] dx. \tag{3.1}$$

Here we assume ν is a positive constant and $f \in L^p(\Omega; \mathbf{R}^3)$ for some $p > \frac{6}{5}$.

In Auchmuty (1988), lemma 9.1, it is shown that (u, p) in $V_1 \times L^2(\Omega)$ is a solution of the stationary Navier-Stokes equations if and only if u obeys

$$-(u.\nabla)u \in \partial\Phi(u). \tag{3.2}$$

The function L corresponding to (1.2) for this problem is $L : V_1 \times V_1 \to \mathbf{R}$ where

$$L(u, v) = \int_\Omega \langle D_s v(x) u(x), u(x) \rangle dx + \Phi(u) - \Phi(v). \tag{3.3}$$

Here $\langle\ ,\ \rangle$ is the usual inner product on $\mathbf{R}^3$ and $D_s v(x)$ is the symmetric part of the Jacobian matrix $Dv(x) = \left(\frac{\partial v_i}{\partial x_j}(x)\right)$. The form (3.3) was obtained from the expression (1.2) after an integration by parts. The preceding analysis leads to the following existence theorem which may be compared with Teman (1977) Chapter 2, theorem 1.5.

Theorem 3 *Assume f, Ω, Φ and L as above, then there is a $\hat{u}$ in V_1 obeying (3.2) and (2.1).*

Proof: To prove this result, we shall verify (A1)-(A4) and use corollary 2. (A1) is easily verified with $K_o = V_1$.

When v is in V_1, then each component of $(v.\nabla)v$ is in $L^q(\Omega)$ for $1 \leq q \leq \frac{3}{2}$ upon using the Sobolev imbedding theorem and Holder's inequality. Thus (A2) holds with $K = V_1$, so $K_1 = V_1$ in this case.

The functional g_v defined in (A3) is given by the first term on the right hand side of (3.3). We shall show it is weakly sequentially continuous, then since V_1 is a reflexive space, it will be weakly continuous and (A3) holds.

Let $\{u^{(m)} : m \geq 1\}$ be a weakly convergent sequence in V_1 with the weak limit $\tilde{u}$. Then by Kondrachov's compactness theorem $u^{(m)}$ converges strongly to $\tilde{u}$ in $L^r(\Omega; \mathbb{R}^3)$ for $1 \leq r < 6$. Each entry of the matrix $D_s v(x)$ is in $L^2(\Omega)$, so Hölder's inequality yields that $g_v(u^{(m)})$ converges to $g_v(\tilde{u})$ as $m \to \infty$. Hence g_v is weakly sequentially continuous and (A3) holds.

To verify (A4), take $v_o = 0$, then $L(u, 0) = \Phi(u)$ is coercive on V_1 and so the set in question is bounded.

Since (A1)-(A4) hold, then Corollary 2 implies that there is a $\hat{u}$ in V_1 obeying (3.2) and (2.1).

It is well-known that the stationary Navier-Stokes equations are not of "potential" type and do not arise in a straight forward manner as the Euler-Lagrange equations of a standard functional of the calculus of variations. Thus the fact that their solutions may be characterized as minimax points of the functional (3.3) which is a simple integral expression is, in itself, of some interest.

REFERENCES

Auchmuty, G. (1988), "Variational Principles for Operator Equations and Initial Value Problems," Nonlinear Anal., T.M.A. 12, 531-564.

Fan, Ky, (1961), "A Generalization of Tychonoff's Fixed Point Theorem", Math. Ann. 142, 305-310.

... (1972), "A Minimax Inequality and Applications," in Inequalities III, A. Shisha. ed. Academic Press, N.Y. 103-113.

Teman, R., (1977) Navier Stokes Equations, North Holland Amsterdam.

Zeidler, E. H. (1985), Nonlinear Functional Analysis and its Applications, III: Variational Methods and Optimization, Springer Verlag, Berlin.

R C BROWN AND D B HINTON
Finding good upper bounds for the best constant in a generalized Hardy Littlewood inequality

In this note we consider the problem of estimating the constant K_p in the inequality

$$\int_0^\infty |y'|^2 \leq K_p \left(\int_0^\infty |y|^p\right)^{1/p} \left(\int_0^\infty |y''|^{p'}\right)^{1/p'} \tag{1}$$

where $1 \leq p \leq \infty$, $1/p + 1/p' = 1$, and y, y' are locally absolutely continuous ("AC_{loc}") real functions such that the right hand integrals are finite. That K_p exists has been shown by Everitt and Giertz [2]. However it seems that K_p is known in only two cases: $p = 2$ is the classic inequality of Hardy and Littlewood [4, Theorem 239] and $K_2 = 2$. $p = \infty$ is a special case of an inequality considered by Arestov; here K_∞ is also 2. (For the latter result as well as a survey of results concerning the best constant of Gabushin's inequality of which (1) is a special case see Kwong and Zettl [5].) For general p and in most cases of Gabushin's inequality the determination of K_p, the existence and determination of extremals, or even questions one might reasonably ask about the behavior of K_p as a function of p—is K_p continuous, differentiable, monotone, etc.?—are difficult and unsolved problems. The situation does not improve much if we restrict ourselves to the less challenging problem of finding good (i.e., small) upper bounds for K_p (a lower bound on K_p is plainly 1). Using numerical methods Everitt and Giertz estimated K_p in the range $1.1 \leq p \leq 2$; examples of the bounds they obtained are $K_{1.1} \leq 13.53$, $K_{1.5} \leq 6.98$, and $K_2 \leq 5$. (Note that the latter bound compares unfavorably with the known result $K_2 = 2$.)

To establish some motivation we remark that estimates of the best constant of this and other types of multiplicative inequalities can be used to establish nonoscillation criteria as well as spectral lower bounds for differential operators (cf.[1], [3], [7]).

Here using a technique derived in part from [2] we will sketch two methods for obtaining improved upper bounds for K_p for all p; moreover we will show that K_p is bounded above by a monotonically decreasing function whose limit as $p \to \infty$ is K_∞. We begin by considering real AC_{loc} functions $\psi_0, \psi_1, \lambda_0, \lambda_1$ defined on $0, \infty)$ such that $\lambda'_0, \lambda'_1 \in AC_{loc}$ and

$$\psi_i(0) = 1 \quad \lim_{t\to\infty} \psi_i(t) = 0, \quad i = 0, 1,$$

$$\lambda_0(0) = 1 \quad \lim_{t\to\infty} \lambda_0(t) = 0$$
$$\lambda'_0(0) = 0 \quad \lim_{t\to\infty} \lambda'_0(t) = 0$$

$$\lambda_1(0) = 0 \quad \lim_{t\to\infty} \lambda_1(t) = 0$$
$$\lambda'_1(0) = 1 \quad \lim_{t\to\infty} \lambda'_1(t) = 0.$$

Further assume that the norms $||\psi_0||_2, ||\psi_0'||_{p'}, ||\psi_1||_p, ||\psi_1'||_2, ||\lambda_0||_p, ||\lambda_0''||_{p'}, ||\lambda_1||_p, ||\lambda_1''||_{p'}$ exist. (All these functions may in particular be of compact support.) Call the classes of such functions respectively $\mathcal{C}_{0,\psi}, \mathcal{C}_{1,\psi}, \mathcal{C}_{0,\lambda}, \mathcal{C}_{1,\lambda}$ and define

$$\Psi_{0,p} := \inf_{\psi_0 \in \mathcal{C}_{0,\psi}} ||\psi_0||_2^{\frac{2}{p+2}} ||\psi_0'||_{p'}^{\frac{p}{p+2}}$$

$$\Psi_{1,p} := \inf_{\psi_1 \in \mathcal{C}_{1,\psi}} ||\psi_1||_p^{\frac{p}{p+2}} ||\psi_1'||_2^{\frac{2}{p+2}}$$

$$\Gamma_{0,p} := \inf_{\lambda_0 \in \mathcal{C}_{0,\lambda}} ||\lambda_0||_p^{\frac{p+1}{p+2}} ||\lambda_0''||_{p'}^{\frac{1}{p+2}}$$

$$\Gamma_{1,p} := \inf_{\lambda_1 \in \mathcal{C}_{1,\lambda}} ||\lambda_1''||_{p'}^{\frac{p+1}{p+2}} ||\lambda_1||_p^{\frac{1}{p+2}}.$$

Then we can show

Theorem. K_p satifies the following inequalities

$$K_p \leq \left(\frac{p+2}{(p+1)^{\frac{p+1}{p+2}}}\right)^2 \Gamma_{0,p}\Gamma_{1,p} + 1 \qquad (2)$$

$$K_p \leq \left(\Psi_{0,p}\Psi_{1,p} \left(\frac{p}{p+2}\right)^{-\frac{4p}{p+2}} \left(\frac{2}{p+2}\right)^{-\frac{8}{p+2}} (K_p)^{\frac{2}{p+2}} + 1\right). \qquad (3)$$

Remark. $\Psi_{0,p}, \Psi_{1,p}, \Gamma_{0,p}$ and $\Gamma_{1,p}$ are themselves solutions of unsolved variational problems which may be of some interest in their own right. Even if we could solve them inequality (2) in particular would not be optimal for all p, since it is shown in [2] via substitution of a test function that estimates of K_2 obtained by the assumptions underlying (2) are bounded below by $1 + 4/3^{\frac{1}{2}}$. Nevertheless suggestive results can be obtained by choosing various candidates for $\psi_0, \psi_1, \lambda_0$, and λ_1 such as splines, interpolating polynomials, $\lambda_0 = (\sin t + 1)e^{-t}$, $\lambda_1 = (\sin t)e^{-t}$, etc.), numerically computing the various norms, substituting directly into (2) or solving the inequality (3), and comparing the results. Generally speaking (2) works best for "low" p, while (3) gives good bounds for K_p for "large"

p. A synthesis of both procedures yields the following table:

p	$K_p \leq$
1	4.1
1.5	3.86
2	3.49
3	3.26
4	3.17
5	3.1
6	3.06
10	2.92
30	2.67
100	2.4
1000	2.05

From this table it appears likely that $K_p \leq 4$ for all $p \geq 1.5$. We suspect in fact that $K_p \leq 4$ for all p, but have not quite been able to verify this numerically at $p = 1$. This however would not be surprising if either $K_1 = 4$ or if the infimum of the upper bounds produced by our method is 4 when $p = 1$. Moreover K_p seems bounded above by a decreasing function of p, whose limit is 2 as $p \to \infty$. We can in fact prove the latter conjecture.

Corollary. There exists a monotonically decreasing function $\Theta(p)$ such that $K_p \leq \Theta(p)$ and $\lim_{p \to \infty} \Theta(p) = 2$. In particular given $\epsilon > 0$ there exists p_ϵ such that for all $p \geq p_\epsilon$

$$K_p \leq \left(\frac{p+2}{(p+1)^{\frac{p+1}{p+2}}}\right)^2 (1+\epsilon) + 1. \tag{4}$$

Thus K_p is upper semicontinuous at ∞.

Proofs. Both the inequalities (2) and (3) are established by bounding $(|y(0)||y'(0)|)(||y||_p ||y''||_{p'})^{-1}$ in the proper way. To derive (2) we start (cf.[2]) by applying integration by parts and Hölder's Inequality to $\int_0^\infty yy''$. This gives the inequality

$$\int_0^\infty |y'|^2 \leq |y'(0)||y(0)| + \left(\int_0^\infty |y|^p\right)^{1/p} \left(\int_0^\infty |y''|^{p'}\right)^{1/p'}. \tag{5}$$

Let $\lambda_0 \in C_{0,\lambda}$, $\lambda_1 \in C_{1,\lambda}$, and $\epsilon > 0$. Define functions $g_{0,\epsilon}$ and $g_{1,\epsilon}$ by

$$g_{0,\epsilon}(t) = \lambda_0(\epsilon t)$$
$$g_{1,\epsilon}(t) = \epsilon^{-1} \lambda_1(\epsilon t).$$

Next integrate $y''g_{0,\epsilon}$ and $y''g_{1,\epsilon}$ by parts. Hölder's inequality and an obvious change of variables yield the estimates

$$|y'(0)| \leq \|y''\|_{p'}\|\lambda_0\|_p \epsilon^{-1/p} + \|y\|_p \|\lambda_0''\|_{p'} \epsilon^{1+1/p}$$

$$|y(0)| \leq \|y''\|_{p'}\|\lambda_1\|_p \epsilon^{-(1+1/p)} + \|y\|_p \|\lambda_1''\|_{p'} \epsilon^{1/p}, \qquad (6)$$

the right side of which may be minimized as functions of ϵ. This in turn gives

$$|y'(0)| \leq (\Gamma_{0,p}/D_p)\|y\|_p^{\frac{1}{p+2}} \|y''\|_{p'}^{\frac{p+1}{p+2}}$$

$$|y(0)| \leq (\Gamma_{1,p}/D_p)\|y\|_p^{\frac{p+1}{p+2}} \|y''\|_{p'}^{\frac{1}{p+2}}$$

where

$$D_p = \left(\frac{p+1}{p+2}\right)^{\frac{p+1}{p+2}} \left(\frac{1}{p+2}\right)^{\frac{1}{p+2}}.$$

(2) follows immediately by combining (5) and (6).

To derive (3) we consider the problem of minimizing $\|y\|_p + \|y'\|_2$ subject to the constraint $y(0) = 1$ over $y \in \mathrm{AC}_{\mathrm{loc}}$. This problem is easily shown equivalent to the problem of minimizing say

$$\{\|e^{-t} + y\|_p + \| - e^{-t} + y'\|_2 : y \in \mathcal{D}\},$$

where $\mathcal{D}$ is the subspace of $\mathrm{AC}_{\mathrm{loc}}$ such that $y(0) = 0$. Call the minimum of this problem "$M(p,2)$". Now using a well known duality principle (cf.[6], pp.119-120) we have that

$$M(p,2) = \sup_z \left\{ \left|\int_0^\infty e^{-t}(z'-z)\right| : \|z'\|_{p'} + \|z\|_2 = 1;\ \int_0^\infty yz' + y'z = 0,\ \forall y \in \mathcal{D} \right\}.$$

Let $f \in \mathcal{C}_{0,\psi}$ such that $f(0) = 1$. Define $f_\alpha(t) := f(\alpha t)$. Then $\|f_\alpha'\|_{p'} = \alpha^{1/p}\|f\|_{p'}$ and $\|f_\alpha\|_2 = \alpha^{-1/2}\|f\|_2$. Taking $z = f_\alpha(\|f_\alpha'\|_{p'} + \|f_\alpha\|_2)^{-1}$ and carrying out the integration yields that

$$M(p,2) \geq \left(\alpha^{1/p}\|f'\|_{p'} + \alpha^{-1/2}\|f\|_2\right)^{-1}.$$

Maximizing the right side of this inequality over α and the definition of $\Psi_{0,p}$ gives

$$\inf_y \{\|y\|_p + \|y'\|_2 : y \in \mathrm{AC}_{\mathrm{loc}}; y(0) = 1\}$$

$$\geq \Psi_{0,p}^{-1} \left(\frac{p}{p+2}\right)^{\frac{p}{p+2}} \left(\frac{2}{p+2}\right)^{\frac{2}{p+2}}.$$

By substituting $\tilde{y}(t) \equiv y(\epsilon t)/y(0)$ and minimizing over ϵ, we finally obtain the inequality

$$|y(0)| \leq \Psi_{0,p} \left(\frac{p}{p+2}\right)^{-\frac{2p}{p+2}} \left(\frac{2}{p+2}\right)^{-\frac{4}{p+2}} \|y\|_p^{\frac{p}{p+2}} \|y'\|_2^{\frac{2}{p+2}}. \qquad (7)$$

By the same method, i.e. minimizing
$$\{\|y\|_2 + \|y'\|_{p'} : y \in AC_{loc}; y(0) = 1\}$$
we can derive the inequality
$$|y(0)| \leq \Psi_{1,p} \left(\frac{p}{p+2}\right)^{-\frac{2p}{p+2}} \left(\frac{2}{p+2}\right)^{-\frac{4}{p+2}} \|y\|_2^{\frac{2}{p+2}} \|y'\|_{p'}^{\frac{p}{p+2}}. \tag{8}$$

The proof is completed by substituting y' for y in (8), multiplying (7) and (8) together, replacing $\|y'\|_2^2$ by $K_p \|y\|_p \|y''\|_{p'}$, and substituting the resulting estimate for $|y(0)| |y'(0)|$ into (5).

Turning to the Corollary we see from (3) that since $K_p > 1$
$$K_p \leq \left(\Psi_{0,p} \Psi_{1,p} \left(\frac{p}{p+2}\right)^{-\frac{4p}{p+2}} \left(\frac{2}{p+2}\right)^{-\frac{8}{p+2}} + 1\right)^{\frac{p+2}{p}}.$$

Taking $\psi_0 = \psi_1 := 1 - t$ on $[0,1]$, we find for example that
$$K_p \leq \Theta(p) := \left(3^{-\frac{1}{p+2}}(p+1)^{-\frac{1}{p+2}} \left(\frac{p}{p+2}\right)^{-\frac{4p}{p+2}} \left(\frac{2}{p+2}\right)^{-\frac{8}{p+2}} + 1\right)^{\frac{p+2}{p}}$$
which is a decreasing function of p having limit 2 at $p = \infty$. Other choices of ψ_0 and ψ_1 will give candidates for $\Theta(p)$ which are smaller near $p = 1$.

The following argument suffices to prove (4). Let λ_1 have exactly one inflection point $c > 0$. Suppose $\lambda'(c) = -\delta$ where $\delta > 0$ and is arbitrarily small. Then $\|\lambda_1''\| = 1 + 2\delta$. Since we can arrange that $\|\lambda_0\|_\infty = 1$, $\Gamma_{0,\infty} = 1$ and $\Gamma_{1,\infty} \leq 1$. Hence $K_\infty \leq 2$. Since the norms of λ_0 and λ_1'' etc. depend continuously on p and p', it follows that given $\epsilon > 0$ there exists p_ϵ such that for all $p \geq p_\epsilon$ (4) is true.

Finally the statement about upper semicontinuity follows from (3) and the fact that $K_\infty = 2$.

The authors would like to acknowledge the assistance of Mr. Obed Matus of Pan American University for some of the numerical calculations in this paper.

REFERENCES

1. M. Ashbaugh, R. C. Brown, and D. B. Hinton *Interpolation inequalities and nonoscillatory differential equations*, to appear in the Proceedings of the Oberwolfach conference "General Inequalities 6".
2. W. N. Everitt and M. Giertz *On the integro-differential inequality* $\|f'\|_2^2 \leq K \|f\|_p \|f''\|_q$, J. Math. Anal. and Appl. **45** (1974), 639-653.
3. I. M. Glazman "Direct Methods of Qualitative Spectral Analysis of Singular Differential Operators", Israel Program for Scientific Translation, Jerusalem, 1965.
4. G. H. Hardy, J. E. Littlewood, and G. Polya "Inequalities", Cambridge University Press, Cambridge, 1959.
5. M. K. Kwong and A. Zettl *Norm inequalities for derivatives*, Lecture Notes in Mathematics, vol. 846, Springer, Berlin, 1981, pp. 229-243.
6. D. G. Luenberger "Optimization by Vector Space Methods", John Wiley & Sons, Inc, New York, 1968.
7. E. J. M. Veling *Optimal lower bounds for the spectrum of a second order linear differential equation with p-integrable coefficient*, Proc. of the Royal Soc. Edinburgh **92A** (1982), 95-101.

T A BURTON
Admissible controls in a PDE of Lurie type

1. Introduction

We consider two forms of a Lurie system, the simplest of which is

$$u_t = \Delta u + b(x)f(\sigma) + \int_{t-h}^{t} p(s, u(s, X), \nabla u(s, X))ds$$
$$\sigma' = \int_{\Omega} c(X)u(t, X)dX - rf(\sigma),$$

$u(t, X) = 0$ on $\partial \Omega$, Ω is a domain with smooth boundary, $\Delta u = u_{xx} + u_{yy} + u_{zz}$, $X = (x, y, z)$, and where

$$\sigma f(\sigma) > 0 \text{ if } \sigma \neq 0. \tag{*}$$

The exact forms are given in (15a), (16a), and (46).

The goal is to give conditions to ensure that all solutions satisfy

$$|\sigma(t)| + \int_{\Omega} u^2(t, X)dX \to 0 \text{ as } t \to \infty$$

without strengthening (*). That is, conditions are sought to ensure that every f satisfying (*) is admissible.

This is a control problem, the first stage of which is to show that all solutions are controllable to zero for every f satisfying (*) (this is absolute stability); the next stage (which we do not consider) is to minimize a cost functional with the assurance that solutions will tend to zero for the control function f which minimizes the cost functional. (c.f. Curtain and Pritchard [4; p. 3] and Nakagiri[10; p. 175].)

Appropriate conditions are given to ensure absolute stability and these use a result in [2] which gives a very surprising differential inequality for a Liapunov functional of the form

$$V'(t) \leq \alpha V(t)$$

derived from a standard inequality relating V' only to space variables.

2. Background

A classical control problem in ordinary differential equations, called the problem of Lurie, has several forms, one of which is

$$\begin{cases} x' = Ax + bf(\sigma) \\ \sigma' = c^T x - rf(\sigma) \end{cases} \tag{1}$$

where b and c are constant column vectors, r is a positive constant, A is an $n \times n$ constant matrix all of whose characteristic roots have negative real parts, f is continuous, and $\sigma f(\sigma) > 0$ if $\sigma \neq 0$. The problem is to give conditions to ensure that every solution tends to zero. Discussion of the problem may be found in [6], [7], [13], for example.

Lurie used the Liapunov function

$$V(x, \sigma) = x^T B x + \int_0^\sigma f(s) ds \tag{2}$$

with $B^T = B$ and $A^T B + BA = -D$ for B and D positive definite. That function has played a major role in the problem, as is seen in the extensive survey book of Lefschetz [7]. The derivative of V along a solution of (1) satisfies

$$V'(x(t), \sigma(t)) = -x^T D x + f(\sigma)[2b^T B + c^T] x - r f^2(\sigma) \tag{3}$$

and Lefschetz [7] showed that this is negative definite if and only if

$$r > (Bb + c/2)^T D^{-1} (Bb + c/2). \tag{4}$$

Still, unless $\int_0^\sigma f(s) ds \to \infty$ as $|\sigma| \to \infty$, one may not immediately conclude that solutions are bounded (and, therefore, tend to zero).

The simple remedy is to ask that admissible controls satisfy

$$f^2(\sigma) \geq \int_0^\sigma f(s) ds \text{ (and so } \int_0^{\pm\infty} f(s) ds = +\infty) \tag{**}$$

so that $V' \leq -\gamma V$, from which we conclude that $V(t) \leq V(0) e^{-\gamma t}$ and that σ is bounded and tends to zero.

But LaSalle [6] showed that solutions of (1) are bounded if and only if

$$C^T A^{-1} b + r > 0, \tag{5}$$

and it is known that (4) implies (5), so (3) and (4) imply that all solutions of (1) tend to zero.

There is a complete parallel between Liapunov functions for (1) and the PDE of interest here, with the same difficulty of proving boundedness occurring. It was hoped that (5) would prove central for the PDE case as well, but LaSalle's idea does not seem to extend to that case. It may be noted that (5) has attracted considerable attention. For example, in [3] we showed that if $f(\sigma)/\sigma \to 0$ as $|\sigma| \to \infty$, then solutions of (1) are uniformly ultimately bounded if and only if (5) holds. Somolinos [12] strengthened that result to $f(\sigma)/\sigma$ bounded as $|\sigma| \to \infty$, among other results.

In [8] (see also [9] and [10]) Nakagiri introduced as a counterpart of (1) the equation

$$u_t = \Delta u + \int_{-h}^0 d\eta(s, X) u(t+s, X) ds + c(X) f(\sigma(t)) \tag{6}$$

where $\sigma(t) = \int_\Omega d(X) u(t, X) dX$ and $u = 0$ on $\partial\Omega$. This formulation is not parallel to (1), but rather to

$$\begin{cases} x' = Ax + bf(\sigma) \\ \sigma = d^T x, \end{cases}$$

as given by Lefschetz [7; pp. 39-40], so that

$$\sigma' = d^T x' = d^T[Ax + bf(\sigma)] = d^T Ax + d^T b f(\sigma)$$

which results in (1) when $d^T A = c^T$ and $d^T b = -r < 0$. We discuss a form of (6) in the last section.

Nakagiri solves his problem using semigroup theory and frequency domain techniques. It is very interesting to note that some of the same questions arise for (6) as for (1). In particular, Nakagiri asked conditions similar to (**) and required that

$$\sigma f(\sigma) \leq k\sigma^2 \text{ and } \int_0^\sigma f(s)ds \to \infty \text{ as } |\sigma| \to \infty \tag{***}$$

in order for a control to be admissible. We will not need such conditions.

To bring into focus the work here and to introduce the main boundedness technique that we use, we summarize the analysis of a one-dimensional problem. Consider

$$\begin{cases} u_t = g(u_x)_x + b(x)f(\sigma) + \int_{t-h}^t p(s, u(s,x), u_x(s,x))ds \\ \sigma' = \int_0^1 c(x)u(t,x)dx - rf(\sigma), \end{cases} \tag{7}$$

$$u(t,0) = u(t,1) = 0, \tag{8}$$

where $0 < \alpha \leq dg(x)/dx$ for some $\alpha > 0$, $\int_0^1 p^2(t, u, u_x)dx \leq \beta \int_0^1 u_x^2 ds$ for some $\beta > 0$, and b and c are at least L^2 functions. Under certain conditions the Liapunov functional

$$V_1(t) = \int_0^1 [(\frac{1}{2})u^2 + k \int_{-h}^0 \int_{t+s}^t p^2(v, u, u_x)dv ds]dx + \int_0^\sigma f(s)ds \tag{9}$$

will satisfy

$$V_1'(t) \leq -\lambda \int_0^1 [u^2 + f^2(\sigma) + \int_{t-h}^t p^2(s, u, u_x)ds]dx \tag{10}$$

If $\int_0^\sigma f(s)ds$ does not diverge with σ, then the question of boundedness raised concerning (1) occurs once more.

And a major point of this paper is to show that $\sigma(t)$ is bounded without any condition on f except (*).

In order to prove that solutions of the PDE are bounded we need a result in [2] which may be stated as follows. Let

$$x' = F(t,x) \tag{11}$$

and suppose that $V_1, P : [0, \infty) \times \mathbb{R}^{n+1} \to [0, \infty)$, $U : [0, \infty) \times \mathbb{R} \to [0, \infty)$, and $Q : [0, \infty) \times [0, \infty) \times \mathbb{R}^{n+1} \to [0, \infty)$ are continuous functions with

$$V_1(t,x) = P(t,x) + U(t,z), \tag{12}$$

where $x = (x_1, \ldots, x_n, z)$, and we also suppose that

$$V_1'(t,x) \leq -Q(P(t,x), t, x) \tag{13a}$$

with $Q(P,t,x) > 0$ if $P > 0$ and Q increasing in P. (In [2], this last condition was inadvertently left out, but was used in the proof.)

Theorem B. Let (12) and (13) hold and suppose there is an $L > 0$ such that if $t_n \to \infty$ and $z_n \to \infty$ then

$$U(t_n, z_n) \to L, \quad U(t, z) < L \text{ for all } (t, z) \text{ with } z > 0. \tag{14}$$

If $V = V_1 - L$ and if $x(t)$ is a solution of (11) on $[t_o, \infty)$ with $\limsup_{t \to \infty} z(t) = \infty$, then

$$V'(t, x) \leq -Q(V(t,x)/2, t, x) \text{ for } z(t) > 0. \tag{13b}$$

Our application will have (13b) as $V' \leq -\phi V$, ϕ constant.

In this result the connection of $x(t)$ to (11) is immaterial. If $x(t)$ is any function on $[t_o, \infty)$ for which (13) holds, then the result is valid.

We have discussed two formulations of the Lurie problem in terms of (1) and (6). There is yet a third standard formulation (c.f., Lefschetz [7]) of the form

$$w_t = \Delta w + \int_{t-h}^{t} p(s, u(s, X), \nabla u(s, X)) ds + b(X)\mu(t),$$

$$\mu'(t) = f(\sigma),$$

$$\sigma = \int_\Omega d(X) w(t, X) dX - r\mu.$$

This formulation can be transformed into a system closely related to (6). We do not consider it here.

3. Absolute stability

In this section we consider two systems, the first of which is

$$u_t = \Delta u + b(X) f(\sigma) + \int_{t-h}^{t} p(s, u(s, X), \nabla u(s, X)) ds, \tag{15a}$$

$$\sigma' = \int_\Omega c(X) u(t, X) dX - r f(\sigma),$$

$$u(t, X) = 0 \text{ for } X \in \partial\Omega, \tag{16a}$$
$$u(0, X) = g^0(X), u(s, X) = g^1(s, X) \text{ a.e. } s \in [-h, 0], X \in \Omega,$$

where f is continuous, (*) holds,

$$\int_\Omega |p(s, u, \nabla u)|^2 dX \leq \beta \int_\Omega |\nabla u|^2 dX \text{ for some } \beta > 0,$$

$c(X)$ and $b(X)$ are at least $L^2(\Omega)$, r and h are positive constants, and p is at least continuous. The Sobolev estimates which we will need also work with the quasi-linear term in

$$\begin{cases} u_t = g(u_x)_x + b(x) f(\sigma) + \int_{t-h}^{t} p(s, u(s, x), u_x(s, x)) ds, \\ \sigma' = \int_0^1 c(x) u(t, x) dx - r f(\sigma), \end{cases} \tag{15b}$$

$$\begin{cases} u(t,0) = u(t,1) = 0 \\ u(0,x) = g^0(x), u(s,x) = g^1(s,x) \text{ a.e. } s \in [-h,0), 0 \le x \le 1, \end{cases} \quad (16b)$$

$\int_0^1 |p(s,u,u_x)|^2 dx \le \beta \int_0^1 u_x^2 dx$ for some $\beta > 0$, $c(x)$ and $b(x)$ are at least $L^2(0,1)$, r and h are positive constants, p is at least continuous, $dg(x)/dx \stackrel{def}{=} g'(x) \ge \alpha$ for some $\alpha > 0$.

Not only will the Sobolev estimates work for this system, but we will conclude absolute stability in the supremum norm for (15b), while we can only prove absolute stability in the L^2-norm for (15a). Nakagiri's stability is also in the L^2-norm. An existence result will be sketched in the next section, but our primary goal here is to establish strong <u>a priori</u> bounds on solution which do not require strenghthening of (*).

Theorem 1. Suppose that $b(X)$ and $c(X)$ are in $L^2(\Omega)$, $b(X) + c(X)$ is bounded, $g'(r) \ge \alpha > 0$, λ_1 is the first eigenvalue of $-\Delta$ on H_0^1 and $k - (\frac{1}{2}) = \gamma > 0$ with

$$\lambda_1(1 - kh\beta) - (h/2) \stackrel{def}{=} \mu > 0, \quad (19)$$

and

$$\int_\Omega (b(X) + c(X))^2 dX < 4\mu r. \quad (20)$$

Then for each solution of (15a), (16a) on $[0,\infty)$ there is an $M > 0$ with

$$\int_\Omega |\nabla u|^2 dX + \int_0^\sigma f(s)ds + \int_0^t \int_\Omega |\Delta u|^2 dX ds < M$$

and

$$|\sigma(t)| + \int_\Omega \left[u^2(t,X) + |\nabla u|^2 + \int_{t-h}^t p^2(s,u,\nabla u)ds\right] dX \to 0 \text{ as } t \to \infty.$$

Also, for each solution of (15b), (16b) on $[0,\infty)$ there is an $M > 0$ with

$$\|u\| + \int_0^1 u_x^2(t,x)dx + \int_0^\sigma f(s)ds + \int_0^t \int_0^1 u_{xx}^2(t,x)dxdt < M$$

(where $\|\cdot\|$ is the supremum norm in x for fixed t) and $|u(t,x)| + |\sigma(t)| \to 0$ as $t \to \infty$.

Proof. Most of the proof of the nonlinear case (b) parallels that for the linear case (a). We focus mainly on (15a). For this solution, define

$$V_1(t) = \int_\Omega \left[(\tfrac{1}{2})u^2(t,X) + k\int_{-h}^0 \int_{t+s}^t p^2(v,u(v,X), \nabla u(v,X))dv ds\right] dX \quad (21)$$
$$+ \int_0^\sigma f(s)ds$$

so that

$$\begin{aligned}V_1'(t) &= \int_\Omega \{u[\Delta u + b(X)f(\sigma) + \int_{t-h}^t p(s,u(s,X), \nabla u(s,X))ds]\\ &\quad + hkp^2(t,u,\nabla u) - k\int_{t-h}^t p^2(s,u(s,X), \nabla u(s,X))ds\\ &\quad + f(\sigma)c(X)u(t,X)\}dX - rf^2(\sigma)\\ &= \int_\Omega \{-|\nabla u|^2 + (b+c)uf(\sigma) + hk\beta|\nabla u|^2 + (h/2)u^2\\ &\quad - (k - (\tfrac{1}{2}))\int_{t-h}^t p^2(s,u(s,X), \nabla u(s,X))ds - (r/|\Omega|)f^2(\sigma)\}dx\end{aligned}$$

(where $|\Omega| = \int_\Omega dX$ and we have used the divergence theorem to say
$$\int_\Omega u\Delta u dX = -\int_\Omega |\nabla u|^2 dX)$$
$$\leq \int_\Omega \{[-\lambda_1(1-kh\beta) + (h/2)]u^2 + (b+c)uf(\sigma)$$
$$-(r/|\Omega|)f^2(\sigma)$$
$$-(k-(\frac{1}{2}))\int_{t-h}^t p^2(s,u,\nabla u)ds\}dX$$
$$= \int_\Omega \{-\mu u^2 + (b+c)uf(\sigma) - (r/|\Omega|)f^2(\sigma)$$
$$-\gamma \int_{t-h}^t p^2(s,u,\nabla u)ds\}dX$$
$$= \int_\Omega \{-\mu\{u^2 + [(b+c)/\mu]uf(\sigma) + [(b+c)/2\mu]^2 f^2(\sigma)$$
$$+ \left((r/|\Omega|\mu) - [(b+c)/2\mu]^2\right) f^2(\sigma)\}$$
$$-\gamma \int_{t-h}^t p^2(s,u,\nabla u)ds\}$$
$$\leq -\bar{\lambda}\int_\Omega \{(u + [(b+c)/2\mu]f(\sigma))^2 + f^2(\sigma)$$
$$+ \int_{t-h}^t p^2(s,u,\nabla u)ds\}dX$$
$$\leq -\lambda \int_\Omega \{u^2 + f^2(\sigma) + \int_{t-h}^t p^2(s,u,\nabla u)ds\}dX$$

(by completing the square again starting with $f^2(\sigma)$ and using $b+c$ bounded) for $\lambda, \bar{\lambda} > 0$. Indeed, we can argue with a different $\lambda, \bar{\lambda} > 0$ that

$$V_1' \leq -\lambda \int_\Omega \{u^2 + |\nabla u|^2 + f^2(\sigma) + \int_{t-h}^t p^2(s,u,\nabla u)ds\}dX \tag{22}$$

by taking a sightly smaller μ when we replaced $|\nabla u|^2$ by u^2. We will also arrive at $|\nabla u|^2$ in another way.

Now define
$$W(t) = \int_\Omega |\nabla u|^2 dX = \int_\Omega (u_x^2 + u_y^2 + u_z^2)dX \tag{23}$$

so that
$$W'(t) = \int_\Omega 2(u_x u_{xt} + u_y u_{yt} + u_z u_{zt})dX$$
$$= -\int_\Omega 2(u_{xx}u_t + u_{yy}u_t + u_{zz}u_t)dX$$

(by the divergence theorem and the boundary conditions)
$$= -2\int_\Omega \Delta u[\Delta u + b(X)f(\sigma) + \int_{t-h}^t p(s,u,\nabla u)ds]dX$$
$$\leq \int_\Omega \{[-2 + (h/K) + (1/K)](\Delta u)^2 + Kb^2(X)f^2(\sigma)$$
$$+ K\int_{t-h}^t p^2(s,u,\nabla u)ds\}dX$$

21

for $K > 0$. Thus, since b is in L^2, if K is large and m is small then for

$$Z(t) = V_1(t) + mW(t) \tag{24}$$

we have (for a new $\lambda > 0$)

$$Z'(t) \leq -\lambda \int_\Omega [(\Delta u)^2 + |\nabla u|^2 + f^2(\sigma) + \int_{t-h}^t p^2(s,u,\nabla u)ds]dX. \tag{25}$$

In fact, under these boundary conditions, Simpson and Spector [11: p. 26] show that there is a $\gamma > 0$ with

$$\gamma \int_\Omega |\nabla u|^2 dX \leq \int_\Omega (\Delta u)^2 dX. \tag{26}$$

(This argument would allow us to delete the condition that $b + c$ is bounded, but would complicate subsequent arguments.) Thus, an integration of (25) will yield the inequality involving M of the theorem.

The case for (15b) uses the same V_1 where Ω is now $(0,1)$ and $W(t) = \int_0^1 \int_0^{u_x} g(s)ds dx$. We will now show that $V_1(t)$ actually tends to zero. Suppose that

$$\limsup_{t \to \infty} \sigma(t) = \infty.$$

Then it is clear that $\int_0^\sigma f(s)ds$ is bounded for $\sigma > 0$ since $V_1'(t) \leq 0$; thus, we can define

$$L = \int_0^\infty f(s)ds$$

and

$$V(t) = V_1(t) - L.$$

Review (12) - (14) and Theorem B, letting

$$P(t,x) = \int_\Omega \{(\frac{1}{2})u^2(t,X) + k\int_{-h}^0 \int_{t+s}^t p^2(v,u,\nabla u)dvds\}dX,$$

$$U(t,z) = \int_0^\sigma f(s)ds,$$

and

$$Q(P(t,x),t,x) = P(t,x)[\lambda/(hk+1)]$$

when we notice that

$$\int_{-h}^0 \int_{t+s}^t p^2(v,u,\nabla u)dvds \leq h \int_{t-h}^t p^2(s,u,\nabla u)ds.$$

Note also that P is a function of t alone; this means that Theorem B will be valid in this setting.

Lemma. If $\limsup_{t \to \infty} \sigma(t) = \infty$, then $\sigma(t) \to \infty$.

Proof. Note that

$$|\int_\Omega c(X)u(t,X)dX|^2 \leq (\int_\Omega c^2(X)dX)^2 + (\int_\Omega u^2(t,X)dX)^2$$

so the term on the left is bounded since $c \in L^2$. As the right-hand-side of σ' in (15a) is bounded for σ bounded, if there is a point $\sigma_0 \neq 0$ and a sequence $\{t_n\} \to \infty$ such that $|\sigma(t_n) - \sigma_0| \to 0$ as $n \to \infty$, then there is a $T > 0$ such that if $0 < \delta < (\frac{1}{2})|\sigma_0|$ then $|\sigma(t) - \sigma_0| \leq \delta$ for $t_n \leq t \leq t_n + T$ and n large. An integration of (22) will then show that $V_1(t) \to -\infty$ as $t \to \infty$, a contradiction. We therefore conclude that $\sigma(t) \to \infty$ as $t \to \infty$ and the lemma is proved. (This argument is part of a general theorem in [1].)

Hence, by Theorem B there is a $t_0 > 0$ and a constant $\phi > 0$ such that $V'(t) \leq -\phi V(t)$ for $t \geq t_0$. But

$$V'(t) = -(|V'|)^{\frac{1}{2}}(|V'|)^{\frac{1}{2}} \leq -(\phi V)^{\frac{1}{2}}(|V_1'|)^{\frac{1}{2}}$$
$$\leq -KV^{\frac{1}{2}}(|\sigma'|^2)^{\frac{1}{2}} = -KV^{\frac{1}{2}}|\sigma'|$$

for some $K > 0$. That is

$$|\sigma'|^2 = (\int_\Omega c(X) u(t,X) dX - rf(\sigma))^2$$
$$\leq 2(\int_\Omega c(X) u(t,X) dX)^2 + 2r^2 f^2(\sigma)$$
$$\leq 2(\int_\Omega c^2(X) dX \int_\Omega u^2(t,X) dX) + 2r^2 f^2(\sigma)$$
$$\leq c_1(\int_\Omega u^2(t,X) dX + f^2(\sigma))$$
$$\leq c_2 |V_1'|$$

for constants c_1 and c_2. We then have

$$V^{-\frac{1}{2}} V' \leq -K|\sigma'|$$

so

$$V^{\frac{1}{2}}(t) \leq V^{\frac{1}{2}}(t_0) - (K/2)|\sigma(t) - \sigma(t_0)|.$$

Thus $\sigma(t)$ is bounded. We can then integrate (22) and conclude that $\sigma(t) \to 0$ so $\int_0^\sigma f(s)ds \to 0$ as $t \to \infty$. (Here, we have used our previous argument with t_n and T.)

We now have

$$V_1'(t) \leq -\lambda \int_\Omega \left[\int_{t-h}^t p^2(s,u,\nabla u)ds + u^2(t,X) \right] dX$$
$$\leq -\lambda V_1(t) + H(t)$$

where $\lambda > 0$ and $H(t) \to 0$ as $t \to \infty$. Hence,

$$V_1(t) \leq V_1(0) e^{-\lambda t} + \int_0^t e^{-\lambda(t-s)} H(s) ds$$

and that integral tends to zero since it is the convolution of an L^1-function with a function tending to zero.

From (25) and (26), together with $V_1(t) \to 0$, it is the case that along any solution we have

$$Z'(t) \leq -\lambda \int_\Omega [(\Delta u)^2 + |\nabla u|^2 + \int_{t-h}^t p^2(s,u,\nabla u) ds] dX$$
$$\leq -\lambda Z(t) + D(t)$$

23

where $D(t) \to 0$ as $t \to \infty$. Just as in the case of V_1 in the last paragraph, we argue that $Z(t) \to 0$ as $t \to \infty$ and so $\int_\Omega [u^2(t,X) + |\nabla u|^2]dX \to 0$ as $t \to \infty$. A parallel argument for (15b) yields $\| u(t,x) \| \leq \int_0^1 u_x^2(t,x)dx \to 0$ as $t \to \infty$. This completes the proof of Theorem 1.

4. Remarks on Existence

As mentioned earlier, our work was motivated by a paper of Nakagiri [8] concerning

$$u_t = \Delta u + \int_{-h}^0 d\eta(s,x)u(t+s,x) + \phi(\sigma(t))c(x), t \geq 0, x \in \Omega, \tag{6}$$

$$\sigma(t) = \int_\Omega d(x)u(t,x)dx \text{ for } t \geq 0, \tag{27}$$

$$u(t,x) = 0 \text{ on } \partial\Omega, t \geq 0, \tag{28}$$

$$u(0,x) = g^0(x), u(s,x) = g^1(s,x) \text{ a.e., } s \in [-h,0), x \in \Omega. \tag{29}$$

The system is written as an abstract ODE

$$u'(t) = Au(t) + \int_{-h}^0 d\eta(s)u(t+s) + \phi(\sigma(t))c, t \geq 0, \tag{30}$$

with (27) and (29) holding. Under certain additional assumptions, if $T(t)$ is the semigroup generated by A and if $W(t)$ is the unique fundamental solution of (c.f., [10; p. 175])

$$W(t) = \begin{cases} T(t) + \int_0^t T(t-s)\int_{-h}^0 d\eta(\xi)W(\xi+s)ds &, t \geq 0 \\ 0 &, t < 0 \end{cases} \tag{31}$$

and if

$$U_t(s) = \int_{-h}^0 W(t-s+\xi)d\eta(\xi) \text{ a.e. } s \in [-h,0], \tag{32}$$

then there is at least one solution $u(t)$ of

$$u(t) = W(t)g^0 + \int_{-h}^0 U_t(s)g^1(s)ds$$

$$+ \int_0^t W(t-s)\phi(\int_\Omega d(x)u(s,x)dx)cds, t \geq 0, \tag{33}$$

and it is the strong solution of (6) when $c \in D(A)$ and

$$g^0 \in D(A), g^1 \in W^{1,2}([-h,0]; L^2(\Omega)), g^1(0) = g^0. \tag{34}$$

There are at least three distinct ways of linking a control and these are discussed in Lafschetz [7]. Nakagiri's formulation will not yield our system. However, existence theory for our system is now readily obtained from his work. Here is a sketch.

The linearity in the delay yields a global solution of (31) and the form of (33) is well suited to contraction mappings. Indeed, the term involving $U_t(s)$ in (33) drops out in the contraction mapping argument and modification of the Nakagiri scheme to fit a linear form of (15a) can be seen by considering the system

$$\begin{cases} u_t = u_{xx} + b(x)f(\sigma) \\ \sigma' = \int_0^1 c(x)u(t,x)dx - rf(\sigma), \end{cases} \tag{35}$$

$$u(t,0) = u(t,1) = 0, \tag{36}$$

$$u(0,x) = g^0(x), \sigma(0) = \sigma_0. \tag{37}$$

Write

$$u_t = u_{xx}, u(t,0) = u(t,1) = 0 \tag{38}$$

as

$$u'(t) + A(u(t)) = 0 \tag{39}$$

and then write the system as

$$\begin{cases} u(t) = e^{-At}g^0(x) + \int_0^t e^{-A(t-s)}b(x)f(\sigma(s))ds \\ \sigma(t) = \sigma_0 + \int_0^t [\int_0^1 c(x)u(s,x)dx - rf(\sigma(s))]ds. \end{cases} \tag{40}$$

When $g^0(x) \in L^2(0,1)$, then $e^{-At}g^0(x) \in D(A)$ for $t > 0$; and when $b(x)f(\sigma(t))$ is Hölder continuous and locally integrable, then the integral in the first equation of (40) is in the $D(A)$ (c.f., Henry [5; p. 50]). If f satisfies a local Lipschitz condition, then (40) will define a contraction mapping with a unique fixed point which will then solve (35).

5. Nakagiri's Equations

We turn now to

$$u_t = \Delta u + \int_{-h}^0 d\eta(s,X)u(t+s,X)ds + f(\sigma(t))c(X), \tag{41}$$

$$\sigma(t) = \int_\Omega d(X)u(t,X)dX, \tag{42}$$

$$u(t,X) = 0 \text{ on } \partial\Omega \text{ and } d(X) = 0 \text{ on } \partial\Omega. \tag{43}$$

Following the lead of Lefschetz [7; pp. 39-40] for systems of ODEs, we write

$$\begin{aligned} \sigma'(t) &= \int_\Omega d(X)u_t(t,X)dX \\ &= \int_\Omega d(X)[\Delta u + \int_{-h}^0 d\eta(s,X)u(t+s,X)ds + f(\sigma(t))c(X)]dX \\ &= \int_\Omega d(X)\Delta u dX + \int_\Omega d(X)\int_{-h}^0 d\eta(s,X)u(t+s,X)dsdX \\ &\quad + \int_\Omega d(X)c(X)dX f(\sigma). \end{aligned}$$

Existence results were stated with (33) and (34). Using the divergence theorem we suppose there is a function $b(X)$ with $\int_\Omega d(X)\Delta u dX = \int_\Omega b(X) \cdot \nabla u dX$ (that is, $b(X) = -\nabla d(X)$) and we ask that

$$\int_\Omega d(X)c(X)dX = -r < 0, b \in L^2, c \text{ and } b \text{ are bounded}. \tag{44}$$

Our work is based on construction of Liapunov functionals and different ones must be used if the delay is discrete. For brevity, then, we replace the delay term by

$$\int_{-h}^0 d\eta(s, X)u(t+s, X)ds \to \int_{t-h}^t p(s, X)u(s, X)ds \tag{45}$$

where p is continuous. Our main concern here is to show absolute stability without strengthening (*).

Our system will now be

$$\begin{cases} u_t = \Delta u + \int_{t-h}^t p(s, X)u(s, X)ds + c(X)f(\sigma) \\ \sigma' = \int_\Omega b(X) \cdot \nabla u dX + \int_\Omega d(X) \int_{t-h}^t p(s, X)u(s, X)dsdX - rf(\sigma), \end{cases} \tag{46}$$

$$u(t, X) = 0 \text{ on } \partial\Omega. \tag{47}$$

Theorem 2. Let λ_1 be the first eigenvalue of $-\Delta$ on H_0^1, α and β positive, $\alpha + \beta = 1$, $k > 1$,

$$\beta\lambda_1 - (h/2) - khp^2(t, X) \geq \gamma > 0, \tag{48}$$

$$-4\int_\Omega [d(X)c(X) + (h/2)d^2(X)]dX > \int_\Omega [(|b(X)|^2/\alpha) + (c^2(X)/\gamma)]dX + \mu \tag{49}$$

for some $\mu > 0$ and let (44) hold. If (u, σ) is any solution of (46) - (47) on $[0, \infty)$, then

$$|\sigma(t)| + \int_\Omega u^2(t, X)dX \to 0 \text{ as } t \to \infty.$$

Proof. The analog of (21) is

$$V_1(t) = \int_\Omega [(\frac{1}{2})u^2(t, X) + k\int_{-h}^0 \int_{t+s}^t p^2(v, X)u^2(v, X)dvds]dX + \int_0^\sigma f(s)ds$$

and its derivative along a solution of (46) satisfies

$$V_1'(t) = \int_\Omega [u\{\Delta u + \int_{t-h}^t p(s, X)u(s, X)ds + c(X)f(\sigma)\} + f(\sigma)b(X) \cdot \nabla u + f(\sigma)d(X)\int_{t-h}^t p(s, X)u(s, X)ds]dX$$

$$-rf^2(\sigma) + kh\int_\Omega p^2(t,X)u^2(t,X)dX$$
$$-k\int_\Omega \int_{t-h}^t p^2(s,X)u^2(s,X)ds\,dX$$
$$\leq \int_\Omega \{-|\nabla u|^2 + (h/2)u^2 + (\frac{1}{2})\int_{t-h}^t p^2(s,X)u^2(s,X)ds$$
$$+c(X)uf(\sigma) + f(\sigma)b(X)\cdot\nabla u + (h/2)f^2(\sigma)d^2(X)$$
$$+(\frac{1}{2})\int_{t-h}^t p^2(s,X)u^2(s,X)ds + khp^2(t,X)u^2(t,X)$$
$$-k\int_{t-h}^t p^2(s,X)u^2(s,X)ds\}dX - rf^2(\sigma)$$

(using the divergence theorem)

$$\leq \int_\Omega \{-\alpha|\nabla u|^2 + (h/2)u^2 - \beta\lambda_1 u^2 + c(X)uf(\sigma) + khp^2(t,X)u^2$$
$$+f(\sigma)b(X)\cdot\nabla u - (k-1)\int_{t-h}^t p^2(s,X)u^2(s,X)ds\}dX$$
$$-[r-(h/2)\int_\Omega d^2(X)dX]f^2(\sigma)$$

(using (48))

$$\leq \int_\Omega \{-\alpha|\nabla u|^2 - \gamma u^2 + c(X)uf(\sigma) + |f(\sigma)||b(X)||\nabla u|$$
$$-(k-1)\int_{t-h}^t p^2(s,X)u^2(s,X)ds\}dX$$
$$-[r-(h/2)\int_\Omega d^2(X)dX]f^2(\sigma)$$
$$\leq \int_\Omega \{-\alpha[|\nabla u|^2 - |f(\sigma)|(|b(X)|/\alpha)|\nabla u|$$
$$+[f^2(\sigma)/4\alpha^2]|b(X)|^2]$$
$$-\gamma[u^2 - (c(X)/\gamma)f(\sigma)u + (c^2(X)/4\gamma^2)f^2(\sigma)]$$
$$-(k-1)\int_{t-h}^t p^2(s,X)u^2(s,X)ds$$
$$-[r-(h/2)\int_\Omega d^2(X)dX - (|b(X)|^2/4\alpha)$$
$$-(c^2(X)/4\gamma)]f^2(\sigma)\}dX.$$

Using (44) and (49), we can find a $\bar{\mu} > 0$ with

$$V_1'(t) \leq -\bar{\mu}\int_\Omega [u^2 + |\nabla u|^2 + f^2(\sigma) + \int_{t-h}^t p^2(s,X)u^2(s,X)ds]dX \tag{50}$$

The identical argument as was given in the proof of Theorem 1 will show that $V_1(t) \to 0$ as $t \to \infty$ and we complete the proof in the same way.

Remark. Theorem 2 does not seem to be comparable to that of Nakagiri. His η is more general than ours and his stability conditions and techniques are different. On the other hand, our class of admissible functions $f(\sigma)$ is much larger than his and includes all those for the ODE counterpart.

References

[1] Burton, T. A., A general stability result of Marachov type, in preparation.

[2] _____, Liapunov functions and boundedness for differential and delay equations, Hiroshima Math. J. 18 (1988), 341-350.

[3] _____, A differential equation of Lurie type, Proc. Amer. Math. Soc. 36 (1972), 491-496.

[4] Curtain, R. F. and Pritchard, A. J., Infinite Dimensional Linear Systems Theory, Springer, New York, 1978.

[5] Henry, D., Geometric Theory of Semilinear Parabolic Equations, Springer, Berlin, 1989.

[6] LaSalle, J. P. Complete stability of a nonlinear control system, Proc. Nat. Acad. Sci. U.S.A. 48 (1962), 600-603.

[7] Lefschetz, S., Stability of Nonlinear Control Systems, Academic Press, Orlando, 1965.

[8] Nakagiri, S., Absolute stability of nonlinear control systems with time delay, in Functional Differential Equations (T. Yoshizawa and J. Kato, eds.) World Scientific, New Jersey (1990), 274-278.

[9] _____, Structural properties of functional differential equations in Banach spaces, Osaka J. Math. 25 (1988), 353-398.

[10] _____, Optimal control of linear retarded systems in Banach spaces, J. Math. Anal. Appl. 120 (1986), 169-210.

[11] Simpson, H. C. and Spector, S. J., On the positivity of the second variation in finite elasticity, Arch. Rat. Mech. Anal. 98 (1987), 1-30.

[12] Somolinos, A., On the condition $c^T A^{-1} b + r > 0$ in the Lurie problem, Proc. Amer. Math. Soc. 47 (1975), 432-441.

[13] _____, Stability of Lurie-type functional equations, J. Differential Equations 26 (1977), 191-199.

T. A. Burton
Department of Mathematics
Southern Illinois University
Carbondale, Illinois 62901

C Y CHAN[1] AND D T FUNG[2]
Quenching for coupled parabolic problems

1. Introduction.
Let
$$H = \frac{\partial^2}{\partial x^2} - \frac{\partial}{\partial t},$$
$$\Omega = (0, a) \times (0, T),$$
$$\Gamma = ([0, a] \times \{0\}) \cup (\{0\} \times (0, T)),$$
$$S = \{a\} \times (0, T),$$

where $T \leq \infty$. Kawarada [13] introduced the concept of quenching through the study of the following first initial-boundary value problem,

$$H\mu = -(1-\mu)^{-1} \text{ in } \Omega, \ \mu = 0 \text{ on } \Gamma \cup S. \tag{1.1}$$

The solution μ is said to quench if there exists a finite time T such that

$$\sup\{|\mu_t(x, t)| : 0 \leq x \leq a\} \to \infty \text{ as } t \to T^-. \tag{1.2}$$

In such a case, T is called the quenching time. Since the solution μ of the problem (1.1) is strictly increasing with respect to t for each $x \in (0, a)$ (cf. Acker and Walter [2]), a necessary condition for quenching is

$$\max\{\mu(x, t) : 0 \leq x \leq a\} \to 1^- \text{ as } t \to T^-. \tag{1.3}$$

Kawarada claimed that (1.3) implied (1.2), and hence the two conditions were equivalent. Chan and Kwong [9] pointed out that there was a gap in Kawarada's proof. Using methods different from his, they established his claim for more general semilinear problems:

$$H\mu = -f(\mu) \text{ in } \Omega, \ \mu = 0 \text{ on } \Gamma \cup S, \tag{1.4}$$

$$H\mu = -g(\mu, \mu_x) \text{ in } \Omega, \ \mu = 0 \text{ on } \Gamma \cup S, \tag{1.5}$$

where $f(\mu)$ tends to infinity as μ approaches to some positive constant c, and $g(\mu, \nu)$ tends to infinity uniformly over all ν in any bounded interval as μ approaches c. Chan and Kaper [8] established the claim for more general parabolic singular operator; they studied the problem:

$$L\mu = -f(\mu) \text{ in } \Omega, \ \mu = 0 \text{ on } \Gamma, \ \mu_x = 0 \text{ on } S, \tag{1.6}$$

---
[1] Department of Mathematics, University of Southwestern Louisiana, Lafayette, LA 70504, U.S.A.
[2] Department of Mathematics, Southeastern Louisiana University, Hammond, LA 70402, U.S.A

where $L\mu \equiv H\mu + b\mu_x/x$ with b a constant less than 1. This includes the problem (1.1) as a special case since the solution of the problem (1.1) is symmetric with respect to the line $x = a/2$. We refer to the papers of Chan and Chen [4] and Chan and Kaper [8] for the significance of the expression $L\mu$. Acker and Kawohl [1] proved Kawarada's claim for the radially symmetric solution of the multidimensional semilinear heat equation, subject to the zero boundary condition and a nonnegative, radially symmetric and nonincreasing initial condition. Such a solution satisfies the singular equation

$$L\mu = -f(\mu) \text{ in } \Omega, \tag{1.7}$$

but with b a positive integer and $\mu_x(0, t) = 0$ for $0 < t < T$. By using an argument based on modifications of that of Friedman and McLeod [12] for blow-up problems, Kawarada's claim was proved by Deng and Levine [11] for a multidimensional semilinear heat equation in a bounded convex domain with a smooth boundary.

Another direction motivated by Kawarada [13] is the study of the critical length a^*, which is the length such that the solution exists globally for $a < a^*$ and quenching occurs for $a > a^*$. He showed that quenching occurs for the problem (1.1) when $a > 2^{3/2}$. Acker and Walter [2] showed that $a^* = 1.5303$ (to 5 significant figures). Furthermore, Acker and Walter [2; 3] proved that there exists a unique critical length a^* for each of the problems (1.4) and (1.5) respectively. Under weaker hypotheses, Chan and Kwong [10] extended the above result to the equation, $H\mu = -h(x, \mu, \mu_x)$. Results on the behavior of the solution of the problem (1.4) with $a = a^*$ were given by Levine and Montgomery [14]. Existence of the critical length a^* and its determination by computational methods were given by Chan and Chen [4] for the problem:

$$L\mu = -(1-\mu)^{-1} \text{ in } \Omega, \ \mu = 0 \text{ on } \Gamma \cup S,$$

with b a constant less than 1. Here, the singular term $b\mu_x/x$ destroys the symmetry of the solution μ about the line $x = a/2$, and shifts the points where μ attains its maxima with respect to x from the line $x = a/2$. Thus, it makes the problem more difficult both theoretically and numerically than those such as problems (1.1) and (1.4). Similar results were given by Chan and Kaper [8] for the problem (1.6), and by Chan and Cobb [6] for the problem (1.7) with b a constant less than 1, subject to

$$\mu = 0 \text{ on } \Gamma, \ \mu_x(a, t) + \delta\mu(a, t) = 0 \text{ on } S,$$

where δ is a positive constant. So far for a coupled parabolic system, only the critical length for global existence of the solution for a system of two equations subject to zero initial-boundary data has been studied by Chan and Chen [5]; no result has been given on the critical length beyond which quenching occurs, and on the blow-up of the time-derivative of the solution.

In the sequel, let $\boldsymbol{u}$ and $\boldsymbol{f}(\boldsymbol{u})$ denote, respectively, the transpose of the vectors (u_1, u_2) and $(f_1(u_1, u_2), f_2(u_1, u_2))$. A solution $\boldsymbol{u}$ is said to quench if there exists a finite time T such that

$$\sup\{|\boldsymbol{u}_t(x, t)| : 0 \le x \le a\} \to \infty \text{ as } t \to T^-.$$

The main purpose here is to study the blow-up of the time-derivative of the solution and the critical length for the following problem,

$$Hu = -f(u) \text{ in } \Omega, \ u = 0 \text{ on } \Gamma \cup S. \tag{1.8}$$

As in the previous papers by Chan and Chen [4, 5], Chan and Cobb [6], and Chan and Kaper [8], we assume that the problem (1.8) has a classical solution before its quenching time. In Section 2, we state without proofs our main results. Their proofs are given in [7]. An example is also given here.

2. Main results.
In the sequel, we assume the following conditions:
(i) $f(0) > 0$.
(ii) There exists a positive constant column vector q with $q_1 \geq q_2$ such that $f(w)$ is of class C^2 in $(-\infty, q_1) \times (-\infty, q_2)$, and $\lim_{w_2 \to q_2} f_1(w) = \infty$.
(iii) to any ϵ with $0 < \epsilon < q_2$, there exists a constant C such that

$$f_2(w) - f_1(w) \geq C(w_2 - w_1) \text{ for } w_i \leq q_i - \epsilon.$$

(iv) the first and second partial derivatives of $f(w)$ are nonnegative, and the first partial derivatives are bounded for $w_i \leq q_i - \epsilon$.

The following result gives the blow-up of the time-derivative of the solution.

Theorem 1. (a) *There exists at most one solution u. If u quenches at some finite time T, then the only quenching point is at $x = a/2$.*
(b) *Suppose $u_2(a/2, t) \to q_2$ as $t \to T^-$ for some finite time T, and there exists a function $p_1(w)$ in $C[0, q_2] \cap C^2(0, q_2)$ such that $p_1 > 0$, $p_1' \leq 0$, $p_1'' \leq 0$, $p_1(q_2) = 0$, and*

$$\frac{\partial(p_1(u_2) f_1(u_1, u_2))}{\partial u_2} \geq 0 \text{ for } (u_1, u_2) \in (0, q_1) \times (0, q_2).$$

Case 1. If $u_1 \equiv u_2$ in Ω, then $u_t(a/2, t) \to \infty$ as $t \to T^-$.
Case 2. Suppose there exists $p_2(w)$ in $C[0, q_1] \cap C^2(0, q_1)$ such that $p_2 > 0$, $p_2' \leq 0$, $p_2'' \leq 0$, and

$$\frac{\partial(p_2(u_1) f_2(u_1, u_2))}{\partial u_1} \geq 0 \text{ for } (u_1, u_2) \in (0, q_1) \times (0, q_2).$$

Then, $u_{1_t}(a/2, t) \to \infty$ as $t \to T^-$. If, in addition, $p_2(q_1) = 0$, and $u_1(a/2, t) \to q_1$ as $t \to T^-$, then $u_t(a/2, t) \to \infty$ as $t \to T^-$.

For illustration, let us consider the following example. In the problem (1.8), let

$$f_1(u_1, u_2) = (1 - u_2)^{-\beta} + k_1 u_1 \tag{2.1}$$

for some nonnegative number k_1 with β a number greater than or equal to 1. We may take $p_1(u_2) = 1 - u_2^n$, where n is a positive integer such that $n \geq 2(k_1 + 1)$. Then,

$$\frac{\partial}{\partial u_2}[p_1(u_2)f_1(u_1,u_2)]$$
$$= (1-u_2)^{-\beta}\left[\beta(1+u_2+u_2^2+\cdots+u_2^{n-1})-nu_2^{n-1}\right]-nk_1u_1u_2^{n-1}$$
$$\geq (1-u_2)^{-1}(1+u_2+u_2^2+\cdots+u_2^{n-1}-nu_2^{n-1})-nk_1u_1u_2^{n-1}$$
$$= 1+2u_2+3u_2^2+\cdots+(n-1)u_2^{n-2}-nk_1u_1u_2^{n-1}$$
$$> \frac{n}{2}u_1u_2^{n-1}(n-1-2k_1)$$
$$> 0 \quad \text{since } 0 < u < 1.$$

If $u_1 \equiv u_2$, then we have Case 1.
To illustrate Case 2, let

$$f_2(u_1, u_2) = (1-u_1)^{-\gamma} + k_2 u_2 \tag{2.2}$$

for some nonnegative number k_2 such that $k_1 < k_2$ with γ a number greater than or equal to β. Here, we may take $p_2(u_1) = 1 - u_1^m$, where m is a positive integer such that $m \geq 2(k_2+1)$. Then, $u_{1_t}(a/2, t) \to \infty$ as $t \to T^-$. If $\beta = \gamma$ and $k_1 = k_2$, then it follows from the maximum principle that $u_1 \equiv u_2$. This illustrates the second part of Case 2 (as well as Case 1) that $\boldsymbol{u}_t(a/2, t) \to \infty$ as $t \to T^-$. We note that the special case of the forcing terms given by (2.1) and (2.2) with $\beta = 1 = \gamma$ was considered by Chan and Chen [5] in illustrating the computation of the critical length for global existence of the solution.

The steady state of the problem (1.8) is given by

$$\boldsymbol{U}'' = -\boldsymbol{f}(\boldsymbol{U}) \text{ for } 0 < x < a, \ \boldsymbol{U}(0) = \boldsymbol{0} = \boldsymbol{U}(a). \tag{2.3}$$

Theorem 2. *(a) If $u_2 \leq q_2 - \epsilon$, then the problem (2.3) has a solution $\boldsymbol{U}$, to which $\boldsymbol{u}$ converges componentwise uniformly from below as t tends to infinity.*
(b) Let $\boldsymbol{u}(x, t; a)$ denote a solution of the problem (1.8). For any positive number η, $\boldsymbol{u}(x,t;a) < \boldsymbol{u}(x,t;a+\eta)$ in Ω.
(c) If $\lim_{t\to\infty} u_2(a/2, t; a) = q_2$, then there exists a finite time T such that

$$\lim_{t\to T^-} u_2((a+\eta)/2, t; a+\eta) = q_2.$$

Theorem 2 (a) and (b) implies that there exists a critical length a^* such that u exists for all $t > 0$ if $a < a^*$. The critical length a^* is determined as the supremum of all positive values a for which a solution of the problem (2.3) exists. Theorem 2(c) implies that quenching occurs for $a > a^*$.

References

[1] A. Acker and B. Kawohl, *Remarks on quenching*, Nonlinear Anal. **13** (1989), 53-61.
[2] A. Acker and W. Walter, *The quenching problem for nonlinear parabolic differential equations*, Lecture Notes in Math. **564** (1976), Springer-Verlag, New York, 1-12.

[3] _____, *On the global existence of solutions of parabolic differential equations with a singular nonlinear term*, Nonlinear Anal. **2** (1978), 499-505.
[4] C. Y. Chan and C. S. Chen, *A numerical method for semilinear singular parabolic quenching problems*, Quart. Appl. Math. **47** (1989), 45-57.
[5] _____, *Critical lengths for global existence of solutions for coupled semilinear singular parabolic problems*, Quart. Appl. Math. **47** (1989), 661-671.
[6] C. Y. Chan and S. S. Cobb, *Critical lengths for semilinear singular parabolic mixed boundary-value problems*, Quart. Appl. Math. **49** (1991), 497-506.
[7] C. Y. Chan and David T. Fung, *Quenching for coupled semilinear reaction-diffusion problems* (to appear).
[8] C. Y. Chan and H. G. Kaper, *Quenching for semilinear singular parabolic problems*, SIAM J. Math. Anal. **20** (1989), 558-566.
[9] C. Y. Chan and M. K. Kwong, *Quenching phenomena for singular nonlinear parabolic equations*, Nonlinear Anal. **12** (1988), 1377-1383.
[10] _____, *Existence results of steady-states of semilinear reaction-diffusion equations and their applications*, J. Differential Equations **77** (1989), 304-321.
[11] K. Deng and H. A. Levine, *On the blowup of u_t at quenching*, Proc. Amer. Math. Soc. **106** (1989), 1049-1056.
[12] A. Friedman and B. McLeod, *Blow-up of positive solutions of semilinear heat equations*, Indiana Univ. Math. J. **34** (1985), 425-447.
[13] H. Kawarada, *On solutions of initial-boundary problem for $u_t = u_{xx} + 1/(1-u)$*, Publ. Res. Inst. Math. Sci. **10** (1975), 729-736.
[14] H. A. Levine and J. T. Montgomery, *The quenching of solutions of some nonlinear parabolic equations*, SIAM J. Math. Anal. **11** (1980), 842-847.

C Y CHAN[†] AND L KE[†]
Existence of solutions and critical lengths for time-periodic semilinear parabolic problems

1. Introduction

Since the introduction of the concept of quenching in 1975 by Kawarada [14], only initial boundary-value problems have been studied. Further references can be found through the work of Acker and Kawohl [1], Chan and Chen [4, 5], Chan and Cobb [6], Chan and Kaper [7], Chan and Kwong [8, 9], Deng [12], and Deng and Levine [13]. The main purpose here is to study quenching phenomena for time-periodic solutions of the following semilinear parabolic problem:

$$Hu = -g(t)f(u) \text{ in } \Omega, \quad u(0,t) = 0 = u(a,t), \tag{1.1}$$

where $Hu = u_{xx} - u_t$, $\Omega \equiv (0,a) \times (-\infty, \infty)$, $g(t)$ is Hölder continuous, nontrivial, nonnegative and T-periodic, and $f(u) > 0$ and $f'(u) \geq 0$ for $-\infty < u < c$ for some positive number c with $\lim_{u \to c^-} f(u) = \infty$. Since a time-periodic solution u either exists for all time t or does not exist, there does not exist a finite time τ such that

$$\sup\{|u_t(x,t)| : 0 \leq x \leq a\} \to \infty \text{ as } t \to \tau^-.$$

A necessary consequence is that there does not exist a finite time τ such that a time-periodic solution reaches c. This is in sharp contrast with that for a parabolic initial-boundary value problem. It is the main purpose here to show that with some modification, the concept of critical length can be carried over from initial-boundary value problems to time-periodic problems although their proofs are quite different. The length a^* is the critical length for the problem (1.1) if a time-periodic solution u (of period T) exists for $a < a^*$, and the problem has no T-periodic solution for $a > a^*$.

2. Critical Length

Let Ω^- denote the closure of Ω. For ease of reference, let us state below the result, which follows from Lemma 4 of Chan and Wong [11].

Lemma 1. Let $u(x,t) \in C^{2,1}(\Omega) \cap C(\Omega^-)$. If $Hu \leq 0$ in Ω, $u(0,t) \geq 0$ and $u(a,t) \geq 0$ for all t, and $u(x,t) = u(x, t+T)$ for $0 < x < a$, then u is nonnegative on Ω^-. Furthermore, if one of the above inequalities is not a strict equality, then u is positive in Ω.

Let $M = \max_{0 \leq t \leq T} g(t)$, and U denote a solution of the nonlinear two-point boundary value problem:

$$U_{xx} = -Mf(U) \text{ for } 0 < x < a, \quad U(0) = 0 = U(a). \tag{2.1}$$

We have the following existence result.

[†]Department of Mathematics, University of Southwestern Louisiana, Lafayette, Louisiana 70504-1010, U.S.A.

Theorem 2. *If the problem (2.1) has a solution which is positive for $0 < x < a$, then the problem (1.1) has at least one T-periodic solution u, and $u > 0$ in Ω.*

Proof. Let us construct a sequence $\{u_n\}$ as follows: $u_0 \equiv 0$, and for $n = 1, 2, 3, \cdots$,

$$Hu_n = -g(t)f(u_{n-1}) \text{ in } \Omega,$$

$$u_n(0,t) = 0 = u_n(a,t), \; u_n(x,t) = u_n(x, t+T).$$

Since $U > 0$ for $0 < x < a$, and $f' \geq 0$, it follows that

$$H(U - u_1) = -g(t)[f(U) - f(0)] \leq 0 \text{ in } \Omega.$$

By Lemma 1, $U(x) \geq u_1(x,t)$ in Ω. By using mathematical induction and Lemma 1, we have $U(x) \geq u_n(x,t)$ in Ω for $n = 1, 2, 3, \cdots$.

From Chan and Wong [10],

$$u_n(x,t) = \int_0^T \int_0^a G(x,t;\xi,\tau)g(\tau)f(u_{n-1}(\xi,\tau))d\xi d\tau, \qquad (2.2)$$

where the T-periodic Green's function $G(x,t;\xi,\tau)$ is given by

$$\frac{2}{a}\sum_{n=1}^\infty E_n(t-\tau)\sin\frac{n\pi}{a}x \sin\frac{n\pi}{a}\xi \text{ for } 0 < t-\tau < T,$$

$$\frac{2}{a}\sum_{n=1}^\infty E_n(T+t-\tau)\sin\frac{n\pi}{a}x \sin\frac{n\pi}{a}\xi \text{ for } -T < t-\tau < 0;$$

here

$$E_n(t) = \exp\left[-\left(\frac{n\pi}{a}\right)^2 t\right] / \left\{1 - \exp\left[-\left(\frac{n\pi}{a}\right)^2 T\right]\right\}.$$

By Lemma 1, $u_1 > 0$ in Ω. Since $f' \geq 0$, one can use mathematical induction and Lemma 1 to conclude $u_{n+1} \geq u_n$ in Ω for $n = 1, 2, 3, \cdots$.

The monotone sequence $\{u_n\}$ is bounded above by U. Let $\underline{u}(x,t)$ denote its limit $\lim_{n\to\infty} u_n(x,t)$. Then, $\underline{u}$ is T-periodic, and positive in Ω. A proof similar to that of Theorem 3.1 on convergence to a solution by Bange [3] shows that $\underline{u}$ is a solution of the problem (1.1).

We also use the notation $u(x,t;a)$ to denote a solution of the problem (1.1).

Lemma 3. *The limit $\underline{u}$ is the minimal T-periodic solution of the problem (1.1). If $a_1 > a$, then $\underline{u}(x,t;a_1) > \underline{u}(x,t;a)$.*

Proof. By Lemma 1, any solution u of the problem (1.1) is positive in Ω. Since

$$H(u - u_1) = -g(t)[f(u) - f(0)] \leq 0 \text{ in } \Omega,$$

it follows from Lemma 1 that $u \geq u_1$. Using $f' \geq 0$ and mathematical induction, we have $u \geq u_n$ for $n = 1, 2, 3, \cdots$, and hence $u \geq \underline{u}$.

Since
$$u_1(a, t; a_1) - u_1(a, t; a) = u_1(a, t; a_1) > 0,$$
it follows from Lemma 1 that
$$u_1(x, t; a_1) > u_1(x, t; a) \text{ in } \Omega.$$

Using $f' \geq 0$ and mathematical induction, we have
$$u_n(x, t; a_1) > u_n(x, t; a) \text{ in } \Omega.$$

Hence,
$$\underline{u}(x, t; a_1) \geq \underline{u}(x, t; a) \text{ in } \Omega.$$

This gives
$$H(\underline{u}(x, t; a_1) - \underline{u}(x, t; a)) \leq 0 \text{ in } \Omega.$$

Since
$$\underline{u}(a, t; a_1) - \underline{u}(a, t; a) = \underline{u}(a, t; a_1) > 0,$$
it follows from Lemma 1 that
$$\underline{u}(x, t; a_1) > \underline{u}(x, t; a) \text{ in } \Omega.$$

Theorem 4. *The problem (1.1) has a unique critical length a^*.*

Proof. By Amann [2], the problem (2.1) has a solution U such that $U > 0$ for $0 < x < a$, provided that a is sufficiently small. Thus by Theorem 2, the problem (1.1) has a T-periodic solution.

By using Lemma 3, when the problem (1.1) has a solution for $a = a_1$, any problem (1.1) with $a = \alpha < a_1$ has a solution. Hence if the theorem is false, then given any length a, the problem (1.1) has a T-periodic solution u. Let
$$m(x) = T^{-1} \int_0^T u(x, t) dt.$$

Then, $m(x) < c$. Integrating the differential equation in (1.1) with respect to t from 0 to T, we have
$$m_{xx} = -T^{-1} \int_0^T g(t) f(u(x, t)) dt.$$

It follows from $u > 0$ in Ω, $f'(u) \geq 0$, and $g(t) \geq 0$ that
$$m_{xx} \leq -\frac{f(0)}{T} \int_0^T g(t) dt.$$

Since $g(t)$ is nontrivial, the right-hand side is a negative constant. Let us denote this by $-\epsilon$. Then, $m_{xx} \leq -\epsilon$. From the boundary condition, $m(0) = 0 = m(a)$. By the maximum principle, a lower bound l of m is given by

$$l_{xx} = -\epsilon, \quad l(0) = 0 = l(a).$$

By direct calculation, $l(x) = \epsilon x(a-x)/2$. Now,

$$\max_{0 \leq x \leq a} m(x) \geq \max_{0 \leq x \leq a} l(x) = \epsilon a^2/8, \tag{2.3}$$

which tends to infinity as a approaches infinity. This contradiction proves the theorem.

We remark that the above construction of l can be used to find an upper bound of a^*. As an illustration, let

$$g(t)f(u) = \frac{\sin^2 t}{1-u}.$$

Thus, $c = 1$, $T = \pi$ and $T^{-1} \int_0^T g(t) dt = 1/2$. Hence, $\epsilon = 1/2$. From (2.3), $\epsilon a^2/8 \geq 1$ implies $a \geq 4$. That is, $a^* < 4$.

3. Numerical example

Let

$$\omega_- = (0, \frac{a}{2}) \times (-\infty, \infty), \text{ and } \omega_+ = (\frac{a}{2}, a) \times (-\infty, \infty).$$

Lemma 5. In ω_-, $\partial u_n(x,t)/\partial x > 0$. In ω_+, $\partial u_n(x,t)/\partial x < 0$.

Proof. Let

$$w_n(x,t) = u_n(x,t) - u_n(2h-x,t) \text{ in } [h, 2h] \times (-\infty, \infty),$$

where h is a constant between 0 and $a/2$. Then,

$$Hw_n = -g(t)[f(u_{n-1}(x,t)) - f(u_{n-1}(2h-x,t))],$$

$$w_n(h,t) = 0, \quad w_n(2h,t) = u_n(2h,t) > 0,$$

$$w_n(x,t) = w_n(x, t+T).$$

Hence, $Hw_1 = 0$ in Ω. By Lemma 1, $w_1 > 0$ in $(h, 2h] \times (-\infty, \infty)$. By the parabolic version of Hopf's lemma, $\partial w_1(h,t)/\partial x > 0$. Hence, $2\partial u_1(h,t)/\partial x > 0$. By varying h from 0 to $a/2$, we obtain $\partial u_1(x,t)/\partial x > 0$ in ω_-. Similarly, $\partial u_1(x,t)/\partial x < 0$ in ω_+. The lemma then follows by using mathematical induction.

From Lemma 5,

$$\max_{0 \leq x \leq a} u_n(x,t) = u_n(\frac{a}{2}, t).$$

A procedure of computing a^* is as follows:

(I). Let $\bar{a}$ be an upper bound (determined by $l(x)$) of a^*. Since 0 is lower bound, we take the average $\bar{a}/2$ as a first approximation to a^*. The sequence $\{u_n\}$ is computed by using the representation formula (2.2).

(II). If $\max_{0\leq t\leq T} u_n(a/2, t) \geq c$, then the problem (1.1) has no solution. Using the method of bisection, we use $\bar{a}/4$ as the new approximation of a^*.

(III). If $\max_{0\leq t\leq T} u_n(a/2, t) < c$, and

$$\max_{\substack{0\leq x\leq a \\ 0\leq t\leq T}} [u_n(x,t) - u_{n-1}(x,t)] < \delta$$

for some pre-assigned tolerance error δ, then u exists and we replace the lower bound 0 of a^* by $\bar{a}/2$. We then use $(\bar{a}/2 + \bar{a})/2 = 3\bar{a}/4$ as the new approximation of a^*.

(IV). The above procedure is repeated until the difference between two successive approximations is within the desired accuracy. Then, their average may be taken as a^*.

For the above example with $g(t) = \sin^2 t$, and $f(u) = (1-u)^{-1}$, we obtain $a^* = 1.8746$ (to 4 decimal points).

REFERENCES

[1] A. Acker and B. Kawohl, *Remarks on quenching*, Nonlinear Anal. **13** (1989), 53-61.
[2] H. Amann, *Supersolutions, monotone iterations, and stability*, J. Differential Equations **21** (1976), 336-377.
[3] D. W. Bange, *Periodic solutions of a quasilinear parabolic differential equation*, J. Differential Equations **17** (1975), 61-72.
[4] C. Y. Chan and C. S. Chen, *A numerical method for semilinear singular parabolic quenching problems*, Quart. Appl. Math. **47** (1989), 45-57.
[5] _____, *Critical lengths for global existence of solutions for coupled semilinear singular parabolic problems*, Quart. Appl. Math. **47** (1989), 661-671.
[6] C. Y. Chan and S. S. Cobb, *Critical lengths for semilinear singular parabolic mixed boundary-value problems*, Quart. Appl. Math. **49** (1991), 497-506.
[7] C. Y. Chan and H. G. Kaper, *Quenching for semilinear singular parabolic problems*, SIAM J. Math. Anal. **20** (1989), 558-566.
[8] C. Y. Chan and M. K. Kwong, *Quenching phenomena for singular nonlinear parabolic equations*, Nonlinear Anal. **12** (1988), 1377-1383.
[9] _____, *Existence results of steady-states of semilinear reaction-diffusion equations and their applications*, J. Differential Equations **77** (1989), 304-321.
[10] C. Y. Chan and B. M. Wong, *Periodic solutions of parabolic problems*, Differential Equations and Applications (1989), Ohio University Press, Athens, Ohio, 142-148.
[11] _____, *Periodic solutions of singular linear and semilinear parabolic problems*, Quart. Appl. Math. **47** (1989), 405-428.
[12] K. Deng, *Quenching for solutions of a plasma type equation*, Nonlinear Anal. (to appear).
[13] K. Deng and H. A. Levine, *On the blowup of u_t at quenching*, Proc. Amer. Math. Soc. **106** (1989), 1049-1056.
[14] H. Kawarada, , *On solutions of initial-boundary problem for $u_t = u_{xx} + (1-u)^{-1}$*, Publ. Res. Inst. Math. Sci. **10** (1975), 729-736.

C Y CHAN[†] AND K K NIP[†]
Quenching for semilinear Euler–Poisson–Darboux equations

1. Introduction.

The concept for quenching was introduced in 1975 by Kawarada [3] through a first initial-boundary value problem for a semilinear heat equation. It was then studied by Chang and Levine [2] in 1981 for a first initial-boundary value problem for a semilinear wave equation. The fact that we do not have as useful a maximum principle for hyperbolic equations as for parabolic equations makes its study quite different. A numerical study on the behavior of the derivatives of solutions was done by Axtell [1]. The effect of nonlinear boundary conditions on the homogeneous wave equation was investigated by Levine [4], and Rammaha [5].

We would like to study the following initial-boundary value problem for the semilinear Euler-Poisson-Darboux equation:

$$u_{tt} + \frac{k}{t}u_t - u_{xx} = f(u) \text{ for } 0 < x < a,\ 0 < t < T, \tag{1.1}$$

$$u(x,0) = u_0(x),\ u_t(x,0) = 0 \text{ for } 0 \le x \le a, \tag{1.2}$$

$$u(0,t) = 0 = u(a,t) \text{ for } 0 < t < T. \tag{1.3}$$

Here, the constant $k \le 1$; $f\colon (-\infty, c) \to (0, \infty)$ for some positive constant c such that f is convex, $f' \ge 0$, and $f(u) \to \infty$ as $u \to c^-$; and $u_0(x)$ is nonnegative and integrable. We show that there exists a length a^* such that for $a \ge a^*$, u quenches (that is, u attains c somewhere) in a finite time T. The case when $k = 0$ and $u_0 \equiv 0$ was studied by Chang and Levine [2]. For illustration, we give an example.

2. Quenching.

The function u is said to be a weak solution of the problem (1.1)-(1.3) if
(i) $u(x,t)$ is continuous on $[0,a] \times [0,T)$ and satisfies (1.2) and (1.3).
(ii) $u_t(x,t)$ is continuous on $(0,a) \times [0,T)$.
(iii) for all $\Phi(x,t) \in C^1([0,a] \times [0,T)) \cap C^2((0,a) \times (0,T))$ satisfying (1.3) and

$$\int_0^T \int_0^a \frac{|\Phi(x,t)|}{t^2} dx\,dt + \int_0^T \int_0^a \frac{|\Phi_t(x,t)|}{t} dx\,dt < \infty, \tag{2.1}$$

we have

$$\int_0^a u_t \Phi\,dx + \frac{k}{t}\int_0^a u\Phi\,dx = \int_0^t \int_0^a \left[\Phi_\tau u_\tau + \Phi_{xx}u + \Phi f + \frac{k}{\tau}\Phi_\tau u - \frac{k}{\tau^2}\Phi u\right] dx\,d\tau. \tag{2.2}$$

[†]Department of Mathematics, University of Southwestern Louisiana, Lafayette, Louisiana 70504-1010, U.S.A.

We remark that (2.2) can be obtained as follows: After multiplying (1.1) by Φ and integrating over the domain $(0, a) \times (\sigma, t)$ for some positive number σ, it follows from Φ and u satisfying the boundary conditions that

$$\int_0^a u_t \Phi dx + \frac{k}{t} \int_0^a u \Phi dx - \int_0^a u_t(x,\sigma)\Phi(x,\sigma)dx - k \int_0^a u(x,\sigma)\frac{\Phi(x,\sigma)}{\sigma}dx$$
$$= \int_\sigma^t \int_0^a \left[\Phi_\tau u_\tau + \Phi_{xx} u + \Phi f + \frac{k}{\tau}\Phi_\tau u - \frac{k}{\tau^2}\Phi u\right] dx d\tau.$$

Using (2.1) and the continuity of $\Phi_t(x,0)$, we have $\Phi_t(x,0) = 0$ for $0 \le x \le a$. By letting $\sigma \to 0^+$, we obtain (2.2).

Let
$$F(t) = \frac{\pi}{2a} \int_0^a u(x,t)\psi(x)dx,$$

$$H(w) = -\frac{\lambda}{2}w^2 + \int_{F(0)}^w f(s)ds,$$

where $\psi(x) = \sin\sqrt{\lambda}x$ with $\lambda = (\pi/a)^2$.

Lemma 1. *If $H'(b) > 0$ for some positive number b ($< c$), and $H'(w) \ge 0$ for $b < w < c$, then*

$$\int_b^c [H(w) - H(b)]^{-1/2} dw < \infty.$$

Proof. Since $H'(b) = -\lambda b + f(b) > 0$, there exists a positive constant δ such that $-\lambda w + f(b) > 0$ for $b \le w \le b + \delta$. For $b < w \le b + \delta$,

$$H(w) - H(b) = -\frac{\lambda}{2}(w^2 - b^2) + \int_b^w f(s)ds$$
$$\ge (w-b)[-\lambda(b+\delta) + f(b)]$$
$$> 0.$$

Therefore,

$$\int_b^{b+\delta} [H(w) - H(b)]^{-1/2} dw \le \int_b^{b+\delta} \{(w-b)[-\lambda(b+\delta) + f(b)]\}^{-1/2} dw,$$

which is bounded. Since $H'(w) \ge 0$ for $b < w < c$, it follows that $H(w) - H(b)$ is bounded away from zero on $(b+\delta, c)$. Hence,

$$\int_{b+\delta}^c [H(w) - H(b)]^{-1/2} dw < \infty.$$

The lemma then follows.

Let
$$\alpha = \inf_{0 < w < c} \frac{f(w)}{w}.$$

Since f is positive, it follows that $\alpha > 0$.

Theorem 2. *Suppose $H'(F(0)) > 0$, and*

$$\lim_{t \to 0+} \left[t^k \int_0^a |u_t(x,t)| dx \right] = 0 \text{ if } k < 0.$$

Then for $a \geq \pi/\sqrt{\alpha}$, every weak solution u must quench in finite time.

Proof. Let us suppose $|u| < c$ for $(x,t) \in [0,a] \times [0,\infty)$. Let $\Phi(x,t) = t^2 \psi(x)$. The left-hand side of (2.1) is

$$3 \int_0^T \int_0^a |\sin \sqrt{\lambda} x| dx dt \leq 3aT,$$

which is bounded for a finite time T. From (2.2), we have for $0 < t < T$,

$$t^2 F'(t) + kt F(t) = \int_0^t [2\tau F'(\tau) + (k - \lambda \tau^2) F(\tau) + \frac{\pi \tau^2}{2a} \int_0^a \psi f(u) dx] d\tau. \quad (2.3)$$

The right-hand side is differentiable with respect to t, and hence the left-hand side is also differentiable. Differentiating (2.3), we have

$$F''(t) + \frac{k}{t} F'(t) = -\lambda F(t) + \frac{\pi}{2a} \int_0^a \psi f(u) dx.$$

By Jensen's inequality,

$$F''(t) + \frac{k}{t} F'(t) \geq -\lambda F(t) + f(F(t)).$$

That is,

$$F''(t) + \frac{k}{t} F'(t) \geq H'(F(t)). \quad (2.4)$$

By assumption, $H'(F(0)) > 0$. It follows from continuity that there exists some positive number σ ($< T$) such that $H'(F(t)) > 0$ for $0 \leq t < \sigma$. From (2.4),

$$\frac{d}{dt}[t^k F'(t)] \geq t^k H'(F(t)) > 0 \text{ for } 0 < t < \sigma,$$

and hence, $t^k F'(t)$ is strictly increasing for $0 < t < \sigma$.

For $0 \leq k \leq 1$, $F'(0) = 0$. For $k < 0$,

$$\lim_{t \to 0+} |t^k F'(t)| = \frac{\pi}{2a} \lim_{t \to 0+} |t^k \int_0^a u_t(x,t) \psi(x) dx|$$

$$\leq \frac{\pi}{2a} \lim_{t \to 0+} \left[t^k \int_0^a |u_t(x,t)| dx \right]$$

$$= 0.$$

Hence, $\lim_{t \to 0+} t^k F'(t) = 0$ for $k \leq 1$. Therefore for $k \leq 1$, $t^k F'(t) > 0$ for $0 < t < \sigma$.

We know that $F'(t) > 0$ on $(0, \sigma)$. Let us suppose that there exists a positive number $t_1 \in [\sigma, T]$ such that $F'(t_1) = 0$ and $F'(t) > 0$ on $(0, t_1)$. Since $F(0) \geq 0$, we have $F(t) > 0$ for $0 < t < t_1$. It follows from $\alpha \geq \lambda$ that for $0 < t < t_1$,

$$H'(F(t)) = -\lambda F(t) + f(F(t))$$
$$\geq -\alpha F(t) + f(F(t))$$
$$\geq 0.$$

From (2.4),

$$t^k F'(t) \frac{d}{dt}[t^k F'(t)] \geq t^{2k} H'(F(t))F'(t) \text{ for } 0 < t < t_1.$$

It follows from $H'(F(0)) > 0$ and $H'(F(t)) \geq 0$ for $0 < t < t_1$ that there exists $t_0 \in (0, \sigma)$ such that $H(w) > H(F(t_0))$ for $F(t_0) < w < F(t_1)$. Thus,

$$\int_{t_0}^{t} s^k F'(s) \frac{d}{ds}[s^k F'(s)] ds \geq \int_{t_0}^{t} s^{2k} H'(F(s))F'(s) ds \text{ for } t_0 < t < t_1. \tag{2.5}$$

For $0 < k \leq 1$,

$$\frac{1}{2}[t^k F'(t)]^2 \geq t_0^{2k}[H(F(t)) - H(F(t_0))] \text{ for } t_0 < t < t_1. \tag{2.6}$$

As $t \to t_1$, the left-hand side approaches zero while the right-hand side is positive. This contradiction shows that $F'(t) > 0$ for $0 < t < T$ and hence $F(t_0) < F(t) < c$ for $t_0 < t < T$. Since $F'(t) > 0$ for $t_0 < t < T$, it follows that (2.6) is true for $t_0 < t < T$, and hence,

$$\int_{t_0}^{t} [H(F(s)) - H(F(t_0))]^{-1/2} F'(s) ds \geq \sqrt{2} \int_{t_0}^{t} \left(\frac{t_0}{s}\right)^k ds.$$

That is, for $t_0 < t < T$,

$$\int_{F(t_0)}^{c} [H(w) - H(F(t_0))]^{-1/2} dw \geq \begin{cases} \dfrac{\sqrt{2}}{1-k} t_0^k (t^{1-k} - t_0^{1-k}) & \text{if } 0 < k < 1, \\ \sqrt{2} t_0 \ln\left(\dfrac{t}{t_0}\right) & \text{if } k = 1. \end{cases}$$

This contradicts Lemma 1 for sufficiently large T.

For $k \leq 0$, it follows from (2.5) that

$$\frac{1}{2}\left\{[t^k F'(t)]^2 - [t_0^k F'(t_0)]^2\right\} \geq t^{2k} \int_{t_0}^{t} H'(F(s))F'(s) ds \text{ for } t_0 < t < t_1.$$

Hence,

$$[F'(t)]^2 \geq 2[H(F(t)) - H(F(t_0))] \text{ for } t_0 < t < t_1. \tag{2.7}$$

An argument as before shows that $F'(t) > 0$ for $0 < t < T$, and $F(t_0) < F(t) < c$ for $t_0 < t < T$. Hence, (2.7) is true for $t_0 < t < T$. Therefore,

$$\int_{t_0}^{t} [H(F(s)) - H(F(t_0))]^{-1/2} F'(s) ds \geq \sqrt{2} \int_{t_0}^{t} ds.$$

That is,

$$\int_{F(t_0)}^{c} [H(w) - H(F(t_0))]^{-1/2} dw \geq \sqrt{2}(t - t_0) \text{ for } t_0 < t < T.$$

This contradicts Lemma 1 for sufficiently large T. Hence, u must quench in finite time.

For illustration, let us consider the problem (1.1)-(1.3) with $0 \leq k \leq 1$, $u(x, 0) = 0$ and $f(u) = (1-u)^{-\beta}$ where β is a positive number. Then, $F(0) = 0$ and $H'(0) = f(0) > 0$. It follows from Theorem 2 that every weak solution must quench in finite time for $a \geq \pi/\sqrt{\alpha}$, that is,

$$a \geq \pi \sqrt{\frac{\beta^\beta}{(\beta+1)^{\beta+1}}}.$$

REFERENCES

[1] J. Axtell, *A numerical study of the derivatives of solutions of the wave equation with a singular forcing term at quenching*, Numer. Methods Partial Differential Equations **1** (1989), 53-76.

[2] P. H. Chang and H. A. Levine, *The quenching of solutions of semilinear hyperbolic equations*, SIAM J. Math. Anal. **12** (1981), 893-903.

[3] H. Kawarada, *On solutions of initial-boundary problem for $u_t = u_{xx} + 1/(1-u)$*, Publ. Res. Inst. Math. Sci. **10** (1975), 729-736.

[4] H. A. Levine, *The quenching of solutions of linear parabolic and hyperbolic equations with nonlinear boundary conditions*, SIAM J. Math. Anal. **14** (1983), 1139-1153.

[5] M. A. Rammaha, *On the quenching of solutions of the wave equation with a nonlinear boundary condition*, J. Reine Angew Mat. **407** (1990), 1-18.

R ESTRADA
Series of Dirac delta functions

1. INTRODUCTION.
In this talk we discuss several properties of some very interesting mathematical objects, the series of Dirac delta functions of the type

$$\sum_{n=0}^{\infty} a_n \delta^{(n)}(x). \qquad (1)$$

Despite the fact that series as (1) diverge in the distributional sense unless only finitely many a_n do not vanish, they have been used as formal tools in several areas of applied mathematics. Krall, Kanwal and Littlejohn [11] studied orthogonal polynomials with the help of these series. Wiener [18] has found series of delta functions as solutions of differential and functional equations. Littlejohn and Kanwal [13] solved the hypergeometric differential equation in terms of such divergent series. Recently, many authors have studied the links between generalized functions and asymptotic analysis [15,17]. As Kanwal and the author [4] have shown, series of Dirac delta functions form the basic block in the asymptotic expansion of generalized functions.

We give a survey of these ideas and some other results on the connection of formal solutions of ordinary differential equations as series of delta functions and the quasiasymptotics of classical solutions. We also study some singular perturbation problems by using this perspective.

2. BASIC NOTIONS.
In this section we introduce the basic definitions needed to handle series of Dirac delta functions of the type $u(x) = \sum_{n=0}^{\infty} a_n \delta^{(n)}(x)$.

An important formula is the ortogonality relation

$$\langle \delta^{(n)}(x), x^m \rangle = \begin{cases} 0 & n \neq m, \\ (-1)^n n! & n = m. \end{cases} \qquad (2)$$

We can say that the sequences $\left\{\frac{x^n}{n!}\right\}$ and $\{(-1)^n \delta^{(n)}(x)\}$ are biortogonal or dual of one another.

In general,(1) diverges in the distributional sense: $u(x)$ is not a distribution. Actually, it is not hard to see that none of the standard summability methods produces a distribution out of the series $\sum_{n=0}^{\infty} a_n \delta^{(n)}(x)$ unless $a_n = 0$ for $n > N$. One could attemp to solve this problem by trying to interpret the series as ultradistributions or, more generally, as hyperfunctions. However, very soon one finds problems whose series of deltas solutions do not belong to any of these spaces. For instance, it follows from the general theory of hyperfunctions [7] that if $\limsup_{n \to \infty}(|a_n|n!)^{\frac{1}{n}} > 0$ then the series (1) does not define a hyperfunction concentrated at $\{0\}$. Naturally, any of these series can be considered as a functional in the space of polinomials, but it is not clear what the relationship of such functionals with classical analytical objects is. The term "dual Taylor series" can be used for them, since in a sense they are the "dual" to the Taylor series $\sum_{n=0}^{\infty} b_n x^n$. But again, this term does not say very much about what they are.

3. SEVERAL PROBLEMS THAT LEAD TO SERIES OF DELTAS.
We shall now mention several problems where dual Taylor series appear naturally.

Let us start with the *problem of moments*. Let $\{\mu_n\}$ be a sequence of real or complex numbers. The problem of finding a function $f(x)$ that satisfies

$$\langle f(x), x^n \rangle = \int_{-\infty}^{\infty} f(x) x^n \, dx = \mu_n \qquad (3)$$

is called the problem of moments. The numbers $\langle f(x), x^n \rangle$ are called the moments of $f(x)$. Classically (3) was studied under the restriction that $f(x)$ is a positive measure, supported in a given closed set [16]. Since the positivity of f is not needed in many situations, the problem of moments was eventually studied for signed measures and more recently in spaces of distributions and ultradistributions. Observe that the series of delta functions provide an inmediate solution to the problem of moments (3). In fact, using (2) it follows that

$$u(x) = \sum_{n=0}^{\infty} \frac{(-1)^n \mu_n \delta^{(n)}(x)}{n!} \tag{4}$$

is a (formal) solution of (3). Writing (4) was very simple and natural, but can a classical solution be obtained out of (4)? and if so, how?. Or put in somewhat different way, if $f(x)$ is a solution of (3) in the ordinary sense, what is the relationship between $f(x)$ and $u(x)$? In this connection, Kanwal and the author [3], when studying the moment problem in the space $\mathcal{E}'(\mathbb{R})$ of compactly supported distributions, observed that if $f \in \mathcal{E}'(\mathbb{R})$ is a solution of (3) then f and u are related in an asymptotic way, namely,

$$f(\lambda x) \sim u(\lambda x) = \sum_{n=0}^{\infty} \frac{(-1)^n \mu_n \delta^{(n)}(x)}{n! \lambda^{n+1}}, \text{ as } \lambda \to \infty. \tag{5}$$

This expansion, the *Moment Asymptotic Expansion* is the starting point in a rich theory for the asymptotic development of generalized functions [2,4]. Observe that this asymptotic approach permits us to work in the distributional framework, since (5) can be expressed equivalently as

$$f(\lambda x) = \sum_{n=0}^{N} \frac{(-1)^n \mu_n \delta^{(n)}(x)}{n! \lambda^{n+1}} + O(\frac{1}{\lambda^{N+2}}), \text{ as } \lambda \to \infty, \tag{6}$$

a relation that involves no infinite series. The moment asymptotic expansion does not hold in $\mathcal{D}'(\mathbb{R}^n)$ or $\mathcal{S}'(\mathbb{R}^n)$. However, it holds in several spaces of distributions, which show suitable decay at infinity. Among others, the moment asymptotic expansion holds in the following spaces: $\mathcal{E}'(\mathbb{R}^n), \mathcal{P}'(\mathbb{R}^n), \mathcal{O}'_M(\mathbb{R}^n), \mathcal{O}'_C(\mathbb{R}^n)$ y $\mathcal{K}'(\mathbb{R}^n)$.

The next problem we would like to mention is the *problem of weights for orthogonal polynomials*. Actually, this one of the first problems where series of Dirac delta functions have been used. Motivated by the failure of classical methods for the study of the weights for the Bessel polynomials [12], various authors [11,14] introduced the divergent series of delta functions in this problem.

Let $\{\mu_n\}$ be a sequence that satisfies the condition

$$\Delta_n = \det [\mu_{i+j}]_{i,j=0}^n \neq 0, n = 0, 1, 2, \ldots \tag{7}$$

Then a sequence of polynomials is introduced by putting $p_0 = 1$ and

$$p_n(x) = \frac{1}{\Delta_{n-1}} \begin{vmatrix} \mu_0 & \cdots & \mu_n \\ \vdots & \ddots & \vdots \\ \mu_{n-1} & \cdots & \mu_{2n-1} \\ 1 & \cdots & x^n \end{vmatrix}, n \geq 1. \tag{8}$$

The polynomials $p_n(x)$ are monic of degree n. The polynomials $\{p_n(x)\}$ can be made orthogonal by a weight $w(x)$ that satisfies $\langle w(x), p_n(x) p_m(x) \rangle = 0, n \neq m, \langle w(x), p_n^2(x) \rangle \neq 0$. Actually, if a dual Taylor series is allowed for the weight $w(x)$ then the solution is inmediate, namely,

$$w(x) = \sum_{n=0}^{\infty} \frac{(-1)^n \mu_n \delta^{(n)}(x)}{n!}. \tag{9}$$

The weight $w(x)$ can also be obtained by solving an ordinary differential equation [11]. If the polynomials satisfy a differential equation of the second order, it should be of the form $p(x)y''(x) + q(x)y'(x) = \lambda_n y(x)$ where $p(x) = ax^2 + bx + c$, $q(x) = dx + e$. If $w(x)$ is a solution of the equation

$$-(p(x)w(x))' + q(x)w(x) = 0 \qquad (10)$$

that vanishes rapidly at $\pm\infty$, then $w(x)$ is a weight for the polynomials $\{p_n(x)\}$.

EXAMPLE For Bessel polynomials the equation is $x^2 y'' + (2x+2)y' + n(n+1)y = 0$. The equation for the weight is $x^2 w' - 2w = 0$, but the classical solution $e^{-\frac{2}{x}}$, $x \neq 0$, cannot be regularized at $x = 0$ to give a distribution. Furthermore, $e^{-\frac{2}{x}}$ does not vanish at $\pm\infty$. On the other hand, substitution of a dual Taylor series $\sum_{n=0}^{\infty} a_n \delta^{(n)}(x)$ gives $w(x) = \sum_{n=0}^{\infty} \frac{2^{n+1} \delta^{(n)}(x)}{(n+1)! n!}$. Although this does not define a distribution, Kim and Kwon [9] established that the series defines a hyperfunction concentrated at $\{0\}$.

The next class of problems we would like to consider is the *solution of ordinary differential equations*. At this point it is convenient to introduce some notation. Given an equation, we denote by S_c the set of classical solutions. Similarly, S_d are the distributional solutions, S_h the hyperfunction solutions and S_δ are the dual Taylor series solutions. Clearly $S_c \subseteq S_d \subseteq S_h$, but the relationship with the space S_δ is not so obvious.

Recall that a normal equation with smooth coefficients does not have distributional solutions other than the classical ones [8]. A similar situation is encountered with hyperfunction solutions, according to Komatsu's theorem [10]. However, the situation with dual Taylor series is more complicated.

EXAMPLE The equation $y' = x^2 y$, has as classical solutions $y = ce^{\frac{x^3}{3}}$. Thus $S_c = S_d = S_h$ has dimension 1. However, if we substitute $y = \sum_{n=0}^{\infty} a_n \delta^{(n)}(x)$ we obtain

$$\sum_{n=0}^{\infty} a_n \delta^{n+1}(x) = \sum_{n=2}^{\infty} n(n-1) a_n \delta^{(n-2)}(x),$$

and thus $a_2 = 0$, while

$$(n+2)(n+1) a_{n+2} = a_{n-1}, \, n \geq 1.$$

Therefore

$$y(x) = a_0 \sum_{m=0}^{\infty} \frac{4 \cdot 7 \cdots (3m-2)}{(3m)!} \delta^{(3m)}(x) + a_1 \sum_{m=0}^{\infty} \frac{2 \cdot 5 \cdots (3m-1)}{(3m+1)!} \delta^{(3m+1)}(x),$$

where a_0 and a_1 are arbitrary. Thus $\dim S_\delta = 2$. Actually for the equation $y' = x^k y$, we have $\dim S_\delta = k$, while $\dim S_c = \dim S_d = \dim S_h = 1$.

Series of Dirac delta functions have also been used in other contexts. They have been employed to solve *functional differential equations*[1,18]. Also series of Dirac delta functions arise in *singular perturbations* [5].

4. DUAL TAYLOR SERIES SOLUTIONS OF O.D.E.'S.

As we have seen, series of Dirac delta functions arise in the solution of equations and also in the asymptotic expansion of generalized functions. Thus, we can expect that there should be a connection between the asymptotics of the distributional solutions and the formal dual Taylor solutions of those equations. The examples we have given show that this is not so evident since, in general, there is no relationship between the dimensions of the spaces S_d and S_δ. L. G. Hernández and myself [6] have obtained several results in this direction, some of wich are highlighted in this section.

THEOREM 1. Let $y(x)$ be a solution of the ordinary differential equation with plynomial coefficients
$$a_k(x)y^{(k)}(x) + \ldots + a_0(x)y(x) = 0. \tag{11}$$
If y belongs to $\mathcal{A}'$, where $\mathcal{A}$ is any of the spaces $\mathcal{E}$, $\mathcal{P}$, $\mathcal{O}_M$, $\mathcal{O}_C$ or another where the moment asymptotic expansion holds,
$$y(\lambda x) \sim \sum_{n=0}^{\infty} \frac{(-1)^n \mu_n \delta^{(n)}(x)}{n! \lambda^{n+1}}, \text{ as } \lambda \to \infty, \tag{12}$$
then the dual Taylor series $\sum_{n=0}^{\infty} \frac{(-1)^n \mu_n \delta^{(n)}(x)}{n!}$ satisfies (10).□

Thus, there is a map from the space $S_d \cap \mathcal{A}'$ to S_δ. For instance, for the equation $y' + 2xy = 0$, we have $S_d = \left\{ ce^{-x^2} : c \in \mathbb{R} \right\} \subseteq \mathcal{P}'$ and the map from S_d to S_δ is a biiection: $ce^{-x^2} \longleftrightarrow c\sqrt{\pi} \sum_{n=0}^{\infty} \frac{\delta^{(2n)}(x)}{4^n n!}$.

For the equation $y'' + y = 0$ we have $S_d = \{c_1 \cos x + c_2 \sin x : c_1, c_2 \in \mathbb{R}\} \subseteq \mathcal{K}'$. But $S_\delta = \{0\}$ and the map of S_d to S_δ is trivial.

The equation $y' = -4x^3 y$ has $S_d = \left\{ ce^{-x^4} : c \in \mathbb{R} \right\}$, of dimension 1, but S_δ has dimension 3. This example shows that not every series of deltas that solves an ordinary differential equation arises from the asymptotic expansion of a distributional solution: the converse of Theorem 1 is not true. However, we have the following result [6]

THEOREM 2. Let $\sum_{n=0}^{\infty} a_n \delta^{(n)}(x)$ be a solution of the ordinary differential equation with polynomial coefficients
$$a_k(x)y^{(k)}(x) + \ldots + a_0(x)y(x) = 0. \tag{13}$$
Then there exist $f \in \mathcal{A}'(\mathbb{R})$, the moment asymptotic expansion holding in $\mathcal{A}'(\mathbb{R})$, such that $f(\lambda x) = o(\lambda^{-n})$, as $\lambda \to \infty$ for every $n \in \mathbb{N}$, and a solution $y \in \mathcal{A}'(\mathbb{R})$ of the equation
$$a_k(x)y^{(k)}(x) + \ldots + a_0(x)y(x) = f(x) \tag{14}$$
such that
$$y(\lambda x) \sim \sum_{n=0}^{\infty} \frac{a_n \delta^{(n)}(x)}{\lambda^{n+1}}, \text{ as } \lambda \to \infty. \tag{15}$$
□

Summarizing, the dual Taylor series solutions of an ordinary differential equation with polynomial coefficients, $Ly = 0$, are the asymptotics of the distributional solutions of the equations $Ly = f$, where $f(\lambda x) = o(\lambda^{-\infty})$ as $\lambda \to \infty$. Observe that for these generalized functions f, the associated dual Taylor series is $0 \cdot \delta(x) + 0 \cdot \delta'(x) + 0 \cdot \delta''(x) + \ldots$ and thus from the point of view of dual Taylor series it is not possible to distinguish the equations $Ly = 0$ and $Ly = f$.

5. SINGULAR PERTURBATIONS.

Series of Dirac delta functions also appear in the solution of singular perturbation problems that do not admit solutions in the form of power series with classical functions as coefficients. Let us start with the initial value problem
$$\varepsilon y'(x) = -y(x), \ x > 0 \ y(0) = 1, \tag{16}$$
where $\varepsilon \ll 1$. Let $z(x) = H(x)y(x)$, where $H(x)$ is the Heaviside function. Then $z' = Hy' + \delta(x)$, thus (15) becomes $\varepsilon z' = -z + \varepsilon \delta(x)$. Next, we look for a solution of the form $z = \sum_{n=0}^{\infty} z_n(x)\varepsilon^n$. Substitution and collection of like powers of ε yields $z_0 = 0, z_1 = -\delta(x), z_n + z'_{n-1} = 0, n \geq 2$. Since it is natural to ask that $z_n(x) = 0$ for $x < 0$, we obtain $z_{n+1}(x) = (-1)^n \delta^{(n)}(x), n \geq 0$, and thus $z(x, \varepsilon) = \sum_{n=0}^{\infty} (-1)^n \varepsilon^{n+1} \delta^{(n)}(x)$.

The solution of (15) is easy to find, however, so that a comparison can be made. The solution is $H(x)\exp(-\frac{x}{\varepsilon})$. The asymptotic relation

$$H(x)\exp\left(-\frac{x}{\varepsilon}\right) \sim \sum_{n=0}^{\infty} (-1)^n \varepsilon^{n+1} \delta^{(n)}(x), \qquad (17)$$

is nothing but the distributional version of Watson's Lemma for the asymptotic expansion of Laplace transforms [4]. More generally, we have the following result.

THEOREM 3. Let $y(x,\varepsilon)$ be the solution of the initial value problem
$$\varepsilon y'(x) = A(x,\varepsilon)y(x),\ x>0, y(x_0) = y_0,$$
where $A(x,\varepsilon)$ is smooth for $x \geq 0$ and $A(x,\varepsilon) \sim A_0(x) + \varepsilon A_1(x) + \varepsilon^2 A_2(x) + \ldots$, as $\varepsilon \to 0$, $A_0(x)$ being negative. Let $z(x,\varepsilon) = H(x)y(x,\varepsilon)$. Then

$$z(x,\varepsilon) \sim z_1(x)\varepsilon + z_2(x)\varepsilon^2 + z_3(x)\varepsilon^3 + \ldots, \text{ as } \varepsilon \to 0, \qquad (18)$$

in $\mathcal{D}'(\mathbb{R})$ where z_i is supported at $\{0\}$ and has order $i-1$ at the most.□

REFERENCES

[1] Cooke, K. L. and Wiener, J., *Distributional and analytical solutions of functional diferential equations*, J. Math. Anal. Appls. **98** (1984), 111-124.

[2] Estrada, R., Gracia-Bondía, J. and Várilly, J., *On asymptotic expansions of twisted products*, J. Math. Phys. **30** (1989), 2789-2796.

[3] Estrada, R. and Kanwal, R. P., *Moment sequences for a class of distributions*, Complex Variables **9** (1987), 31-39.

[4] Estrada, R. and Kanwal, R. P., *A distributional theory for asymptotic expansions*, Proc. Roy. Soc. London A. **428** (1990), 399-430.

[5] Glizer, V. J. and Dimitriev, M. G., *Singular pertubation and generalized functions*, Soviet Math. Dokl. **20** (1979), 1360-1364.

[6] Hernández, L. G. and Estrada, R., *Solution of ordinary differential equations by series of Dirac delta dunctions*, to appear.

[7] Kaneko, A., *Introduction to Hyperfunctions*, Klewer Press, Boston, 1989.

[8] Kanwal, R. P., *Generalized Functions: Theory and Technique*, Academic Press, New York, 1983.

[9] Kim, S. S. and Kwon, K. H., *Generalized weights for orthogonal polynomials*, Diff. Int. Eqns. **4** (1991), 601-608.

[10] Komatsu, H., *On the index of ordinary differential operators*, J. Fac. Sci. Univ. Tokio Sect. 1A **18** (1971), 379-398.

[11] Krall, A. M., Kanwall, R. P. and Littlejohn, L. L., *Distributional solutions of ordinary differential equations*, Canad. Math. Soc. Conf. Proceed. **8** (1987), 227-246.

[12] Krall, H. L. and Frink, O., *A new class of orthogonal polynomials: the Bessel polynomials*, Trans. Amer. Math. Soc. **65** (1949), 100-115.

[13] Littlejohn, L. L. and Kanwal, R. P., *Distributional solutions of the hypergeometric differential equation*, J. Math. Anal. Appls. **122** (1987), 325-345.

[14] Morton, R. D. and Krall, A. M., *Distributional weight function for orthogonal polynomials*, SIAM J. Math. Anal. **9** (1978), 604-626.

[15] Pilipovíc, S., *Quasiasymptotic expansion and the Laplace transform*, Applicable Analysis **35** (1990), 247-261.

[16] Shoat, J. A. and Tamarkin, J. D., *The Problem of Moments*, A.M.S., Providence, 1943.

[17] Vladimirov, V. S., Drozhinov, Y. N. and Zav'alov, B. I., *Multidimensional Tauberian Theorems for Generalized Functions* Nauka, Moscow, 1986.

[18] Wiener, J., *Generalized function solution of differential and functional differential equations*, J. Math. Anal. Appls. **88** (1982), 170-182. Address: Ricardo Estrada, Escuela de Matemática, Universidad de Costa Rica, San José, Costa Rica.

H GINGOLD AND V TRUTZER
The large time pointwise behaviour of linear degenerate hyperbolic systems

1. Introduction. The asymptotic behaviour of solutions of systems of conservation laws, as $t \to \infty$,

$$u_t + \sum_{j=1}^{m} \underline{F}_j(u) u_{x_j} = B(u), \quad x = (x_1, \ldots, x_m), \quad t > 0, \quad u(x,t) \in \mathbb{R}^n$$

was studied, in different contexts, by various authors. See e.g. Glimm & Lax [5], for genuinely nonlinear hyperbolic systems with $m=1$, $n=2$, and Vainberg [9], J. Rauch and M. Taylor [7] and J. Rauch [8], for linear systems in connection with scattering problems. Absent in the literature are discussions of the decay of solutions of systems – even linear ones –

$$u_t + \sum_{j=1}^{m} A_j(x,t,u) u_{x_j} = 0 \quad \text{for the case when multiplicities of the}$$

eigenvalues of the matrix $\tilde{A} = \sum_{j=1}^{m} A_j(x,t,u) \xi_j$, vary with (x,t) or the eigenvalues of $\tilde{A}$ depend on t and decay to zero as $t \to \infty$, even for $m = 1$.

Coalescing eigenvalues could be studied as a limiting case of the 'close' propagating speeds case. They also occur in the case of equations with type change. Also note that Lax [6] showed that in a three dimensional vector space of nxn symmetric matrices the eigenvalues must cross if $n \equiv 2 \pmod 4$.

In the sequel we examine two model cases, both with $m = 1$.

I. We examine systems of the form

$$u_t + A(t) u_x = B(t) u, \quad u(x,0) = \varphi(x) \tag{1.1}$$

with symmetric A, where eigenvalue multiplicity varies with t and with no requirement that $B(t)$ be dissipative.

We show that if all speeds of propagation – eigenvalues of A – decay to zero at a certain rate and φ decays at $\pm\infty$ then the solution $u(x,t)$ decays as $t \to \infty$. (We point out that previous results of this kind, assumed that the speeds of propagation were bounded away from zero.)

II. We find, without using Fourier transforms, a __uniform__ representation for the solution $u(x,t,\epsilon)$ of systems with constant coefficients which depend solely on a parameter ϵ:
$$u_t + A(\epsilon)u_x = 0, \qquad u(x,0) = \varphi(x) \tag{1.2}$$
where $A(\epsilon)$ has real eigenvalues which are allowed to coalesce in any possible manner. This representation is used to show uniform decay of the solutions u and it may be useful in implementing a "triangular" form of a numerical method of characteristics.

2. Time Dependent Symmetric Case. We have

Theorem 2.1. *Consider the system* (1.1). *Assume*

i. $A(t) = A^*(t)$, $A(t)$, $B(t)$ *are analytic for* $0 \le t \le \infty$.

ii. $\int_0^\infty |B(s)|ds < \infty$ *(or* $B(t) = \dfrac{B_2}{t^2} + \dfrac{B_3}{t^3} + \cdots$ *as* $t \to \infty$*)*

iii. $\varphi \in C^1(R) \cap L^1(R)$, $\quad x\varphi \in L^1(R)$

iv. *The real analytic eigenvalues* $\lambda_1(t), \lambda_2(t), \cdots, \lambda_n(t)$ *of* $A(t)$ *satisfy*
$$\int_0^t \lambda_j(s)ds \sim c_j \ln(1+t) \quad \text{as} \quad t \to \infty, \; j=1,2\ldots n,$$

c_j *some constants. They are allowed to coalesce in any manner but no two are assumed to be identical. Then, for any* $b > 0$, $|x| \le \lim_{t \to \infty} |u(x,t)| = 0$.

Sketch of Proof. The Fourier transform (in x) of u, $u(\xi,t)$ satisfies the system
$$\hat{u}_t = [-i\xi A(t) + B(t)]\hat{u}, \qquad \hat{u}(\xi,0) = \hat{\varphi}(\xi)ds \tag{2.1}$$

It follows from Gingold & Hsieh [2,3] that
$$\hat{u}(\xi,t) = U(t)\left\{\exp(-i\xi)\int_0^t \begin{bmatrix} \lambda_1(s) & & 0 \\ & \ddots & \\ 0 & & \lambda_n(s) \end{bmatrix} ds\right\}\phi_2(\xi,t)U^{-1}(0)\hat{\varphi}(\xi) \tag{2.2}$$

where $U(t)$ is a matrix function which is unitary and analytic on $[0,\infty]$ $\lambda_1(t), \lambda_2(t), \cdots, \lambda_n(t)$ are the eigenvalues of $A(t)$ and $\phi_2(\xi,t)$ is a certain differentiable matrix function. Assume that (2.2) holds and let I be the identity matrix and let $D(t)$, M be given by

$$D(t) := \int_0^t \begin{bmatrix} \lambda_1(s) & & 0 \\ & \ddots & \\ 0 & & \lambda_n(s) \end{bmatrix} ds \quad , \quad M := (xI - D).$$

Then using integration by parts it can be shown that

$$u(x,t) = \frac{1}{\sqrt{2\pi}} \int_{-\infty}^{\infty} U(t)(iM)^{-1} e^{i\xi M} \left[\frac{\partial \phi_2}{\partial \xi} U^{-1}(0)\hat{\varphi}(\xi) + \phi_2 U^{-1}(0)\hat{\varphi}'(\xi) \right] d\xi =: I_1 + I_2$$

A lengthy calculation reveals that there exist constants K_1, K_2 such that $|\phi_2| \leq K_1$ and $|\frac{\partial \phi_2}{\partial \xi}| \leq K_2$ uniformly for $0 \leq \xi \leq \infty$, $0 \leq t < \infty$. Further analysis shows that

$$|I_1| + |I_2| = 0(\max_{1 \leq j \leq n} (\frac{1}{|\int_0^t \lambda_j(s) ds|})) \quad \text{as } t \to \infty.$$

REMARK: It is possible to show that the conclusion of theorem 2.1 holds if the condition iv in the theorem is replaced by the simple condition

$$|\int_0^t \lambda_j(s) ds| \to \infty \quad \text{as } t \to \infty .$$

However, the proof then is more complicated.

3. **Second Case.** We consider a system (1.2). First we need

Theorem 3.1. (Gingold & Hsieh [3]). *Let $A(\epsilon)$ be a matrix valued function analytic on $[0, \epsilon_0]$ with real eigenvalues. Then there exists a labeling of the eigenvalues such that $\lambda_j(\epsilon)$, $j = 1, 2, \cdots, n$ are analytic and there exists a unitary, analytic matrix function $U(\epsilon)$ such that*

$$\tilde{A}(\epsilon) := U^*(\epsilon) A(\epsilon) U(\epsilon) \quad , \quad \psi(x) := U^*(\epsilon)\varphi(x) \quad , \quad 0 \leq \epsilon \leq \epsilon_0 \tag{3.1}$$

is lower triangular.

If we let $u = Uv$, we reduce the system (1.2) to the triangular system

$$v_t + \tilde{A}(\epsilon) v_x = 0 \, , \, v(x,0) = \psi(x) \, , \, \tilde{A} = (a_{jk}) \, , \, a_{jj} = \lambda_j \, , \, a_{jk} = 0 \text{ if } k > j. \tag{3.2}$$

In the sequel we employ the notations and the machinery of the calculus of finite differences (see e.g. [1]). Denote

$$\tilde{\varphi}_{x,t}(\lambda) = \tilde{\varphi}(\lambda) = \varphi(x - \lambda t) \quad , \quad h[\lambda,\mu] = \frac{h(\lambda) - h(\mu)}{\lambda - \mu} \tag{3.3}$$

$$\tilde{\varphi}[\lambda_1, \cdots, \lambda_k] =$$

$$= (-1)^{k-1} \int_0^t ds_{k-1} \int_0^{s_{k-1}} ds_{k-2} \cdots \int_0^{s_2} \varphi^{(k-1)} (x - \lambda_k t + (\lambda_k - \lambda_{k-1})s_{k-1} + \cdots +$$

$$+ (\lambda_2 - \lambda_1)s_1) \, ds_1 \tag{3.4}$$

With this notation, it follows that the solution of the system (3.2) is given, uniformly in ϵ, by

$$v_m(x,t) = \tilde{\psi}_m[\lambda_m] +$$

$$+ \sum_{k=1,\ I_{m,k}}^{n-1} \sum a_{mm_2} a_{m_2 m_3} \cdots a_{m_j k} \tilde{\psi}_k [\lambda_m, \lambda_{m_2}, \cdots, \lambda_{m_j}, \lambda_k] \tag{3.5}$$

where $I_{m,k} = \{(m_2, m_3, \cdots, m_j): m > m_2 > \cdots > m_j > k\}$, $m=1\ldots n$.
The following theorem can be proven.

THEOREM 3.2. *Let* $A(\epsilon)$ *be an* n *by* n *analytic matrix function on* $[0,\epsilon_0]$, *with real eigenvalues different from* 0 *on* $[0,\epsilon_0]$. *If the initial vector* $\varphi(x)$ *in (1.2) is such that* $\lim_{|x|\to\infty} x^\nu \varphi^{(\nu)}(x) = 0$, $\nu = 0,1,\ldots n-1$. *Then, uniformly for* $|x|$ *in bounded sets and* $0 \le \epsilon \le \epsilon_0$, *the solution of (1.2), satisfies* $\lim_{t\to\infty} u(x,t,\epsilon) = 0$.

REMARKS: We have not come across an *explicit uniform* representation like (3.5) in the literature although the idea of triangularization is commonly used.

Degeneracy in a hyperbolic system (1.2) could have drastic effects. The representation (3.5) informs us that each component of the solution vector to (1.2) could be a superposition of up to $\frac{n(n+1)}{2}$ "distinct traveling waves". Moreover, the coefficients in the linear combination of the traveling waves could be polynomials in t of maximal degree $n-1$. Thus, *resonance without source terms* could occur. If (1.2) would have been a strictly hyperbolic system, then each component of the solution vector would be a linear combination of n traveling waves and no resonance would occur.

In principle one could utilize a Fourier transform to find a representation for solutions of (1.2). However, one must then supplement the

Fourier representation by an asymptotic analysis. The uniform representation (3.5) makes the conclusions of theorem 3.2 transparent.

The emphasis in this work is on *construction* of solutions.

References

[1]   Gelfond, A. O., *Calculus of finite differences*. Nauka, Moscow 1967. (Russ.), English Trans.: Hindustan Publishing Corp., 1971.

[2]   Gingold, H. and Hsieh, P.F., Global approximation of perturbed Hamiltonian Differential Equations with Several Turning Points. Siam. J.Math. Anal., 18, No.5, (1987), 1275 –1293.

[3]   Gingold, H. and Hsieh, P.F., An algorithm for globally analytic tri angularization of a matrix function. Linear Alg. Applic. (1991) (to appear).

[4]   Gingold, H. and Trutzer, V., On resonance in linear conservation laws without source terms, Utilitas Mathematica, 37, (1990), 60–78.

[5]   Glimm, J. and Lax, P.D., Decay of solutions of systems of Nonlinear Hyperbolic Conservation Laws. Memoirs AMS, No. 101, (1970).

[6]   Lax, P.D., The multiplicity of eigenvalues. Bull. AMS 6, (1982), 213–214.

[7]   Rauch, J. and Taylor, M., Decay of Solutions to Nondissipative Hyperbolic Systems on Compact Manifolds. Comm. on Pure and Applied Mathematics. Vol. XXVIII, (1975), 501–523.

[8]   Rauch, J., Asymptotic Behaviour of Solutions to Hyperbolic Partial Differential Equations with Zero Speeds. Comm. on Pure and Applied Mathematics. Vol. XXXI, 430–448 (1978).

[9]   Vainberg, B.R., On the short wave asymptotic behaviour of solutions of stationary problems and the asymptotic behaviour as $t \to \infty$ of solutions of non-stationary problems. Russian Math. Surveys 30:2, (1975), 1 – 58. (Uspeki Mat. Nauk. 30:2, (1975), 3–55.)

J-P GOSSEZ
Nonresonance in some semilinear Neumann problems

1. Let Ω be a bounded open subset in $\mathbf{R}^N$ with smooth boundary $\partial\Omega$. Let f be a continuous function from $\mathbf{R}$ to $\mathbf{R}$. We consider the semilinear Neumann problem

(1) $$\begin{cases} -\Delta u = f(u) + h(x) \text{ in } \Omega, \\ \frac{\partial u}{\partial \nu} = 0 \text{ on } \partial\Omega, \end{cases}$$

where ν denotes the unit exterior normal. We are interested in the conditions to be imposed on the nonlinearity f in order that problem (1) admits at least one solution $u(x)$ for any given $h(x)$. Such conditions are usually called <u>nonresonance conditions</u>.

In the linear case $f(u) = \lambda u$, the equation in (1) reads $-\Delta u = \lambda u + h(x)$. A necessary and sufficient condition for (1) to be solvable for any $h(x)$ (say in $L^2(\Omega)$) is then clearly that λ be different from the (distinct) eigenvalues $\lambda_1 = 0 < \lambda_2 < \lambda_3 < \ldots$ of $-\Delta$ with homogenous Neumann boundary conditions.

In the nonlinear case, the equation in (1) formally reads $-\Delta u = (f(u)/u)u + h(x)$ and we see that it is the quotient $f(u)/u$ which plays the role of the above number λ. One can then guess that if this quotient does not interfere too much with the spectrum $\lambda_1 < \lambda_2 < \ldots$, then nonresonance should occur.

It is our purpose in this talk to review some recent results in that direction and to mention some related open questions.

2. Our starting point will be the classical result of Dolph [Do]: if, for some k,

(2) $$\lambda_k < \liminf_{s \to \pm\infty} \frac{f(s)}{s} \leq \limsup_{s \to \pm\infty} \frac{f(s)}{s} < \lambda_{k+1},$$

then (1) is solvable for any h in $L^2(\Omega)$. The proof goes by degree theory. In fact (2) implies an a priori bound in $H^1(\Omega)$ for all solutions of (1).

Condition (2) of Dolph was weakened by Costa-Oliveira into a condition which involves the primitive of the nonlinearity: $F(s) = \int_0^s f$. Solvability of (1) for any h in $L^2(\Omega)$ is derived in [Co-Ol] under the following two assumptions:

(3) $$\lambda_k \leq \liminf_{s \to \pm\infty} \frac{f(s)}{s} \leq \limsup_{s \to \pm\infty} \frac{f(s)}{s} \leq \lambda_{k+1},$$

(4) $$\lambda_k < \liminf_{s \to \pm\infty} \frac{2F(s)}{s^2} \leq \limsup_{s \to \pm\infty} \frac{2F(s)}{s^2} < \lambda_{k+1}.$$

The proof here is variational. Observe that there is no a priori bound in general (think of $k = 0$, $h = 0$, with f having an unbounded set of zeros, which yield constant solutions of (1)). Condition (3) and (4) are used to prove that the corresponding functional

$$\Phi(u) = \frac{1}{2}\int_\Omega |\nabla u|^2 - \int_\Omega F(u) - \int_\Omega hu, \quad u \in H^1(\Omega)$$

satisfies the (P.S.) condition and has the right shape in order to apply the Rabinowitz saddle point theorem.

The case where equality holds in the extreme left or right inequalities of (4) was also considered recently. Various sufficient conditions for nonresonance in such cases can be found in [Co], [Si], [Ra], [Mo].

3. We wish now to describe a result near the first eigenvalue $\lambda_1 = 0$ which provides a <u>necessary and sufficient condition</u> for nonresonance. Let us assume that no interference occurs with the higher part of the spectrum, in the following sense (inspired from (3), (4)):

(5) $$\limsup_{s \to \pm\infty} \frac{f(s)}{s} \leq \lambda_2,$$

(6) $$\limsup_{s \to \pm\infty} \frac{2F(s)}{s^2} < \lambda_2.$$

Then a necessary and sufficient condition for (1) to be solvable for any $h \in L^\infty(\Omega)$ is that $f : \mathbf{R} \to \mathbf{R}$ be unbounded from above and from below (cf. [Go-Om$_2$]).

Necessity follows immediately by integrating the equation in (1). Sufficiency is proved by using degree theory, by constructing a bounded open set $\mathcal{O}$ in $C^1(\bar{\Omega})$ which contains o and which is such that no solution of the family of homotopic problems

$$\begin{cases} -\Delta u = (1-\mu)\theta u + \mu[f(u) + h(x)] \text{ in } \Omega, \\ \frac{\partial u}{\partial \nu} = 0 \text{ on } \partial\Omega \end{cases}$$

arises on the boundary $\partial\mathcal{O}$. (Here θ is fixed with $\lambda_1 = 0 < \theta < \lambda_2$ and μ varies with $0 \leq \mu < 1$). One should observe here that no a priori bound holds in general. Actually even the (P.S.) condition fails in general (think of $h = 0$, with F uniformly bounded and f having an unbounded set of zeros, which provide an unbounded (P.S.) sequence). In this respect obtaining a variational proof of the above result would certainly be of interest.

Condition (6) on the primitive F of the nonlinearity f is used through its equivalence with a positive density condition. Precisely, assuming (5) and that f has at most linear growth, then (6) holds if and only if there exists $\eta > 0$ such that the set

$$E = \left\{ s \in \mathbf{R}_0; \frac{f(s)}{s} \leq \lambda_2 - \eta \right\}$$

has a positive density at $+\infty$ and at $-\infty$. This means that

$$\liminf_{r\to+\infty} \frac{|E\cap[0,r]|}{|[0,r]|} > 0, \liminf_{r\to-\infty} \frac{|E\cap[r,0]|}{|[r,o]|} > 0,$$

where $|\ |$ denotes Lebesgue measure. This notion of positive density was introduced in [DF-Go] and the above equivalence was proved in [Go-Om$_1$].

4. The results of [Do], [Co-Ol], [Co], [Si], [Ra], [Mo] mentionned above also hold for the Dirichlet problem (and actually were originally derived in that setting). This is not so however for the result of [Go-Om$_2$]. Consider the problem

(7)
$$\begin{cases} -\Delta u = \lambda_1 u + f(u) + h(x) \text{ in } \Omega, \\ u = 0 \text{ on } \partial\Omega, \end{cases}$$

where $0 < \lambda_1 < \lambda_2 < \ldots$ now denotes the eigenvalues of $-\Delta$ on $H_0^1(\Omega)$. Multiplying by the first (positive) eigenfunction and integrating, one immediately sees that a necessary condition for nonresonance is again that $f : \mathbf{R} \to \mathbf{R}$ be unbounded from above and from below. This condition however is no more sufficient (assuming of course no interference with the higher part of the spectrum), even in the ODE case, as was shown recently by Njoku [Nj]. No necessary and sufficient condition seems to be known in the case of the Dirichlet problem.

5. The proof of the result discussed in §3 involves the consideration of constant lower and upper solutions which may not be well ordered. We will now state a general result in that direction.

Let us consider the problem

(8)
$$\begin{cases} -\Delta u = f(x,u) \text{ in } \Omega, \\ \text{Dirichlet or Neumann homogeneous condition on } \partial\Omega, \end{cases}$$

where f is a L^p Caratheodory function for some $p > N$. A classical result says that if (8) admits a lower solution $\alpha(x)$ and an upper solution $\beta(x)$, with $\alpha(x) \leq \beta(x)$ in Ω, then (8) admits a solution $u(x)$, with $\alpha(x) \leq u(x) \leq \beta(x)$. See e.g. [Am]. Assume now that

(9)
$$\lambda_1 \leq \liminf_{s\to\pm\infty} \frac{f(x,s)}{s} \leq \limsup_{s\to\pm\infty} \frac{f(x,s)}{s} \leq\!\!\!\!\!/\ \lambda_2,$$

where $\lambda_1 < \lambda_2 < \ldots$ denote the eigenvalues of $-\Delta$ under the corresponding boundary conditions. Then it is shown in [Go-Om$_3$] that the sole existence of a lower solution $\alpha(x)$ and of an upper solution $\beta(x)$ (with possibly no ordering relation between them) implies the existence of a solution $u(x)$.

This allows to recover in a different way the characterization of nonresonance discussed in §3, under however a noninterference condition with respect to λ_2 which is a little bit stronger than (5), (6). Weakening the right inequality in (9) so as to reach a condition of the type (5), (6) remains unclear at this moment. Suppressing the restriction with respect to λ_1 in (9) also remains unclear at this moment (this is possible in the case of the Neumann problem when $f(x,u)$ splits as $f(u) + h(x)$ with $h \in L^\infty(\Omega)$).

Another question concerns the possibility of considering in §3 forcing terms in $L^2(\Omega)$ instead of $L^\infty(\Omega)$. The technical difficulty is related to the use of constant lower and upper solutions.

REFERENCES

[Am] H.AMANN, *Fixed point equations and nonlinear eigenvalue problems in ordered Banach spaces,* Siam Review, 18 (1976), 620-709.

[Co] D.COSTA, *A note on unbounded perturbations of linear resonant problems,* to appear.

[Co-Ol] D.COSTA and A.OLIVEIRA, *Existence of solution for a class of semilinear elliptic problems at double resonance,* Boll. Soc. Bras. Mat., 19 (1988), 21-37.

[DF-Go] D.DE FIGUEIREDO and J.-P.GOSSEZ, *Conditions de non résonance pour certains problèmes elliptiques semi linéaires,* C.R. Ac. Sc. Paris, 302 (1986), 543-545.

[Do] C.DOLPH, *Nonlinear integral equations of Hammerstein type,* Trans. Am. Math. Soc., 66 (1949), 289-307.

[Go-Om$_1$] J.-P.GOSSEZ and P.OMARI, *Periodic solutions of a second order ordinary differential equation: a necessary and sufficient condition for nonresonance,* J. Diff. Equat., to appear.

[Go-Om$_2$] J.-P.GOSSEZ and P.OMARI, *A necessary and sufficient condition of nonresonance for a semilinear Neumann problem,* Proc. Am. Math. Soc., to appear.

[Go-Om$_3$] J.-P.GOSSEZ and P.OMARI, *Nonordered lower and upper solutions in semilinear elliptic problems,* to appear.

[Mo] M.MOUSSAOUI, *Questions d'existence dans les problèmes semi-linéaires elliptiques,* Thèse Doct., Université Libre de Bruxelles, 1991.

[Ra] M.RAMOS, *Remarks on resonance problems with unbounded perturbations,* to appear.

[Si] E.A.SILVA, *Critical point theorems and applications to differential equations,* Ph. D. thesis, Univ. Wisconsin, 1988.

Département de Mathématique
Campus Plaine C.P.214
Université Libre de Bruxelles
1050 Bruxelles - Belgique

Q DU, M GUNZBURGER AND J PETERSON
On the Ginzburg–Landau equations of superconductivity

A main goal of our ongoing work is to develop robust and efficient codes that can be used to help determine, through comparisons with experimental observations, the extent to which Ginzburg-Landau models can be applied to high-temperature superconductors and also to use these codes to study electromagnetic phenomena in these materials. Necessary to achieving this goal is a thorough understanding, from both the analytic and numerical points of view, of low-temperature Ginzburg-Landau models. Here, we summarize the results of our work in these directions. Space limitations preclude a presentation of details or proofs; these may be found in [3]-[5]. A reader interested in a good introduction to the subject of superconductivity may consult [6].

1. The Ginzburg-Landau Model on Bounded Regions

We first focus on the Ginzburg-Landau model for bounded regions. This model is based on the principle that the material is in a state such that the Gibbs free energy, given by

$$\mathcal{G}(\psi, \mathbf{A}) = \int_\Omega \left(f_n - |\psi|^2 + \frac{1}{2}|\psi|^4 + \left|\left(\frac{i}{\kappa}\nabla + \mathbf{A}\right)\psi\right|^2 + |\mathbf{h}|^2 - 2\mathbf{h} \cdot \mathbf{H} \right) d\Omega, \quad (1)$$

is minimized. Here ψ, $\mathbf{A}$, and $\mathbf{h} = \text{curl } \mathbf{A}$ denote the non-dimensionalized complex order parameter, magnetic potential, and magnetic field, respectively; $\mathbf{H}$ is the applied field, f_n is the constant free energy density of the normal state in the absence of a magnetic field, and κ, known as the Ginzburg-Landau parameter, is a material constant. For a type-I superconductor, we have $\kappa < 1/\sqrt{2}$, while $\kappa > 1/\sqrt{2}$ for a type-II superconductors.

The Ginzburg-Landau free energy functional (1) has a very important property, namely, *gauge invariance*. If, for some $\phi \in H^2(\Omega)$, we have that

$$\zeta = \psi e^{i\kappa\phi} \quad \text{and} \quad \mathbf{Q} = \mathbf{A} + \nabla\phi,$$

then $(\psi, \mathbf{A})$ and $(\zeta, \mathbf{Q})$ are said to be gauge equivalent; it is easily shown that $\mathcal{G}(\zeta, \mathbf{Q}) = \mathcal{G}(\psi, \mathbf{A})$, i.e., $\mathcal{G}$ is gauge invariant. Based on this property, the existence of minimizers of the Ginzburg-Landau functional can be established via standard variational arguments. In fact, we can show that any minimizer of the Ginzburg-Landau functional is gauge equivalent to a solution which has a divergence free magnetic potential with vanishing normal component on the boundary. This corresponds to the "Coulomb gauge". The Ginzburg-Landau equations and natural boundary conditions in the Coulomb gauge are then given by

$$\left(\frac{i}{\kappa}\nabla + \mathbf{A}\right)^2 \psi - \psi + |\psi|^2\psi = 0 \quad \text{in } \Omega \tag{2}$$

$$\operatorname{curl}\operatorname{curl}\mathbf{A} = -\frac{i}{2\kappa}(\psi^*\nabla\psi - \psi\nabla\psi^*) - |\psi|^2\mathbf{A} + \operatorname{curl}\mathbf{H} \quad \text{and} \quad \operatorname{div}\mathbf{A} = 0 \quad \text{in } \Omega \tag{3}$$

$$\mathbf{A}\cdot\mathbf{n} = 0, \quad \text{and} \quad (\frac{i}{\kappa}\nabla\psi + \mathbf{A}\psi)\cdot\mathbf{n} = 0, \quad \text{and} \quad \operatorname{curl}\mathbf{A}\times\mathbf{n} = \mathbf{H}\times\mathbf{n} \quad \text{on } \Gamma, \tag{4}$$

where Γ denotes the boundary of Ω. We have also conidered other possible boundary conditions.

Various mathematical properties of the solutions of the Ginzburg-Landau equations have veen established. These include the proof of non-existence of the local maxima for the Ginzburg-Landau functional and the boundedness of the order parameter, i.e., appropriately non-dimensionalized, $|\psi| \leq 1$ almost everywhere. An examination of simple analytic solutions or trivial solutions has been made. Many of the properties may be used to interpret certain physical phenomenon such as the perfect Meissner effect in the absence of an applied field and the existence of a mixed state (the existence of vortices or filaments) in the presence of an applied field below a critical value.

We have also developed and analyzed finite element algorithms for approximating solutions to the model. Finite element subspaces $\mathcal{S}^h$ and $\mathbf{V}^h$ are constructed in a standard way, from partitions of Ω into finite elements; h is then some measure of the size of the finite elements in a partition. The normal component of functions belonging to $\mathbf{V}^h$ is set to be zero while no boundary conditions are required of functions belonging to $\mathcal{S}^h$. These subspaces are assumed to satisfy the usual approximation properties. Finite element approximations are then defined as follows: seek $\psi^h \in \mathcal{S}^h$ and $\mathbf{A}^h \in \mathbf{V}^h$ such that

$$\int_\Omega \Big[(\frac{i}{\kappa}\nabla\psi^h + \mathbf{A}^h\psi^h)\cdot(-\frac{i}{\kappa}\nabla(\tilde{\psi}^h)^* + \mathbf{A}^h(\tilde{\psi}^h)^*)$$
$$+ (\frac{i}{\kappa}\nabla\tilde{\psi}^h + \mathbf{A}^h\tilde{\psi}^h)\cdot(-\frac{i}{\kappa}\nabla(\psi^h)^* + \mathbf{A}^h(\psi^h)^*) \tag{5}$$
$$+ (|\psi^h|^2 - 1)(\psi^h(\tilde{\psi}^h)^* + (\psi^h)^*\tilde{\psi}^h)\Big]d\Omega = 0 \quad \forall\,\tilde{\psi}^h \in \mathcal{S}^h$$

and

$$\int_\Omega [\operatorname{div}\mathbf{A}^h\operatorname{div}\tilde{\mathbf{A}}^h + \operatorname{curl}\mathbf{A}^h\cdot\operatorname{curl}\tilde{\mathbf{A}}^h + |\psi^h|^2\mathbf{A}^h\cdot\tilde{\mathbf{A}}^h$$
$$+ \frac{i}{2\kappa}((\psi^h)^*\nabla\psi^h - \psi^h\nabla(\psi^h)^*)\cdot\tilde{\mathbf{A}}^h]\,d\Omega = \int_\Omega \mathbf{H}\cdot\operatorname{curl}\tilde{\mathbf{A}}^h\,d\Omega \quad \forall\,\tilde{\mathbf{A}}^h \in \mathbf{V}^h. \tag{6}$$

Note that the above is a discretization of a weak formulation of the Ginzburg-Landau equations with the gauge div $\mathbf{A} = 0$ in Ω and $\mathbf{A} \cdot \mathbf{n} = 0$ on Γ.

The convergence of finite element approximations for a linear problem corresponding to the principal part of the Ginzburq-Landau equations can be obtained by standard approaches similar to those discussed in [2]. Under the framework of [1], we can then establish the convergence of the finite element solutions of (5)-(6) to a branch of regular solutions of the nonlinear Ginzburg-Landau equations (2)-(4) and the convergence was shown to be uniform for κ in a compact interval. Optimal error estimates were also derived under the usual regularity assumptions.

2. The Periodic Ginzburg-Landau Model

While our analysis of the Ginzburg-Landau model on bounded regions is valid in both type-I and type-II regime, it is of use mostly for type-I superconductors. It is well-known that, for type-II superductors, solutions exhibit much more complicated structures. The grid size necessary to resolve the fine structures based on the bounded region model, in any computation of practical utility, would be prohibitively small. For this reason we also studied a periodic Ginzburg-Landau model wherein the effect due to boundaries is neglected.

A important assumption is that away from bounding surfaces, certain physical variables, such as the magnetic field, the current, and the density of superconducting charge carriers, exhibit periodic behavior with respect to a two-dimensional lattice. This lattice is non-orthogonal and is not necessarily aligned with the coordinate axes. A function f is *periodic with respect to the lattice vectors* $\mathbf{t}_1$ *and* $\mathbf{t}_2$ if $f(\mathbf{x}+\mathbf{t}_k) = f(\mathbf{x})$ for $k = 1, 2$ and $\forall \, \mathbf{x} \in \mathbb{R}^2$.

A very important issue is the choice of gauge, about which there has been conflicting discussions in the literature. We have made a rigorous and detailed derivation of consistent gauge choices. In short, given a cell of the lattice cell, we showed that the vector magnetic potential is gauge equivalent to a potential of the form $\mathbf{Q}-\mathbf{A}_0$ where $\mathbf{A}_0 = \bar{B}/2(x_2, -x_1)^T$ and $\mathbf{Q}$ is divergence free, periodic, and has mean zero in a single cell. Here $\bar{B}$ denotes the average value of the magnetic field. To insure the periodicity of the physical variables, the order parameter should satisfy the following boundary condition:

$$\psi(\mathbf{x}+\mathbf{t}_k) = \psi(\mathbf{x})e^{i\kappa g_k(\mathbf{x})} \quad \forall \, \mathbf{x} \in \Gamma_{-k}, \; k = 1,2, \tag{7}$$

where $g_k(\mathbf{x}) = -\bar{B}(\mathbf{x} \times \mathbf{t}_k)/2$, $k = 1, 2$, and Γ_{-k}, $k = 1, 2$ denote the bottom and left boundaries of a lattice cell, respectively.

The equivalence of the different forms of the Ginzburg-Landau functional in the periodic setting has been established as has the existence of a minimizer belonging to $\mathcal{H}_P^1(\Omega_P) \times \mathbf{H}_{\text{per}}^1(\text{div}\,;\Omega_P)$ where

$$\mathcal{H}_P^1(\Omega_P) = \left\{ \psi \in \mathcal{H}^1(\Omega_P) \mid \psi(\mathbf{x} + \mathbf{t}_k) = \psi(\mathbf{x})e^{i\kappa g_k} \quad \forall\, \mathbf{x} \in \Gamma_{-k} \text{ a.e.}, \quad k = 1,2 \right\}$$

and

$$\mathbf{H}_{\text{per}}^1(\text{div}\,;\Omega_P) = \Big\{ \mathbf{Q} \in \mathbf{H}^1(\Omega_P) \mid \text{div}\,\mathbf{Q} = 0 \quad \text{in} \quad \Omega_P \text{ a.e.}, \quad \int_{\Omega_P} \mathbf{Q}\,d\Omega = 0$$
$$\text{and} \quad \mathbf{Q}(\mathbf{x} + \mathbf{t}_k) = \mathbf{Q}(\mathbf{x}) \quad \forall\, \mathbf{x} \in \Gamma_{-k} \text{ a.e.}, \quad k = 1,2 \Big\}.$$

In this gauge, the minimizers satisfy the Ginzburg-Landau equations in the following form:

$$\mathbf{curl}\,\mathbf{curl}\,\mathbf{Q} + |\psi|^2(\mathbf{Q} - \mathbf{A}_0) - \frac{1}{\kappa}\Im(\psi^*\mathbf{grad}\,\psi) = 0 \quad \text{in}\ \Omega_P$$

$$-\frac{1}{\kappa^2}\Delta\psi + (|\mathbf{Q} - \mathbf{A}_0|^2 + |\psi|^2 - 1)\psi + \frac{2i}{\kappa}\mathbf{grad}\,\psi \cdot (\mathbf{Q} - \mathbf{A}_0) = 0 \quad \text{in}\ \Omega_P$$

together with natural boundary conditions

$$\mathbf{curl}\,(\mathbf{Q} - \mathbf{A}_0)|_{\mathbf{x}+\mathbf{t}_k} = \mathbf{curl}\,(\mathbf{A} - \mathbf{A}_0)|_{\mathbf{x}} \quad \forall\, \mathbf{x} \in \Gamma_{-k},\ k = 1,2$$

$$\Big(\mathbf{grad}\,|\psi|\Big)|_{\mathbf{x}+\mathbf{t}_k} = \Big(-\mathbf{grad}\,|\psi|\Big)|_{\mathbf{x}} \quad \forall\, \mathbf{x} \in \Gamma_{-k},\ k = 1,2$$

$$|\psi|\Big(\mathbf{grad}\,\omega - \kappa(\mathbf{Q} - \mathbf{A}_0)\Big)|_{\mathbf{x}+\mathbf{t}_k}$$
$$= |\psi|\Big(\mathbf{grad}\,\omega - \kappa(\mathbf{Q} - \mathbf{A}_0)\Big)|_{\mathbf{x}} \quad \forall\, \mathbf{x} \in \Gamma_{-k},\ k = 1,2.$$

We have established various properties of the solutions of the periodic Ginzburg-Landau model. Some of the results, such as the boundedness of the order parameter and the discussion on simple analytic solutions, are silmilar to that for bounded regions. Unlike the latter case, for the periodic case we have been able to obtain a complete regularity theory. In fact we have shown that solutions of the nonlinear Ginzburg-Landau equations may be extended to $\mathbb{R}^2$ periodicically for the reduced magnetic potential $\mathbf{Q}$ and "quasi"-periodically for the order parameter ψ and that these extentions are infinitely differentiable funtions that satisfy the Ginzburg-Landau equations in all $\mathbb{R}^2$ in the classical sense.

We have also studied finite element approximations of the periodic Ginzburg-Landau model. The periodicity of the physical variables implies non-standard, in the context of periodic problems, relations for the primary dependent variables employed in the model, e.g., see (7). As far as boundary conditions are concerned, functions belonging to the finite element space are required to satisfy (7) only at the nodes on the boundary, i.e.,

$$\psi^h(\mathbf{x}_j + \mathbf{t}_k) = \psi^h(\mathbf{x}_j)e^{i\kappa g_k(\mathbf{x}_j)} \quad \forall\, \mathbf{x}_j \in \Gamma_{-k},\, k = 1, 2,$$

where $\mathbf{x}_j$ may be any boundary vertex of a triangle for piecewise linear and quadratic elements and may also be the midpoint of any triangular edge on the boundary for quadratic elements. The periodicity of the finite element solutions with respect to the lattice is also defined only at interpolating nodes. Thus, the finite element spaces are not subspaces of those used for the exact solution space. The analysis of approximations was centered around this issue.

Our analysis of finite element approximations resulted in optimal error estimates, i.e.,

$$\|\psi - \psi^h\|_s \leq Ch^{p+1-s}\|\psi\|_{p+1} \quad \text{and} \quad \|\mathbf{Q} - \mathbf{Q}^h\|_s \leq Ch^{p+1-s}\|\mathbf{Q}\|_{p+1}, \quad s = 0, 1,$$

where $p = 1$ for continuous piecewise linear elements and $p = 2$ for continuous piecewise linear quadratic elements.

The key to the analysis of finite element errors is an estimate for a boundary integral term which would disappear if the finite element spaces were subspaces of the exact solution spaces $\mathcal{H}_P^1(\Omega_P) \times \mathbf{H}_{\text{per}}^1(\text{div}\,;\Omega_P)$. We showed that this integral, although not zero in general, is of higher order error compared to approximation theoretic errors and thus the optimal convergence rates are retained. The idea may well be generalized to higher-order elements with suitable choices of interpolation procedures.

References

1. F. Brezzi, J. Rappaz and P.-A. Raviart, Finite-dimensional approximation of nonlinear problems. Part I: branches of nonsingular solutions, *Numer. Math.* **36**, 1980, 1-25.
2. P. Ciarlet, *The Finite Element Method for Elliptic Problems*, North-Holland, Amsterdam, 1978.
3. Q. Du, M. Gunzburger and J. Peterson, Analysis and approximation of Ginzburg-Landau models for superconductivity, *SIAM Review*, to appear.
4. Q. Du, M. Gunzburger and J. Peterson, Modeling and analysis of a periodic Ginzburg-Landau model for type-II superconductors, to appear.
5. Q. Du, M. Gunzburger and J. Peterson, Finite element approximation of a periodic Ginzburg-Landau model for type-II superconductors, to appear.
6. M. Tinkham, *Introduction to Superconductivity*, McGraw-Hill, New York, 1975.

Qiang Du
Department of Mathematics
Michigan State University
East Lansing, MI 48224, USA

Max Guznburger and Janet Peterson
Department of Mathematics and
Interdisciplinary Center for Applied Mathematics
Virginia Tech
Blacksburg, VA 24061, USA

J H KIM AND K S HA
Comparison theorems for solutions of one dimensional stochastic differential and integral equations of Ito's type

Lakshmikantham-Zhang [5] considered a comparison theorem for solutions of the stochastic differential equation of the form

$$u'(t,\omega) = f(t,u(t,\omega),\omega)$$

$$u(0,\omega) = u_0(\omega)$$

in a probability space.

In this talk, we investigate comparison theorems for solutions of the stochastic differential equation

$$du(t) = \sigma(t,u(t))dB(t) + f(t,u(t))dt$$

$$u(0) = u_0,$$

equivalently

$$u(t) = u_0 + \int_0^t \sigma(s,u(s))dB(s) + \int_0^t f(s,u(s))ds \tag{1}$$

and the stochastic integral equation

$$u(t) = u_0 + \int_0^t \sigma(t,s,u(s))dB(s) + \int_0^t f(t,s,u(s))ds \tag{2}$$

of Ito's type under some conditions for the coefficient functions. Here $B(t)$, $t \geq 0$, is one dimensional standard Brownian motion in a probability space (Ω, F, P) with reference family F_t, $t \geq 0$.

Now we prepare the useful functions in this talk which appeared in [2], [6]. Let $a_0 = 1$, $a_n = 2/(n(n+1)+2)$ and let $\phi_n(x)$ be a continuous function such that its support is contained in (a_n, a_{n-1}),

$$0 \leq \phi_n(x) \leq 1/nx^2 \text{ and } \int_{a_n}^{a_{n-1}} \phi_n(x)dx = 1 \text{ for } n=1,2,\cdots.$$

Set

$$\psi_n(x) = \int_0^x dy \int_0^y \phi_n(z)dz \text{ if } x>0, \quad \psi_n(x) = 0 \text{ if } x \le 0 \qquad (3)$$

for $n=1,2,\cdots$. Then $a_0 = 1 > a_1 > a_2 > \cdots > a_n \to 0$ as $n \to \infty$, $\psi_n \in C^2(R)$,

$$0 \le \psi_n'(x) \le 1, \quad 0 \le \psi_n''(x) \le 1/nx^2 \text{ and } \psi_n(x) \uparrow x^+ \text{ as } n \to \infty,$$

where $x^+ = \max(0,x)$ for $x \in R$.

Using Ito's formula, we have the following result.

Lemma 1. Let $u(t)$ be a solution of (2) and $\psi \in C^2(R)$. Then

$$\psi(u(t)) = \psi(u_0) + \int_0^t \psi'(u(s))\sigma(s,s,u(s))dB(s)$$

$$+ \int_0^t \psi'(u(s))f(s,s,u(s))ds + \int_0^t \psi'(u(s))(\int_0^s \frac{\partial \sigma}{\partial s}(s,r,u(r))dB(r))ds$$

$$+ \int_0^t \psi'(u(s))(\int_0^s \frac{\partial \sigma}{\partial s}(s,r,u(s))dr)ds + \frac{1}{2}\int_0^s \psi''(u(s))\sigma^2(s,s,u(s))ds.$$

We now discuss a comparison theorem for solutions of the stochastic differential equation (1).

Theorem 2. Suppose that $\sigma(t,x)$, $f(t,x)$, $g(t,x) : [0,\infty) \times R \to R$ is continuous functions satisfying

$$|\sigma(t,x) - \sigma(t,y)| \le |x - y|$$

for every $x, y \in R$, $t \ge 0$,

$$f(t,x) - f(t,y) \le M(x - y)$$

for every $x, y \in R$ with $x \ge y$, $t \ge 0$ and some $M > 0$. Let $u(t)$, $v(t)$ be solutions of

$$u(t) = u_0 + \int_0^t \sigma(s,u(s))dB(s) + \int_0^t f(s,u(s))ds \qquad (4)$$

$$v(t) = v_0 + \int_0^t \sigma(s,v(s))dB(s) + \int_0^t g(s,v(s))ds \qquad (5)$$

respectively. Assume that $f(t,x) \le g(t,x)$ for every $t \ge 0$ and $x \in R$.

If $u_0 \le v_0$, then $u(t) \le v(t)$ for every $t \ge 0$ a.s..

Sketch of proof. From (3),(4),(5) and Ito's formula, we have

$$\psi_n(u(t) - v(t)) = I_0(n) + I_1(n) + I_2(n) + I_3(n),$$

where

$$I_0(n) = \psi_n(u_0 - v_0),$$

$$I_1(n) = \int_0^t \psi_n'(u(s) - v(s))(\sigma(s,u(s)) - \sigma(s,v(s)))dB(s),$$

$$I_2(n) = \int_0^t \psi_n'(u(s) - v(s))(f(s,u(s)) - g(s,v(s)))ds,$$

$$I_3(n) = \int_0^t \psi_n''(u(s) - v(s))(\sigma(s,u(s)) - \sigma(s,v(s)))^2 ds.$$

Taking the expectation, since $E(I_0(n)) = E(I_1(n)) = 0$,

$$E(\psi_n(u(t) - v(t))) \le M \int_0^t E((u(s) - v(s))^+)ds + \frac{t}{n}.$$

Hence, as $n \to \infty$, we have

$$E((u(t) - v(t))^+) \le M \int_0^t E((u(s) - v(s))^+)ds.$$

From Gronwall's lemma, $E((u(t) - v(t))^+) = 0$ and thus $u(t) \le v(t)$ for every $t \ge 0$ a.s.. (cf. [1],[2],[6]).

We next investigate a comparison theorem for solutions of stochastic integral equation (2).

Theorem 3. Set $I_{t \ge s} = \{(t,s) \in [0,\infty) \times [0,\infty) | t \ge s\}$. Suppose that $\sigma(t,s,x)$: $I_{t \ge s} \times C([0,\infty)) \to R$ is a continuous function satisfying that

(i) for every $x, y \in C([0,\infty))$, there exists a nonnegative continuous function $k_1(t)$ on $[0,\infty)$ such that

$$|\sigma(t,t,x) - \sigma(t,t,y)| \le k_1(t)|x(t) - y(t)|$$

and

$$|\sigma(t,s,x)|^2 \le k_1(t)k_1(s)(1 + \sup_{r \le s}|x(r)|^2),$$

(ii) $\frac{\partial \sigma}{\partial t}(t,s,x)$ exists and for every $x, y \in C([0,\infty))$ there exists a nonnegative continuous function $k_2(t)$ on $[0,\infty)$ such that

$$\sup_{s \leq t} \left| \frac{\partial \sigma}{\partial t}(t,s,x) - \frac{\partial \sigma}{\partial t}(t,s,y) \right| \leq k_2(t) |x(t) - y(t)|$$

and

$$\left| \frac{\partial \sigma}{\partial t}(t,s,x) \right|^2 \leq k_2(t) k_2(s) (1 + \sup_{r \leq s} |x(r)|^2).$$

And also let $f(t,s,x)$, $g(t,s,x) : I_{t \leq s} \times C([0,\infty)) \to R$ be continuous functions satisfying conditions (i) and (ii) as the function $\sigma(t,s,x)$. Let $u(t)$ and $v(t)$ be solutions of

$$u(t) = u_0 + \int_0^t \sigma(t,s,u(s)) dB(s) + \int_0^t f(t,s,u(s)) ds \tag{6}$$

$$v(t) = v_0 + \int_0^t \sigma(t,s,v(s)) dB(s) + \int_0^t g(t,s,v(s)) ds \tag{7}$$

respectively. Assume that $f(t,t,x) \leq g(t,t,x)$ for every $t \geq 0$ and $x \in C([0,\infty))$, $\frac{\partial f}{\partial t}(t,s,x) \leq \frac{\partial g}{\partial t}(t,s,x)$ for every $(t,s) \in I_{t \geq s}$ and $x \in C([0,\infty))$. If $u_0 \leq v_0$, then $u(t) \leq v(t)$ for every $t \geq 0$ a.s..

<u>Sketch of proof.</u> From (3), (6), (7) and Lemma 1, we have

$$\psi_n(u(t) - v(t)) = J_0(n) + J_1(n) + J_2(n) + J_3(n) + J_4(n) + J_5(n),$$

where

$$J_0(n) = \psi_n(u_0 - v_0),$$

$$J_1(n) = \int_0^t \psi_n'(u(s) - v(s))(\sigma(s,s,u(s)) - \sigma(s,s,v(s))) dB(s),$$

$$J_2(n) = \int_0^t \psi_n'(u(s) - v(s))(f(s,s,u(s)) - g(s,s,v(s))) ds,$$

$$J_3(n) = \int_0^t \psi_n'(u(s) - v(s)) \left(\int_0^s \left(\frac{\partial \sigma}{\partial s}(s,r,u(r)) - \frac{\partial \sigma}{\partial s}(s,r,v(r)) \right) dB(r) \right) ds,$$

$$J_4(n) = \int_0^t \psi_n'(u(s) - v(s))(\int_0^s (\frac{\partial f}{\partial s}(s,r,u(r)) - \frac{\partial g}{\partial s}(s,r,v(s))dr)ds$$

and

$$J_5(n) = \frac{1}{2}\int_0^t \psi_n''(u(s) - v(s))(\sigma(s,s,u(s)) - \sigma(s,s,v(s)))^2 ds.$$

Taking the expectation, since $E(J_0(n)) = E(J_1(n)) = E(J_3(n)) = 0$,

$$E(\psi_n(u(t) - v(t)))$$

$$\leq (\max_{0\leq s\leq t} k_1(s) + t \max_{0\leq s\leq t} k_2(s))\int_0^t E((u(s) - v(s))^+)ds + \frac{t}{2n}\max_{0\leq s\leq t} k_1^2(s).$$

Hence, as $n \to \infty$, we have

$$E((u(t) - v(t))^+) \leq (\max_{0\leq s\leq t} k_1(s) + t\max_{0\leq s\leq t} k_2(s))\int_0^t E((u(s) - v(s))^+)ds.$$

From Gronwall's lemma, $E((u(t) - v(t))^+) = 0$ and thus $u(t) \leq v(t)$ for every $t \geq 0$ a.s.. (cf. [3],[4]).

References

[1] K.S. Ha and J.H. Kim, Monotone iterative technique for 1-dimensional stochastic differential equations, Stoc. Anal. Appl. 6 (1988), 191-203.

[2] N. Ikeda and S. Watanabe, Stochastic differential equations and diffusion processes, Kodansha, Tokyo, 1981.

[3] J.H. Kim, On comparison theorem for solutions of Ito-Volterra equations, J. Sci. PNU 51 (1991), 47-50.

[4] J.H. Kim and K.S. Ha, Monotone iterative technique for 1-dimensional Ito-Volterra integral equations, to appear in Stoc. Anal. Appl., 1991.

[5] V. Lakshmikantham and B.G. Zhang, On the method of sample upper and lower solutions for stochastic differential equations, Stoc. Anal. Appl. 3 (1985), 341-347.

[6] T. Yamada, On a comparison theorem for solutions of stochastic differential equations and its applications, J. Math. Kyoto Math. 13 (1973), 497-512.

Jai Heui KIM and Ki Sik HA   Department of Mathematics
Pusan National University   Pusan, 609-735   Republic of Korea

P W HAMMER
Applications of quasilinearization to parameter identification in partial differential equations

ABSTRACT We develop a technique for identifying unknown coefficients in a nonlinear parabolic partial differential equation. The identification scheme is based on quasilinearization and is applied to equations where the unknown coefficients may be spatially varying.

1. INTRODUCTION

In this paper we wish to summarize our efforts to date involving the application of quasilinearization to parameter identification in nonlinear partial differential equations. Quasilinearization has been successfully applied to parameter identification in linear partial differential equations. In [4], we considered the heat equation where the unknown parameter was a spatially varying diffusion coefficient and established both theoretical and numerical convergence of a quasilinearization based parameter identification algorithm. We believe the framework presented in [4], may be extended for use in parameter identification for a much broader class of partial differential equations. The numerical results presented here are quite encouraging.

The nonlinear parabolic equation we consider is Burgers' equation, given by

$$u_t = (qu_x)_x - uu_x + g.$$

This equation was introduced by Burgers as a simple model for turbulence where $q(x) > 0$ represents the viscosity coefficient. Other applications include supersonic flow about airfoils, traffic flows, and acoustic transmission as described in [6].

The parameter identification problem can be stated as follows. Consider the model

$$u_t = (qu_x)_x - uu_x + g \tag{1.1}$$
$$u(t,0) = u(t,1) = 0 \tag{1.2}$$
$$u(0,x) = u_0(x) \tag{1.3}$$

for $x \in [0,1], t > 0$. Given data u_{ij} at discrete times $t_i, i = 1, 2, \ldots m$ and discrete locations $x_j, j = 1, 2, \ldots r$, we seek to identify $q^*(x)$ in some admissible parameter space **Q** that minimizes

$$J(q) = \frac{1}{2}\sum_{i=1}^{m}\sum_{j=1}^{r}(u(t_i, x_j; q) - u_{ij})^2$$

where $u(t,x,q)$ denotes the solution of (1.1)–(1.3) corresponding to q.

---
*The work of this author was supported in part by the Air Force Office of Scientific Research under grant AFOSR-88-0074 and Defense Advanced Research Projects, ACMP, under contract F49620-87-C-0016.

2. The Quasilinearization Algorithm

The first step in defining the qusilinearization based algorithm is to rewrite (1.1)–(1.3) in an abstract setting. Define $X = L^2(0,1)$ and for each $q(x) \in H^1(0,1)$ define the operator $A(q): D \subset X \to X$

$$A(q)\Psi = q(D_x^2 \Psi) + (D_x q)(D_x \Psi)$$

for all $\Psi \in D = H^2(0,1) \cap H_0^1(0,1)$. It is well established in the literature that (1.1)–(1.3) can be rewritten as an abstract Cauchy problem in X. That is, $u(t,x;q)$ is the unique solution of (1.1)–(1.3) corresponding to q iff $U(t;q) = u(t,\cdot;q)$ is the unique solutions of the following problem in X

$$\dot{U}(t) = A(q)U(t) + F(t, U(t)) + G(t) \qquad (2.1)$$
$$U(0) = u_0 \qquad (2.2)$$

where $F(t,U) = -UU_x$ and $G(t) = g(t,\cdot)$.

In this setting the parameter identification problem can be restated as follows. Given data u_{ij}, we must determine $q^*(x)$ in some admissible parameter space **Q** that minimizes

$$J(q) = \frac{1}{2} \sum_{i=1}^{m} |CU(t_i; q) - Y_i|_{\mathbf{R}^r}^2 \qquad (2.3)$$

where $U(t;q)$ denotes the solution of (2.1)–(2.2) corresponding to q, C is the linear operator from X into $\mathbf{R}^r$ defined by $Cf = (f(x_1), f(x_2) \ldots f(x_r))$ and $Y_i = (u_{i1}, u_{i2}, \ldots u_{ir})$ for $i = 1, 2 \ldots m$.

As seen in [4], the restrictions on the admissable parameter space **Q** are those that ensure the solution of (2.1)–(2.2) is continuously Fréchet differentiable with respect to q. For the purpose of this paper, we will assume each $q(x)$ is smooth enough to ensure that the Fréchet derivative, denoted $D_q U(t;q)$, exists and that $[D_q U(t,q_0)h](x) = v(t,x)$ satisfies the following sensitivity equation

$$v_t = (q_0 v_x)_x - uv_x - u_x v + (hu_x)_x$$
$$v(t, 0) = v(t, 1) = 0$$
$$v(0, x) = 0$$

for each $q_0, h \in \mathbf{Q}$ where $u = u(t, x; q_0)$ is the solution of (1.1)–(1.3) corresponding to q_0. The precise definition of the admissable parameter space **Q** is the subject of ongoing investigation.

We parallel the quasilinearization approach taken in [4] and attempt to determine the minimizer $q^*(x)$ in some admissable parameter space **Q** with orthonormal basis $\{h_i(x) | i = 1, 2, \ldots\}$. We reformulate the parameter identification problem over **Q** as an equivalent parameter identification problem over l^2. For each $\alpha = (\alpha_1, \alpha_2, \ldots) \in l^2$, we define $q_\alpha = \sum_{i=1}^{\infty} \alpha_i h_i(x)$. Then starting with some initial estimate $\alpha^0 = (\alpha_1^0, \alpha_2^0, \ldots) \in l^2$ ($q_{\alpha^0} = \sum_{i=1}^{\infty} \alpha_i^0 h_i(x)$) the quasilinearization algorithm takes the following iterative form in l^2:

$$\alpha^{k+1} = \alpha^k - [D(\alpha^k)]^{-1} \sum_{i=1}^{m} M^T(t_i; q_{\alpha^k})[CU(t_i; q_{\alpha^k}) - Y_i]$$
$$= H(\alpha^k)$$

where $M(t;q) = [CD_qU(t;q)h_1 \quad CD_qU(t;q)h_2 \ldots]$ and $D(\alpha) = \sum_{i=1}^{m} M^T(t_i;q_\alpha)M(t_i;q_\alpha)$. Local convergence of the algorithm depends primarily on continuous Fréchet differentiability of the state with respect to the parameter, as seen in the following theorem.

THEOREM 2.1. *Assume* $q \to U(t;q)$ *is continuously Fréchet differentiable for each* $q \in \mathbf{Q}$. *Moreover, assume* $D(\alpha^*)^{-1}$ *exists,* $\alpha^* = H(\alpha^*)$, *and* $J(q_{\alpha^*}) = 0$. *Then* α^* *is a point of attraction of the iterative scheme* $\alpha^{k+1} = H(\alpha^k)$.

Details of the proof may be found in [3] and [4]. We refer the reader to [4] for a complete derivation of the quasilinearization based algorithm.

We generally expect the admissable parameter space $\mathbf{Q}$ to be infinite dimensional and clearly, this leads to an algorithm that is not implementable due to the infinite dimensional matrix $M(t;q)$. We can only implement the algorithm on finite dimensional subspaces $\mathbf{Q}^s \subset \mathbf{Q}$ and so we choose the spaces $\mathbf{Q}^s$ which approximate $\mathbf{Q}$. For example, $\mathbf{Q}^s$ may be a space of nth degree polynomials, of linear splines, or of cubic splines. Then the algorithm takes a similar iterative form on $\mathbf{R}^s$.

3. NUMERICAL IMPLEMENTATION AND EXAMPLES

The iterative scheme developed in the previous section may be summarized in the following steps.

1. Start with $\alpha^k = (\alpha_1^k, \alpha_2^k, \ldots \alpha_s^k) \in \mathbf{R}^s$ and define $q_{\alpha^k} = \sum_{i=1}^{s} \alpha_i^k g_i(x)$ for $\{g_i(x) : i = 1, 2, \ldots s\}$ a basis for $\mathbf{Q}^s$.
2. Solve

$$u_t = (q_{\alpha^k} u_x)_x - u u_x + g$$
$$u(t,0) = u(t,1) = 0$$
$$u(0,x) = u_0(x).$$

3. For each $i = 1, 2 \ldots s$, solve

$$v_t^i = (q_{\alpha^k} v_x^i)_x - u v_x - u_x v + (g_i u_x)_x$$
$$v^i(t,0) = v^i(t,1) = 0$$
$$v^i(0,x) \equiv 0$$

where $u \equiv u(t,x;q_{\alpha^k})$, the solution found in step 2.

4. For $i = 1, 2, \ldots m$, construct the matrices

$$M(t_i;q_{\alpha^k}) = \begin{bmatrix} v^1(t_i,x_1) & v^2(t_i,x_1) & \cdots & v^s(t_i,x_1) \\ v^1(t_i,x_2) & & & \vdots \\ \vdots & & & \\ v^1(t_i,x_r) & \cdots & \cdots & v^s(t_i,x_r) \end{bmatrix}.$$

5. Solve $\alpha^{k+1} = H(\alpha^k)$.
6. Replace α^k with α^{k+1} and repeat algorithm.

In order to solve the state and sensitivity equations, we use a standard finite element scheme to approximate each partial differential equation by a system of ordinary differential equations. Then, we implement a fourth order Runge-Kutta routine with fixed step size. Standard software packages were used only to calculate integrals, to invert matrices, and to generate random numbers (IMSL routines DQDAGS, DLFSRG, DLFTRG, RNUN).

The first example we consider is a very simple one in which we assume it is known that the viscosity coefficient is constant. The model is given by

$$u_t = qu_{xx} - uu_x$$
$$u(t,0) = u(t,1) = 0$$
$$u(0,x) = \sin\pi x.$$

We choose the true parameter value q^* and then use a standard finite element scheme to determine the approximate solution corresponding to q^*. We generate data at times $t_i = \frac{1}{2}, 1, \frac{3}{2}$ and $x_j = \frac{1}{17}, \frac{2}{17}, \ldots \frac{16}{17}$ according to this approximate solution and then add random noise.

Example 3.1. We choose $q^* = \frac{1}{60}$ and start with an initial estimate $q^0 = 0.1$. The iterative results are summarized in Table 3.1 with a fit to data curve at t_1, presented in Figure 3.1.

We turn now to more complicated examples where we allow for a spatially varying viscosity coefficient. All of the remaining examples were formulated in the following manner. We consider the model

$$u_t = (qu_x)_x - uu_x + g \tag{3.1}$$
$$u(t,0) = u(t,1) = 0 \tag{3.2}$$
$$u(0,x) = u_0(x). \tag{3.3}$$

A true parameter function $q^*(x)$ and true solution $u^*(t,x)$ were chosen first. For each choice of q^*, u^* we chose

$$g(t,x) = u_t^* - (q^*u_x^*)_x + u^*u_x^* \tag{3.4}$$
$$u_0(x) = u^*(0,x). \tag{3.5}$$

Data $u_{ij} = u^*(t_i, x_j)$ was generated for $t_i = \frac{1}{2}, 1, \frac{3}{2}$ and $x_j = \frac{1}{10}, \frac{2}{10}, \ldots \frac{9}{10}$ and based on this data, we determined $q^{s*}(x) \in \mathbf{Q}^s = \{\alpha_0 + \alpha_1 x + \alpha_2 x^2 + \alpha_3 x^3 + \alpha_4 x^4 : \alpha_0, \alpha_1, \alpha_2, \alpha_3, \alpha_4 \in \mathbf{R}\}$ that minimizes (2.3) where $u(t,x;q)$ is the solution of (3.1)–(3.3) corresponding to q with $g(t,x)$ and $u_0(x)$ fixed as in (3.4) and (3.5). A variety of choices were made for $q^*(x)$ with $u^*(t,x) = (x - x^2)(t^2 + t + 1)$ in each example.

Example 3.2. We choose $q^*(x) = 2 - x^2$ and start the iterative scheme with an initial estimate $\alpha^0 = (.2, .2, .2, .2, .2)$, $(q^0(x) = .2 + .2x + .2x^2 + .2x^3 + .2x^4)$. Since $q^* \in \mathbf{Q}^s$, we expect accurate results. The computational findings are presented in Table 3.2 with corresponding $q_{\alpha k}(x)$ presented in Figure 3.2.

Example 3.3. We choose $q^*(x) = \frac{1}{x+\frac{1}{4}}$ and start the iterative scheme with an initial estimate $\alpha^0 = (.2, .2, .2, .2, .2)$. We observe that $q^*(x)$ does not belong to $\mathbf{Q}^s$. However, since $q^*(x)$ does have a power series representation, we still expect q^{s*} to provide a very accurate approximation to q^*. The results are summarized in Table 3.3 and Figure 3.3.

Example 3.4. We choose a discontinuous function for $q^*(x)$, $q^*(x) = \frac{1}{2}$ on $[0, \frac{1}{4}]$ and $\frac{1}{4}$ on $(\frac{1}{4}, 1]$. We start with an initial estimate $\alpha^0 = (.0, .1, .1, .1, .1)$. The results are summarized in Table 3.4 and Figure 3.4

4. Concluding Remarks

In this paper we have presented a quasilinearization based algorithm that successfully identifies unknown parameters in a nonlinear parabolic partial differential equation. The numerical examples of section 3 demonstrate the rapid convergence and accuracy of the algorithm. We believe that theoretical convergence can be established as well as rates of convergence. As seen in [3] and [4], theoretical convergence depends primarily on continuous Fréchet differentiability of the state with respect to the unknown parameter. Thus, the admissable parameter space must be determined in such a way to ensure this differentiability. The proper choice for the admissable parameter space is the main focus of our current research efforts and will be the subject of future papers.

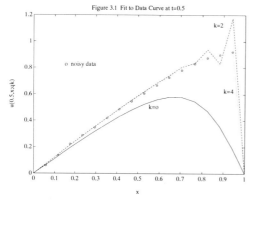

Figure 3.1 Fit to Data Curve at t=0.5

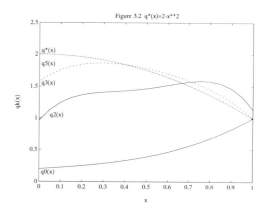

Figure 3.2 $q^*(x)=2-x^{**}2$

k	q^k	$J(q^k)$
0	0.1000	1.22858
1	0.00582	0.32470
2	0.01055	0.06665
3	0.01529	0.00596
4	0.01706	0.00207
5	0.01722	0.00205
6	0.01723	0.00205
7	0.01723	0.00205

Table 3.1

k	α_0^k	α_1^k	α_2^k	α_3^k	α_4^k	$J(q_{\alpha k})$
0	0.200	0.200	0.200	0.200	0.200	2.13295570
1	0.495	2.593	−8.601	13.345	−6.422	0.32500487
2	0.963	3.991	−13.531	20.497	−10.784	0.03588246
3	1.576	2.379	−6.325	6.158	−2.783	0.00205491
4	1.935	0.600	−2.621	1.720	−0.636	0.00002654
5	1.997	0.050	−1.213	0.307	−0.144	0.00000036
6	2.000	0.000	−1.002	0.001	−0.001	0.00000034

Table 3.2

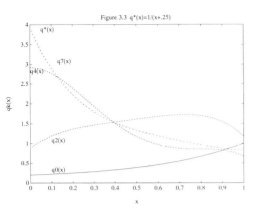

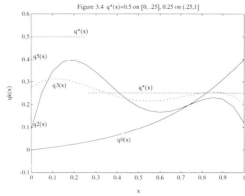

k	α_0^k	α_1^k	α_2^k	α_3^k	α_4^k	$J(q_{\alpha k})$
0	.1000	.1000	.1000	.1000	.1000	241.060655
1	.3592	−.9493	1.4264	.0905	.2227	51.477515
2	.3084	−2.8407	3.4551	0.1854	.4561	10.174342
3	6.6950	−16.5298	10.3804	−0.0546	.9171	1.780307
4	13.1743	−20.9273	5.2363	1.8847	1.6815	.279966
5	−23.1152	50.7312	−33.4424	3.7576	2.6851	.049967
6	0.1017	−4.2464	10.0112	−8.7447	3.6568	.001242
7	8.5001	−23.1571	23.9896	−12.4156	3.9157	.000008
8	7.7101	−21.4753	22.8379	−12.1525	3.9054	.000005
9	7.6639	−21.3755	22.7684	−12.1362	3.9046	.000005

Table 3.3

k	α_0^k	α_1^k	α_2^k	α_3^k	α_4^k	$J(q_{\alpha k})$
0	0.000	0.100	0.100	0.100	0.100	0.47108943
1	−0.182	4.283	−19.284	31.143	−16.153	0.13197294
2	0.091	3.969	−16.346	22.787	−10.384	0.02170193
3	0.275	0.740	−4.365	7.037	−3.482	−0.00257036
4	0.374	−0.394	−0.213	1.306	−0.834	0.00006586
5	0.395	−0.656	0.732	0.036	−0.264	0.00001777
6	0.396	−0.666	0.765	−0.004	−0.246	0.00001770

Table 3.4

References

[1]. H.T. Banks and G.M. Groome, Jr.,, *Convergence Theorems for Parameter Estimation by Quasilinearization*, J. Math Anal. Appl. **42** (1973), 91–109.

[2]. D.W. Brewer, *The Differentiability with Respect to a Parameter of the Solution of a Linear Abstract Cauchy Problem*, SIAM J. Math. Anal. **13** (1982), 607–620.

[3]. D.W. Brewer, J.A. Burns and E.M. Cliff, *Parameter Identification for an Abstract Cauchy Problem by Quasilinearization*, Quart. J. Math. Anal. (submitted).

[4]. P.W. Hammer, *Parameter Identification in Parabolic Partial Differential Equations using Quasilinearization*, (submitted).

[5]. P.W. Hammer, *Parameter Identification in Parabolic Partial Differential Equations using Quasilinearization*, Virginia Tech Ph.D. dissertation (August 1990).

[6]. S. Kang and J.A. Burns, *A Control Problem for Burgers' Equation with Bounded Input/Output*, ICASE Report No. 90-45 (June 1990).

Department of Mathematics
Hollins College
Roanoke, VA 24020

D W BREWER† AND T L HERDMAN‡
A parameter dependence problem in neutral functional differential equations

Abstract. A parameter dependence problem for a class of neutral functional differential equations is considered. Our investigation is in the context of semigroup theory. The differentiability with respect to a parameter of the semigroup for the corresponding abstract Cauchy problem is established. The results are applicable to a class of singular integro-differential equations that occur in aeroelasticity.

1. Introduction. In [4] a complete dynamic model for the elastic motions of a three-degree-of-freedom "typical" airfoil section, with flap, in a two-dimensional, incompressible flow was formulated. An evolution equation for the circulation on the airfoil was derived and coupled to the rigid-body dynamics of the airfoil to obtain a complete set of functional differential equations that describe the composite system. The resulting model for the aeroelastic system including a forcing term f has the form

$$\frac{d}{dt}\left[Ax(t) + \int_{-\infty}^{0} A(s)x(t+s)ds\right] = Bx(t) + \int_{-\infty}^{0} B(s)x(t+s)ds + f(t) \qquad (1.1)$$

for $t > 0$, where $x(t) = col(h(t), \theta(t), \beta(t), \dot{h}(t), \dot{\theta}(t), \dot{\beta}(t), \Gamma(t), \dot{\Gamma}(t))$. The functions h, θ, β denote the plunge, pitch angle and flap angle respectively. The function Γ denotes the total airfoil circulation and is defined for all $t \in (-\infty, \infty)$, (see [4]). The 8×8 matrix A is singular (each entry of the last row is zero) while the 8×8 matrix function $A(s)$ is weakly singular ($A_{88}(s) = (Us-2)/Us)^{1/2}$, U denotes the undisturbed stream velocity). It is to be noted that the state of the system (1.1) includes the past history of $\dot{\Gamma}$, (see [4]). A "finite delay" version of (1.1), $r > 0$ a fixed constant,

$$\frac{d}{dt}\left[Ax(t) + \int_{-r}^{0} A(s)x(t+s)ds\right] = Bx(t) + \int_{-r}^{0} B(s)x(t+s)ds + f(t) \qquad (1.2)$$

has been studied extensively (see [4],[5],[9] for well-posedness; [7], [8] for approximation results and [6] for approximation of (1.1) using (1.2)).

We shall restrict our attention to the finite delay system (1.2) with the forcing term f being zero, and allow an unknown parameter q to be present in the right hand side of (1.2). Our investigation is motivated by the aeroelastic system described above. As an example of a possible parameter that one may need to identify, consider the 5th row 5th column entry for B in the aeroelastic system,

$$B_{55} = U^2\left[c\cos^{-1}c - \frac{1}{3}(2+c^2)\sqrt{1-c^2} - (\frac{1}{2}+a+a^2)\cos^{-1}c + (\frac{1}{6}+2a^2-2ac)\sqrt{1-c^2}\right],$$

---

†The work of this author was supported in part by the Air Force Office of Scientific Research under Grant AFOSR-89-0472.
‡The work of this author was supported in part by the Defense Advanced Research Projects Agency under Grant F49620-87-C-116.

where U, stream velocity, c, joint point of flap, and a, the point on the airfoil where the plunge $h(t)$ is measured, are constants. The corresponding 5th entry of the vector $x(t)$ is $\dot{\theta}(t)$. Consequently the 5th equation in system (1.2) contains the nonstatic term $B_{55}\dot{\theta}(t)$ where B_{55} is given above. The coefficient B_{55} is derived using thin airfoil theory and may provide only an approximation for the true coefficient of $\dot{\theta}(t)$. One could use the approximation together with a parameter identification scheme to obtain a better approximation for the coefficient B_{55}. Differentiability results of the type found in Section 3 have been used to establish the convergence of a parameter estimation algorithm for an abstract Cauchy problem in [2]. An application of this method to a Volterra integro-differential equation may be found in [3].

2. Abstract Setting. We consider the linear functional differential equation of neutral-type

$$\frac{d}{dt}Dx_t = Lx_t, \quad t > 0 \tag{2.1}$$

together with the initial data

$$\lim_{t \to 0^+} Dx_t = \eta, \quad x_0 = \phi \tag{2.2}$$

where D and L are linear $\mathbb{R}^n$-valued operators with domains $\mathcal{D}(D)$ and $\mathcal{D}(L)$ contained in the space of Lebesque-measurable $\mathbb{R}^n$-valued functions on $[-r, 0]$. It is assumed that $\mathcal{D}(D) \cap \mathcal{D}(L)$ contains $W^{1,p} = W^{1,p}([-r, 0]; \mathbb{R}^n)$, the standard Sobolev space of $\mathbb{R}^n$-valued functions with $1 \le p < \infty$. The initial data (η, ϕ) belongs to $\mathbb{R}^n \times L_p$ where $L_p = L_p([-r, 0]; \mathbb{R}^n)$ denotes the standard Lebesque space of $\mathbb{R}^n$-valued functions. Our objective here is to present sufficient conditions for D and L that yield the well-posedness of (2.1)–(2.2) on the product space $\mathbb{R}^n \times L_p$.

The operator D is said to satisfy *Condition $\mathcal{H}$ with index p*, $1 \le p < \infty$, provided that there exists an $n \times n$ matrix function $H(\lambda) = (h_{ij}(\lambda))$ such that $h_{ij}(\lambda) \to 0$ as $\lambda \to \infty$, $\lambda \in \mathbb{R}$, $i, j = 1, 2, \ldots, n$ and $D(e^{\lambda \cdot}I)\lambda^{1/p}H(\lambda) \to I$ as $\lambda \to \infty$, $\lambda \in \mathbb{R}$. For convenience we denote D satisfying Condition $\mathcal{H}$ with index p by $D \in \mathcal{H}(p)$. We denote the space of bounded linear operators from $W^{1,p}$ into $\mathbb{R}^n$ by $\mathcal{B}(W^{1,p}; \mathbb{R}^n)$.

Define the operator $\mathcal{A}$ with domain

$$\mathcal{D}(\mathcal{A}) = \{(\eta, \phi) \in \mathbb{R}^n \times L_p : \phi \in W^{1,p}, \quad D\phi = \eta\} \tag{2.5}$$

by

$$\mathcal{A}(\eta, \phi) = (L\phi, \dot{\phi}). \tag{2.6}$$

The abstract Cauchy problem corresponding to the neutral functional differential system (2.1)–(2.2) is given by

$$\dot{z}(t) = \mathcal{A}z(t), \quad z(0) = z_0 = (\eta, \phi) \in \mathcal{D}(\mathcal{A}). \tag{2.7}$$

Burns, Herdman and Turi (see [5, Theorem 3.4]) have established the following sufficiently results for $\mathcal{A}$ to generate a C_0-semigroup on $\mathbb{R}^n \times L_p$.

THEOREM 2.1. *Let $\mathcal{A}$ be defined by (2.5)–(2.6) where L and D belong to $\mathcal{B}(W^{1,p}; \mathbb{R}^n)$ with $D \in \mathcal{H}(p)$. If the abstract Cauchy problem (2.7) has a unique continuously differentiable solution on $[0, \infty)$, then $\mathcal{A}$ generates a C_0-semigroup on $\mathbb{R}^n \times L_p$.*

Throughout the remainder of this paper we will direct our attention to a class of neutral functional differential equations where D and L have a special structure that guarantees the sufficient conditions of Theorem 2.1. This class of equations will include the aeroelastic system (finite delay) proposed in [4]. Concerning the linear operators D and L we assume

(C1) $W^{1,p} \subseteq \mathcal{D}(D) \cap \mathcal{D}(L)$;
(C2) D and L belong to $\mathcal{B}(W^{1,p}; \mathbb{R}^n)$;
(C3) $D\phi = A\phi(0) + \int_{-r}^{0} A(s)\phi(s)ds$;
(C4) $L\phi = B\phi(0) + \int_{-r}^{0} B(s)\phi(s)ds$;
(C5) $A = diag(1, 1, \ldots, 1, 0)$ is a $n \times n$ diagonal matrix; $A(\cdot)$ is a $n \times n$ matrix function with $A_{ij}(s) = 0$ for $i = 1, 2, \ldots, n$, $j = 1, 2, \ldots, n-1$, $A_{nn}(s) = a|s|^{-\alpha} + \psi(s)$, $0 < \alpha < 1$, with $\psi(s)$ of bounded variation on $[-r, 0]$, $\psi(0) = 0$; $A_{in}(\cdot)$ is continuous on $[-r, 0]$, $i = 1, 2, \ldots, n-1$.
(C6) B is a constant $n \times n$ matrix with $B_{nn} = 0$; $B_{ij}(\cdot)$ is a $n \times n$ matrix function with $B_{ij}(s) = 0$ for $i = 1, 2, \ldots, n$, $j = 1, 2, \ldots, n-1$; $B_{in}(\cdot)$ is continuous on $[-r, 0]$, $j = 1, 2, \ldots, n-1$ and $B_{nn}(s) = 0$.

We employ results found in [5], (see Theorem 3.7 together with Example 3.8) to obtain the following result which yields the well-posedness of the system (2.1)–(2.2).

THEOREM 2.2. *If Conditions (C1)–(C6) hold then the operator $\mathcal{A}$ defined by (2.5)–(2.6) is the infinitesimal generator of a C_0-semigroup on $\mathbb{R}^n \times L_p$ for $p < 1/(1 - \alpha)$.*

The neutral functional differential equation describing the aeroelastic system (finite delay) satisfies the above conditions (C1)–(C6) with $n = 8$ and $\alpha = 1/2$, (see [5, page 389] for details). Consequently, we have the well-posedness for the aeroelastic system on the product space $\mathbb{R}^8 \times L_p$ for $1 \leq p < 2$.

We conclude this section by giving the representation of the C_0-semigroup $S(t)$ corresponding to the generator $\mathcal{A}$ defined by (2.5)–(2.6). For $(\eta, \phi) \in \mathbb{R}^n \times L_p$ the semigroup $S(t)$ is defined by

$$S(t)(\eta, \phi) = (Dx_t, x_t), \qquad t \geq 0$$

where $x(\cdot)$ is the unique solution of (2.1)–(2.2).

3. **Differentiability.** Motivated by the possibility of an unknown parameter in the L operator for the aeroelastic problem we shall now consider the neutral system discussed in Section 2 with $L = L(q)$, q an unknown parameter. In particular we shall consider the parameter dependent problem

$$\frac{d}{dt} Dx_t = L(q)x_t, \qquad t \geq 0 \tag{3.1}$$

$$\lim_{t \to 0^+} Dx_t = \eta, \qquad x_0 = \phi \tag{3.2}$$

where $(\eta, \phi) \in \mathbb{R}^n \times L(-r, 0) \equiv X$, (note we let $p = 1$) and $q \in Q$ a closed subspace of a normed linear space Q. We shall assume that D has representation (C3) while $L = L(q)$ will depend on q but be restricted to having representation

$$L(q)\phi = B(q)\phi(0) + \int_{-r}^{0} B(s, q)\phi(s)ds.$$

In addition to requiring that conditions (C1)–(C6) of Section 2 hold for D and $L = L(q)$ for each $q \in Q$ we will assume that the value α of condition (C5) is fixed, $\alpha = 1/2$. Consequently the system (3.1)–(3.2) with $q \in Q$ is well-posed in the sense that the system has a unique solution $x(\cdot)$ for every $(\eta, \phi) \in X = \mathbb{R}^n \times L(-r, 0)$ that is given by the semigroup

$$S(t; q)(\eta, \phi) = (Dx_t, x_t), \qquad t \geq 0 \tag{3.3}$$

where for each $q \in Q$, the semigroup $S(t; q)$ is generated by the operator $\mathcal{A}(q)$ defined on

$$\mathcal{D}(\mathcal{A}(q)) = \{(\eta, \phi) \in X : \phi \in W^{1,1}, \ D\phi = \eta\} \tag{3.4}$$

by
$$\mathcal{A}(q)(\eta,\phi) = (L(q)\phi, \dot\phi). \tag{3.5}$$

As an immediate consequence of semigroup theory for each $q \in Q$ we have the existence of a pair of constants $M(q)$ and $\omega(q)$ such that

$$\|S(t;q)(\eta,\phi)\|_X \leq M(q) e^{\omega(q)t} \|(\eta,\phi)\|_X, \quad t \geq 0 \tag{3.6}$$

for $(\eta,\phi) \in X$. We shall assume that constants M and ω independent of $q \in Q$ exist so that

C7) $\quad \|S(t,q)(\eta,\phi)\|_X \leq M e^{\omega t} \|(\eta,\phi)\|_X, \quad t \geq 0, \quad q \in Q, \quad (\eta,\phi) \in X.$

Condition (C7) yields the following estimate for the unique solution $x(\cdot)$ of (3.1)–(3.2)

$$\begin{aligned}\int_{t-r}^{t} |x(s)|ds = \int_{-r}^{0} |x_t(s)|ds &\leq |Dx_t| + \int_{-r}^{0} |x_t(s)|ds \\ &\leq \|S(t;q)(\eta,\phi)\|_X \\ &\leq M e^{\omega t} \|(\eta,\phi)\|_X, \quad t \geq 0, \quad q \in Q, \quad (\eta,\phi) \in X.\end{aligned} \tag{3.7}$$

We now turn our attention to the problem of differentiating $S(t;q)(\eta,\phi)$ with respect to q for any fixed $(\eta,\phi) \in X$. Differentiation will be in the sense of Fréchet. We shall state and prove three lemmas which will establish the conditions of hypothesis (H4)–(H6) found in Brewer [1]. The desired differentiability result will follow from Theorem 1 of [1]. It is to be noted that the problem of differentiating the semigroup $S(t;q)$ with respect to q is associated with the problem of differentiating with respect to q the solution of the Cauchy problem

$$\dot{z}(t) = \mathcal{A}(q)z(t), \quad z(0) = (\eta,\phi). \tag{3.8}$$

We are interested in the behavior of $\mathcal{A}(q)$ as q varies. Let $q, q_0 \in Q$ with q_0 fixed. For $(\eta,\phi) \in X$ define $\mathcal{B}(q)$ on X by

$$\mathcal{B}(q)(\eta,\phi) = (L(q)\phi, 0). \tag{3.9}$$

It now follows from (3.9) that the perturbation $\mathcal{A}(q) - \mathcal{A}(q_0)$ has the representation

$$\mathcal{A}(q) - \mathcal{A}(q_0) = \mathcal{B}(q) - \mathcal{B}(q_0).$$

LEMMA 3.1. *Let (C1)–(C7) hold. For every $q \in Q$, $T > 0$ there exists a constant K so that*

$$\int_0^T \|\mathcal{B}(q)S(t;q)(\eta,\phi)\|_X dt \leq K\|(\eta,\phi)\|_X \tag{3.10}$$

for $(\eta,\phi) \in X$.

PROOF: We employ (3.3) together with (3.9) to obtain the estimate

$$\begin{aligned}\int_0^T \|\mathcal{B}(q)S(t;q)(\eta,\phi)\|dt &= \int_0^T \|\mathcal{B}(q)(Dx_t, x_t)\|dt \\ &= \int_0^T |L(q)x_t|dt \\ &\leq |B(q)| \int_0^T |x(t)|dt + \int_0^T \int_{-r}^0 |B(s,q)|\,|x(t+s)|ds\,dt,\end{aligned}$$

where x denotes the solution of (3.1)–(3.2). Let k be the smallest integer such that $T \leq kr$. Inequality (3.7) yields

$$\int_0^T |x(t)|dt \leq \sum_{i=1}^k \int_{(i-1)r}^{ir} |x(t)|dt \leq \sum_{i=1}^k M e^{\omega ir}\|(\eta,\phi)\|$$
$$\leq kMe^{\omega kr}\|(\eta,\phi)\| \leq \left(\frac{T}{r}+1\right) Me^{\omega(T+r)}\|(\eta,\phi)\|.$$

Therefore, we obtain

$$\int_0^T \|\mathcal{B}(q)S(t;q)\,(\eta,\phi)\|dt \leq |B(q)|\left(\frac{T}{r}+1\right) Me^{\omega(T+r)}\|(\eta,\phi)\|$$
$$+ \|B(\cdot,q)\|_\infty \int_0^T \int_{-r}^0 |x_t(s)|ds\,dt$$
$$\leq |B(q)|\left(\frac{T}{r}+1\right) Me^{\omega(t+r)}\|(\eta,\phi)\|$$
$$+ \|B(\cdot,q)\|_\infty TMe^{\omega T}\|(\eta,\phi)\|$$
$$\leq K\|(\eta,\phi)\|$$

where the constant K depends on q and T since r is fixed and M, ω are independent of q. ∎

We now introduce the mapping $\hat{B}: Q \to L^\infty((-r,0); \mathbb{R}^{n\times n})$ defined by

$$\hat{B}(q) = B(\cdot,q) \in L^\infty((-r,0); \mathbb{R}^{n\times n}). \tag{3.11}$$

Before stating our next two lemmas we shall give two conditions concerning $\hat{B}$ and B as functions of $q \in Q$ that we will need. Let $q_0 \in Q$ be fixed and suppose B and $\hat{B}$ satisfy

(C8) $B(q)$ and $\hat{B}(q)$ are continuous in q at $q = q_0$, and

(C9) $B(q)$ and $\hat{B}(q)$ are Fréchet differentiable with respect to q at q_0.

LEMMA 3.2. Assume conditions (C1)–(C8). For every $\epsilon > 0$ there exists a $\delta > 0$ such that

$$\int_0^T |L(q_0+h)x_t - L(q_0)x_t|dt \leq \epsilon|\eta|$$

for all h satisfying $q_0 + h \in Q$, $|h| < \delta$, where x is the unique solution of (3.1)–(3.2) with $(\eta,\phi) = (\eta,0)$ belonging to $Y = \{(\eta,\phi) \in X : \phi = 0\}$ and $q = q_0$.

PROOF: Using estimates as those employed in obtaining inequality (3.10) in Lemma 1 we have

$$\int_0^T |L(q_0+h)x_t - L(q_0)x_t|dt \leq |B(q_0+h) - B(q_0)|\left(\frac{T}{r}+1\right) Me^{\omega(T+r)}|\eta|$$
$$+ \|B(\cdot,q_0+h) - B(\cdot,q_0)\|TMe^{\omega T}|\eta|.$$

The desired result now follows from the continuity of $B(q)$ and $\hat{B}(q)$ at $q = q_0$. ∎

LEMMA 3.3. Assume that conditions (C1)–(C9) above hold. Then the mapping $q \to (L(q)x_t, 0) \in L^1((0,T);$ where x is the solution of (3.1)–(3.2) with $q = q_0$, is Fréchet differentiable with respect to q at q_0. Moreover, $L'(q_0)x_t = B'(q_0)x(t) + \int_{-r}^0 B'(s,q_0)x_t(s)ds$.

PROOF: Let B' and $\hat{B}'$ denote the Fréchet derivatives of B and $\hat{B}$. Pick $h \in Q$. Then we employ the same estimates used earlier to obtain (3.10) to yield

$$|L(q_0+h)x_t - L(q_0)x_t - B'(q_0)hx(t) + \int_{-r}^{0} B'(s,q_0)hx_t(s)ds|$$

$$\leq \left(|B(q_0+h) - B(q_0) - B'(q_0)h| \left(\frac{T}{r}+1\right) Me^{\omega(T+r)} \right.$$

$$\left. + \|B(\cdot,q_0+h) - B(\cdot,q_0) - \hat{B}'(q_0)h\| TMe^{\omega T} \right) \|(\eta,\phi)\|.$$

The desired result follows from the differentiability of $B(q)$ and $\hat{B}(q)$ at $q = q_0$. ∎

We now state the differentiability result for the semigroup $S(t;q)$.

THEOREM 3.4. *Let Conditions (C1)–(C9) and hold. If $x(t)$ denotes the unique solution of (3.1)–(3.2), $T > 0$ and $q_0 \in Q$ then the semigroup $S(t;q)$ $(\eta,\phi) = (Dx_t, x_t)$ is Fréchet differentiable with respect to q at q_0 for $0 \leq t \leq T$.*

PROOF: Lemma 3.1, 3.2 and 3.3 imply that hypotheses (H4), (H5) and (H6) found in Brewer [1] hold, consequently the differentiability of $S(t,q)$ with respect to q follows from Theorem 1 in [1]. ∎

We conclude this paper with the remark that if y denotes the derivative of the solution x of (3.1)–(3.2) with respect to q, then y satisfies the "sensitivity equation"

$$\frac{d}{dt}Dy_t = L(q)y_t = L'(q)x_t, \quad t \geq 0$$

$$\lim_{t \to 0^+} Dy_t = 0, \quad y_0 = 0.$$

REFERENCES

1. D. Brewer, *The differentiability with respect to a parameter of the solution of a Linear Abstract Cauchy Problem*, SIAM J. Math Anal. **13** (1982), 607–620.
2. D.W. Brewer, J.A. Burns and E. M. Cliff, *Parameter identification for an abstract Cauchy problem by quasilinearization*, Quarterly Appl. Math. (to appear).
3. D.W. Brewer and R.K. Powers, *Parameter identification in a Volterra equation with weakly singular kernel*, J. Integral Equations **2** (1990), 353–373.
4. J.A. Burns, E.M. Cliff and T.L. Herdman, *A state-space model for an aeroelastic system*, 22nd IEEE Conference on Decision and Control **3** (1983), 1074–1077.
5. J.A. Burns, T.L. Herdman and J. Turi, *The Neutral functional integro-differential equations with weakly singular kernals*, J.M.A.A. **145** (January 1990), 371–401.
6. E.M. Cliff, T.L. Herdman and J. Turi, *Finite memory approximations for a singular Neutral System Arising in Aeroelasticity*, 30th IEEE Conference on Decision and Control (1991 to appear).
7. T.L. Herdman and J. Turi, *On the solutions of a class of integral equations arising in unsteady aerodynamics*, Differential Equations: Stability and Control, Lecture Notes in Pure and Applied Mathematics, S. Elaydi, Ed., Marcel Dekker **127** (1990), 241–248.
8. K. Ito and J. Turi, *A Numerical method for a class of singular integro-differential equations based on semi-group approximation*, Siam J. Num. Anal. (to appear).
9. F. Kappel and Kang Pei Zhang, *Equivalence of functional equations of neutral type and abstract Cauchy problems*, Monatsch. Math **101** (1986), 115–133.

Dennis W. Brewer
Department of Mathematics
University of Arkansas
Fayetteville, Arkansas 72701

Terry L. Herdman
Interdisciplinary Center for Applied Mathematics
Virginia Polytechnic Institute and State University
Blacksburg, VA 24061

J H JAROMA, V Lj KOCIĆ[1] AND G LADAS
Global asymptotic stability of a second-order difference equation

Recently, Jaroma, Kuruklis and Ladas [3] investigated the nonlinear difference equation

$$x_{n+1} = \frac{\alpha x_n}{1 + \beta x_n + \gamma x_{n-1}}, \quad n = 0, 1, \ldots \quad (1)$$

and showed that when

$$\alpha \in (1, \infty) \text{ and } \beta, \gamma \in (0, \infty)$$

the positive equilibrium of Eq. (1) is a global attractor of all positive solutions.

Our aim in this paper is to present the following generalization of the above result.

Theorem 1 *Assume that $f : (0, \infty) \times (0, \infty) \to (0, \infty)$ is continuously differentiable, and that for all $(u, v) \in (0, \infty) \times (0, \infty)$,*

$$f_u(u, v) < 0, \, f_v(u, v) < 0, \text{ and } \frac{d}{du}[uf(u, u)] \leq 0.$$

Suppose also that the equation

$$x_{n+1} = x_n f(x_n, x_{n-1}), \quad n = 0, 1, \ldots \quad (2)$$

has a unique positive equilibrium $\bar{x}$. Then $\bar{x}$ is globally asymptotically stable.

Our interest in Eq. (2) has been influenced by the discrete delay logistic model

$$N_{t+1} = \frac{\alpha N_t}{1 + \beta N_{t-k}} \quad (3)$$

where

$$\alpha \in (1, \infty), \, \beta \in (0, \infty) \text{ and } k \in \{0, 1, 2, \ldots\}$$

which was proposed by E. C. Pielou in her books [7, p.22] and [8, p.79], as a discrete analog of the delay logistic equation

$$\dot{N}(t) = rN(t)[1 - \frac{N(t-\tau)}{P}].$$

The local asymptotic stability of the linearized equation associated with Eq. (3) is completely determined by the result of Levin and May [6] on the stability of the linear delay difference equation

$$y_{n+1} - y_n + py_{n-k} = 0.$$

The global attractivity of the positive equilibrium of Eq. (3) has been investigated in [5] for $k = 0$ and $k = 1$, and in [4] for $k \geq 2$. See also [2].

---

[1]On leave from the Deparment of Mathematics, Faculty of Electrical Engineering, University of Belgrade, Belgrade, Yugoslavia

1. The Proof of Theorem 1

Let $\bar{x}$ denote the unique positive equilibrium of Eq. (2) and let $\{x_n\}$ be any solution of Eq. (2) with initial conditions $x_{-1}, x_0 \in (0, \infty)$. We must prove that $\bar{x}$ is locally asymptotically stable and that
$$\lim_{n \to \infty} x_n = \bar{x}. \tag{4}$$
The linearized equation associated with Eq. (2) about $\bar{x}$ is
$$y_{n+1} + p y_n + q y_{n-1} = 0, \quad n = 0, 1, \ldots \tag{5}$$
where
$$p = -1 - \bar{x} f_u(\bar{x}, \bar{x}) \text{ and } q = -\bar{x} f_v(\bar{x}, \bar{x}).$$
It is well known, see for example Finizio and Ladas [1], that the trivial solution of Eq. (5) is asymptotically stable if and only if
$$|p| < 1 + q < 2.$$
This is clearly satisfied because $f(\bar{x}, \bar{x}) = 1$ and from the hypotheses on f,
$$f_u(\bar{x}, \bar{x}) < 0, \ f_v(\bar{x}, \bar{x}) < 0 \text{ and } 1 + \bar{x} f_u(\bar{x}, \bar{x}) + \bar{x} f_v(\bar{x}, \bar{x}) \geq 0.$$

It remains to show that (4) holds. This will be accomplished by means of a thorough study of the positive and negative semicycles of the solution $\{x_n\}$. A **positive semicycle** of $\{x_n\}$ consists of a "string" of terms $\{x_{l+1}, x_{l+2}, \ldots, x_m\}$ with $l \geq -2$ and $m \leq \infty$, all greater than or equal to $\bar{x}$ and such that

$$\text{either} \quad l = -2 \quad \text{or} \quad l \geq -1 \text{ and } x_l < \bar{x}$$
and
$$\text{either} \quad m = \infty \quad \text{or} \quad m < \infty \text{ and } x_{m+1} < \bar{x}.$$

A **negative semicycle** of $\{x_n\}$ consists of a "string" of terms $\{x_{k+1}, x_{k+2}, \ldots, x_l\}$ with $k \geq -2$ and $l \leq \infty$, all less than $\bar{x}$ and such that

$$\text{either} \quad k = -2 \quad \text{or} \quad k \geq -1 \text{ and } x_k \geq \bar{x}$$
and
$$\text{either} \quad l = \infty \quad \text{or} \quad l < \infty \text{ and } x_{l+1} \geq \bar{x}.$$

The first semicycle of a solution starts with the term x_{-1} and is positive if $x_{-1} \geq \bar{x}$ and negative if $x_{-1} < \bar{x}$. A solution may have a finite number of semicycles or infinitely many. A solution with infinitely many semicycles is called **oscillatory**. Otherwise the solution is called **nonoscillatory**.

The following technical lemma which is a consequence of the hypotheses imposed on f will be useful in the sequel.

Lemma 1 *Assume that $0 < a < \bar{x} < b$. Then the following inequalities are true:*

$$af(a,a) < \bar{x}; \quad bf(b,b) > \bar{x}. \tag{6}$$

$$\bar{x}f(\frac{\bar{x}^2}{a},\frac{\bar{x}^2}{a}) > a; \quad \bar{x}f(\frac{\bar{x}^2}{b},\frac{\bar{x}^2}{b}) < b. \tag{7}$$

$$\bar{x}f(\bar{x}f(a,a),\bar{x}f(a,a)) > a; \quad \bar{x}f(\bar{x}f(b,b),\bar{x}f(b,b)) < b. \tag{8}$$

Proof. Inequalities (6) are a simple consequence of the increasing character of the function $uf(u,u)$ and the fact that $f(\bar{x},\bar{x}) = 1$.

Since $\bar{x}^2/a > \bar{x}$, the second inequality in (6) yields, $\frac{\bar{x}^2}{a}f(\frac{\bar{x}^2}{a},\frac{\bar{x}^2}{a}) > \bar{x}$ which proves the first inequality in (7). The second inequality in (7) is proved in a similar way.

From (6), $\bar{x}f(a,a) < \bar{x}^2/a$ and so from (7) and the fact that $f(u,u)$ decreases in u we have $\bar{x}f(\bar{x}f(a,a),\bar{x}f(a,a)) > a$. The second inequality in (8) is proved similarly. □

The next lemma presents a detailed description of the semicycles of $\{x_n\}$.

Lemma 2 *(a) If x_{-1} and x_0 are not both equal to $\bar{x}$, a positive semicycle cannot have two consecutive terms equal to $\bar{x}$.*

(b) Every semicycle of $\{x_n\}$, except perhaps for the first one, has at least two terms.

(c) The extreme in a semicycle is equal to the first or to the second term of the semicycle. More precisely, after the first term, the remaining terms in a positive semicycle decrease and the remaining terms in a negative semicycle increase.

(d) Except perhaps for the first semicycle of a solution, in a semicycle with finitely many terms, the extreme of the semicycle cannot be equal to the last term.

(e) The maxima in successive positive semicycles are decreasing and the minima in successive negative semicycles are increasing.

Proof. The proof of (a) is simple and will be omitted. For the remaining statements we will only give the proof for positive semicycles whose terms are not equal to $\bar{x}$. The proof for negative semicycles is similar and will be omitted. The proof for the trivial semicycle where $x_{-1} = x_0 = \bar{x}$ is obvious.

(b) If x_n is the first term in a positive semicycle (other than the first semicycle), then

$$x_{n+1} = x_n f(x_n, x_{n-1}) > x_n f(x_n, x_n) \geq \bar{x} f(\bar{x}, \bar{x}) = \bar{x}$$

so x_{n+1} is also in the same semicycle.

(c) If x_n, x_{n+1} are two consecutive terms in a positive semicycle, then

$$x_{n+2} = x_{n+1} f(x_{n+1}, x_n) \leq x_{n+1} f(\bar{x}, \bar{x}) = x_{n+1}. \tag{9}$$

(d) This follows from the observation that unless $x_n = x_{n+1} = \bar{x}$, the inequality in (9) is strict. On the other hand, if $x_n = x_{n+1} = \bar{x}$ then by (a) the entire solution reduces to a positive semicycle which contradicts the hypothesis that the semicycle has finitely many terms.

(e) Consider four consecutive semicycles

$$\begin{aligned} C_{r-1} &= \{x_{k+1}, x_{k+2}, \ldots, x_l\} & &- \text{ negative semicycle,} \\ C_r &= \{x_{l+1}, x_{l+2}, \ldots, x_m\} & &- \text{ positive semicycle,} \\ C_{r+1} &= \{x_{m+1}, x_{m+2}, \ldots, x_n\} & &- \text{ negative semicycle,} \\ C_{r+2} &= \{x_{n+1}, x_{n+2}, \ldots, x_p\} & &- \text{ positive semicycle.} \end{aligned}$$

If b_{r-1}, b_r, b_{r+1} and b_{r+2} denote the extreme values in these four semicycles, respectively, we must prove that

$$b_{r+2} < b_r \text{ and } b_{r-1} < b_{r+1}. \tag{10}$$

It follows from (c) that either $b_r = x_{l+1}$ or that $b_r = x_{l+2}$. In the first case

$$b_r = x_l f(x_l, x_{l-1}) < \bar{x} f(b_{r-1}, b_{r-1}).$$

In the second case

$$\begin{aligned} b_r &= x_{l+1} f(x_{l+1}, x_l) = x_l f(x_l, x_{l-1}) f(x_{l+1}, x_l) \\ &\leq x_l f(x_l, x_l) f(x_l, x_{l-1}) < \bar{x} f(b_{r-1}, b_{r-1}). \end{aligned}$$

Thus, in either case

$$b_r < \bar{x} f(b_{r-1}, b_{r-1}).$$

In a similar way we obtain

$$b_{r+1} > \bar{x} f(b_r, b_r).$$

Hence

$$b_{r+2} < \bar{x} f(b_{r+1}, b_{r+1}) < \bar{x} f(\bar{x} f(b_r, b_r), \bar{x} f(b_r, b_r)) \tag{11}$$

and

$$b_{r+1} > \bar{x} f(b_r, b_r) > \bar{x} f(\bar{x} f(b_{r-1}, b_{r-1}), \bar{x} f(b_{r-1}, b_{r-1})). \tag{12}$$

From these and (8), it follows that (10) holds and the proof of Lemma 2 is complete. □

We are now ready to establish (4). This is a simple consequence of Lemma 2(c), if the solution is nonoscillatory. So assume that $\{x_n\}$ is oscillatory and set

$$\lambda = \liminf_{n \to \infty} x_n \text{ and } \Lambda = \limsup_{n \to \infty} x_n$$

which by Lemma 2(e) both exist and satisfy

$$0 < \lambda \leq \bar{x} \leq \Lambda.$$

It follows from (11) and (12) that

$$\Lambda \leq \bar{x} f(\bar{x} f(\Lambda, \Lambda), \bar{x} f(\Lambda, \Lambda)) \text{ and } \lambda \geq \bar{x} f(\bar{x} f(\lambda, \lambda), \bar{x} f(\lambda, \lambda)).$$

In view of (8), these imply that $\lambda = \bar{x} = \Lambda$ and the proof of the theorem is complete. □

References

[1] N. Finizio, and G. Ladas, *An Introduction to Differential Equations*, Wadsworth Publishing Co., 1982.

[2] I. Györi and G. Ladas, *Oscillation Theory of Delay Differential Equations with Applications*, Oxford University Press, 1991.

[3] J. H. Jaroma, S. A. Kuruklis and G. Ladas, *Oscillations and stability of a discrete delay logistic model* (to appear).

[4] V. Lj. Kocić and G. Ladas, *Global attractivity in nonlinear delay difference equations* Proc. Amer. Math. Soc. (to appear).

[5] S. A. Kuruklis and G. Ladas, *Oscillation and global attractivity in a discrete delay logistic model*, Quarterly Appl. Math. (to appear).

[6] S. Levin and R. May, *A note on difference-delay equations*, Theor. Pop. Biol. **9** (1976), 178–187.

[7] E. C. Pielou, *An Introduction to Mathematical Ecology*, Wiley Interscience, 1969.

[8] E. C. Pielou, *Population and Community Ecology*, Gordon and Breach, New York, 1974.

Department of Mathematics
The University of Rhode Island
Kingston, RI 02881, U. S. A.

B S LALLI AND B G ZHANG
Existence of nonoscillatory solutions for odd order neutral equations

An existence criterion for nonoscillatory solutions for an odd order neutral differential equation is provided. Some sufficient conditions are also given for the existence of a positive solution of an nth order equation with nonlinearity in the neutral term.

I. Introduction

In this paper we consider the first order neutral differential equation of the form

$$\bigl(x(t) - cx(t-\tau)\bigr)' - p(t)x\bigl(g(t)\bigr) = 0, \quad t \geq t_0 \tag{1.1}$$

and a nonlinear neutral differential equation of nth order of the form

$$\Bigl(x(t) - h(t)f_1\bigl(x(\tau(t))\bigr)\Bigr)^{(n)} + p(t)f_2\bigl(x(g(t))\bigr) = 0 \quad t \geq t_0 \tag{1.2}$$

Most of the works on oscillation theory for neutral equations deal with stable type equations. There are a few papers (see e.g. [1], [5], [6]) where nonoscillation of unstable type equations of order larger than one is discussed. We establish a result for the existence of an unbounded solution of Eq. (1.1) which tends to infinity exponentially. As pointed out by Hale [1] it is useful to study neutral nonlinear differential equations of the form

$$\bigl(x(t) - G(t, x(t-\tau))\bigr)' = H(t, x(t-\tau)).$$

As usual a solution $x(t)$ of Eq. (1.j), $j = 1, 2$, is said to be oscillatory on $[t_0, \infty)$ if the set of zeros of $x(t)$ is unbounded, otherwise it is called nonoscillatory. In the sequel we need the following lemma:

Lemma (See [8]) *Let X be a Banach space, Γ a bounded closed and convex subset of X, A, B be maps on Γ to X such that $Ax + By \in \Gamma$ for every pair x, $y \in \Gamma$. If A is a contraction and B is completely continuous then the equation $Ax + Bx = x$ has a solution in Γ.*

2. Results for Equation (1.1)

We assume that

$$c \geq 0, \quad \tau > 0 \quad g(t) \leq t, \quad \lim_{t \to \infty} g(t) = \infty \quad p(t) \geq 0,$$

and the functions p, g are continuous on $[t_0, \infty)$, $t_0 \geq 0$. In case

$$p(t) \equiv p, \quad g(t) = t - \sigma, \quad \sigma > 0,$$

from the analysis of the characteristic equation of Eq. (1.1) we know that Eq. (1.1) has always an unbounded solution $x(t) = Ae^{\alpha t}$, $\alpha > 0$. The question arises whether Eq. (1.1) has always an unbounded solution $x(t)$ which tends to infinity exponentially as t tends to infinity. We explore that possibility. For $c \geq 0$, let $x(t)$ be a positive solution of Eq. (1.1). We put $z(t) = x(t) - cx(t - \tau)$. Then $z'(t) \geq 0$, and therefore two possibilities exist: (i) $z(t) > 0$, or (ii) $z(t) \leq 0$, eventually.

Consequently, the nonoscillatory solution $x(t)$ must satisfy one of the following types of asymptotic behavior:

(a) $\lim_{t \to \infty} x(t) = 0$; (b) $\lim_{t \to \infty} x(t) = \ell \neq 0$, (c) $\lim_{t \to \infty} x(t) = \infty$.

By using Lemma one can prove the following:

Theorem 2.1. *Based on the value of c we have the conclusions:*

(i) *If $c \geq 0$, Eq. (1.1) has always a positive solution $x(t)$ satisfying (b) or (c);*

(ii) *If $c = 1$, Eq. (1.1) has always an unbounded solution $x(t)$ satisfying (c);*

(iii) *If $c > 1$, Eq. (1.1) has always an unbounded solution $x(t)$ which tends to infinity exponentially;*

(iv) *If $0 \leq c < 1$, and $\int_T^\infty p(t)dt = \infty$, $T \geq t_0$ Eq. (1.1) has always an unbounded positive solution, and every bounded solution of Eq. (1.1) either oscillates or tends to zero as t tends to infinity.*

Example 2.1. Consider the equation

$$\left(x(t) - \frac{1}{2}x(t-1)\right)' = \frac{t-2}{2t^2} x(t-1), \quad \text{for} \quad t \geq 2 \tag{2.1}$$

has a bounded solution $x(t) = 1 + \frac{1}{t}$.

Example 2.2. The equation

$$\left(x(t) - x(t-1)\right)' = (1 - e^{-1})x(t), \quad \text{for} \quad t \geq 2 \tag{2.2}$$

has an unbounded solution $x(t) = e^t$.

Example 2.3. The equation

$$(x(t) - 2x(t-1))' = \frac{(e-2)t + e}{t-1} x(t-1), \quad \text{for} \quad t \geq 2 \tag{2.3}$$

satisfies the assumptions of thereom 2.1 (iii). It has a solution $x(t) = te^t$.

3. Nonlinear Equation (1.2)

We prove the following:

Theorem 3.1. *Assume that n is an odd integer and that*

(i) $h \in C(R_+, R)$;

(ii) $f_i \in C(R, R)$, $xf_i(x) > 0$, $i = 1, 2$, as $x \neq 0$

$$|f_1(x) - f_1(y)| \leq L|x - y|, \quad x, y \in [0, 1]$$

f_2 is a nondecreasing function, L is a positive constant;

(iii) $\tau, g \in C(R_+, R)$, $0 \leq t - \tau(t) \leq M$, M is a constant

$$\lim_{t \to \infty} \tau(t) = \infty, \quad \lim_{t \to \infty} g(t) = \infty$$

(iv) *there exists $\alpha > 0$ such that $Lh(t)e^{\alpha(t-\tau(t))} \leq c < 1$, and*

$$Lh(t)e^{\alpha(t-\tau(t))} + \frac{e^{\alpha t}}{(n-1)!} \int_t^\infty (s-t)^{n-1} p(s) f_2(e^{-\alpha g(s)}) ds \leq 1 .$$

Then Eq. (1.2) has an eventually positive solution $x(t)$ which tends to zero exponentially.

Proof. Let t_0 be sufficiently large so that

$$T = \min\{\inf_{t \geq t_0} \tau(t), \inf_{t \geq T_0} g(t)\} .$$

Denote by $BC([T, \infty))$ the Banach space of all bounded and continuous real valued functions defined on $[T, \infty)$. Let Ω be a subset of BC defined as follows:

$$\Omega = \{y \in BC : 0 \leq y(t) \leq 1, \quad t_0 \leq t < \infty\} .$$

Define operators S_1 and S_2 on Ω as follows:

$$(S_1y)(t) = \begin{cases} h(t)e^{\alpha t}f_1\Big(y(\tau(t))e^{-\alpha\tau(t)}\Big), & \text{if } t \geq t_0 \\ \frac{t}{t_0}(S_1y)(t_0) + (1 - \frac{t}{t_0}), & \text{for } T \leq t \leq t_0, \end{cases}$$

$$(S_2y)(t) = \begin{cases} \frac{e^{\alpha t}}{(n-1)!}\int_t^\infty (s-t)^{(n-1)} p(s) f_2\Big(y(g(s))e^{-\alpha g(s)}\Big) ds & \text{if } t \geq t_0 \\ \frac{t}{t_0}(S_2y)(t_0) & \text{for } T \leq t \leq t_0. \end{cases}$$

By (iv), for every $x, y \in \Omega$ we have $S_1x + S_2y \in \Omega$. Condition (iv) implies that S_1 is a contraction on Ω. It is easy to see that $|\frac{d}{dt}(S_2y)(t)| \leq M_1$ for $y \in \Omega$ where, M_1 is a positive constant. From this it follows that S_2 is completely continuous. By Lemma there exists a $y \in \Omega$ such that

$$(S_1 + S_2)y = y.$$

That is

$$y(t) = \begin{cases} h(t)e^{\alpha t}f_1\Big(y(\tau(t))e^{-\alpha\tau(t)}\Big) \\ + \frac{e^{\alpha t}}{(n-1)!}\int_t^\infty (s-t) p(s) f_2\Big(y(g(s))e^{-\alpha g(s)}\Big) ds, & t \geq t_0 \\ \frac{t}{t_0}y(t_0) + (1 - \frac{t}{t_0}) & T \leq t \leq t_0 \end{cases}$$

It is easy to see that $y(t) > 0$ for $t \geq T$. Set

$$x(t) = y(t)e^{-\alpha t}.$$

Then

$$x(t) = h(t)f_1\Big(x(\tau(t))\Big) + \frac{1}{(n-1)!}\int_t^\infty (s-t)^{n-1} p(s) f_2\Big(x(g(s))\Big) ds, \quad t \geq t_0$$

or

$$\Big(x(t) - h(t)f_1\big(x(\tau(t))\big)\Big)^{(n)} + p(t)f_2\big(x(g(t))\big) = 0, \quad t \geq t_0.$$

This completes the proof.

Example 3.1. Consider a nonlinear neutral equation of the form

$$\Big(x(t) - \frac{1}{4}x^3(t-1)\Big)' + p(t)x^{\frac{1}{3}}(t) = 0, \tag{3.1}$$

where

$$p(t) = e^{-\frac{2}{3}t} - \frac{3}{4}e^3 e^{-\frac{8}{3}t} > 0$$

for all large values of t. In our notation

$$h(t) = \frac{1}{4}, \quad f_1(x) = x^3, \quad L = 3, \quad f_2(x) = x^{\frac{1}{3}}$$

Obviously the hypotheses of theorem 3.1 are satisfied. Therefore Eq. (3.1) has a solution $x(t)$ which tends to zero exponentially as $t \to \infty$. In fact, $x(t) = e^{-t}$ is such a solution of (3.1).

In recent years many results in oscillation theory of functional differential equations of neutral type have appeared. However, there are so far only a few results about the existence of nonoscillatory solutions. A systematic investigation of nonoscillation of higher order neutral differential equations has been done by Jaros and Kusano ([5], [6]). In their study it is shown that the nonoscillatory solutions of

$$\left(x(t) - h(t)x(\tau(t))\right)^{(n)} + \sum_{i=1}^{N} p_i(t)x(g_i(t)) = 0, \tag{3.2}$$

fall into four possible categories with regard to their asymptotic behavior. Thus if $x(t)$ is nonoscillatory solution of (3.2) then one of the following holds:

(A) $\quad 0 < \liminf\limits_{t \to \infty} \dfrac{|x(t)|}{t^k} \le \limsup\limits_{t \to \infty} \dfrac{|x(t)|}{t^k} < \infty \quad$ for some $\quad k \in \{1, 2, \cdots, n-1\}$;

(B) $\quad \lim\limits_{t \to \infty} \dfrac{x(t)}{t^l} = 0 \quad$ and $\quad \lim\limits_{t \to \infty} \dfrac{|x(t)|}{t^{l-1}} = \infty$,

$\qquad\qquad$ for some $\quad l \in \{1, 2, \cdots, n-1\} \quad$ with $\quad (-1)^{n-l-1} = 1$;

(C) $\quad \lim\limits_{t \to \infty} \dfrac{|x(t)|}{t^{n-1}} = \infty$, and $\quad$ (D) $\quad \lim\limits_{t \to \infty} x(t) = 0$,

The existence criteria for nonoscillation of solutions of the type (A) and (B) are given in [5]. However, the question of existence of nonoscillatory solutions of the type (D) is unanswered.

In this note we establish an existence criterion for nonoscillatory solutions of the type (D) for Eq. (3.2).

Theorem 3.2. *Let n be odd and assume that*

(i) $\;h : [t_0, \infty) \to R = (-\infty, \infty)$ *is continuous,*

(ii) $\;\tau$ *is continuous on* $[t_0, \infty)$, $\tau(t) < t \;$ *for* $\; t \ge t_0 \;$ *and* $\; \lim\limits_{t \to \infty} \tau(t) = \infty$,

(iii) $\;p_i, \; g_i : [t_0, \infty) \to R \;$ *are continuous,* $p_i(t) \ge 0 \;$ *and* $\; \lim\limits_{t \to \infty} g_i(t) = \infty, \;$ *for* $1 \le i \le N$

(iv) there exists a positive number α such that

$$0 \leq h(t)e^{\alpha(t-\tau(t))} \leq C < 1 \tag{3.3}$$

$$h(t)e^{\alpha(t-\tau(t))} + e^{\alpha t}\int_t^\infty (s-t)^{n-1}\sum_{i=1}^N p_i(s)e^{-\alpha g_i(s)}ds \leq 1, \tag{3.4}$$

holds eventually. Then Eq. (3.2) has an eventually positive solution $x(t)$ which approaches zero exponentially.

Remark. The complete text with proofs of the results will be published elsewhere.

References

1. Chuanxi Q. and Ladas G. S., Existence of positive solutions of neutral equations (to appear).
2. Chuanxi Q., Ladas G., Zhang B. and Zhao T. (1990) Sufficient conditions for oscillation and existence of positive solutions. Appli. Analysis. 35, 187-194.
3. Grove E. A., Kulenovic M.R.S. and Ladas G. S. (1987) Sufficient conditions for oscillation and nonoscillation of neutral equations, J. Differential Equations, 68, 373-382.
4. Hale J. K. (1977) Theory of functional differential equations, Springer Verlag.
5. Jaros J. and Kusanso T. (1988) Oscillation theory of higher order linear functional differential equations of neutral type. Hiroshima Math, J., 18, 509-531.
6. Jaros J. and Kusano T. (1989) Asymptotic behavior of nonoscillatory solutions of nonlinear differential equations of neutral type, Funcialaj Eckvacioj, 32, 251-263.
7. Ladas G. S., Lakshmikantham V. and Zhang B. G. (1987) Oscillation Theory of Differential Equations with Deviating Arguments, Marc Dekker, New York.
8. Nashed M. and Wong J. S. W. (1969) Some variations of a fixed point theorem of Krasnoselskii and applications to nonlinear integral equations, J. Math. Mech., 18, 666-677.
9. Nato Y. (1990) Nonoscillatory solutions of neutral differential equations, Hiroshima Math. J., 20, 231-258.
10. Tarski A. (1955) A lattice theoretic fixed point theorem and its applications. Pacific J. Math. 5, 285-309.

B. G. Zhang
Department of Applied Mathematics
Ocean University of Qingdao
Shandong, P. R. of China

B. S. Lalli
Department of Mathematics
University of Saskatchewan
Saskatoon, SK, S7N 0W0, Canada

Y LATUSHKIN
Exact Lyapunov exponents and exponentially separated cocycles

1. Introduction.

Starting from the classic series of the papers of R. Sacker and G. Sell [7,8] a number of works are devoted to studying the exponential dichotomies (hyperbolicity) of linear extensions of dynamical systems (or linear skew-product flows, LSPF). Recently the hyperbolicity of LSPF has been considered in connection (see [2,3]) with the multiplicative ergodic theorem. Dynamical spectrum and continuous spectral subbundles were described in terms of the multiplicative ergodic theorem [2] and the spectrum and spectral subspaces of a special class of operators: weighted composition operators (WCO), associated with the linear extension of dynamical system [4,5].

At the same time the investigation of other, and in some sense, more delicate characteristics of the LSPF was continued: the fact of their being exponentially separated. These investigations are being provided both in differential equations language and in the language of LSPF [1]. In [1], in particular, the question when the cocycle is exponentially separated was reduced to the question of the hyperbolicity of another cocycle which differs by the scalar factor from the given one. This fact gives a simple possibility to describe when the cocycle is exponentially separated in the terms of the multiplicative ergodic theorem and the spectral theory of WCO.

This description is given in this note. We give the algorithm of the description of exponential separability of cocylce starting from the set of exact Lyapunov exponents and measurable subbundles, generated by all ergodic (relatively to the dynamical system in base) measures.

This work continues papers [3-5], written together with A.M. Stepin. It is the author's pleasure to thank Prof. A. Stepin for a remarkable opportunity to think together on exponentially separated cocycles, hyperbolicity, etc.

2. Preliminaries.

Let $\alpha : X \to X$ be a homeomorphism of a compact metric space X, $A : X \times \mathbf{Z} \to GL(\mathbf{R}^m)$ - be a continuous cocycle:

$$A(x, n+k) = A(\alpha^n x, k) A(x, n), \quad A(x, 0) = I, \quad k, n \epsilon \mathbf{Z}, \quad x \epsilon X$$

over α and

$$\hat{\alpha} : X \times H \to X \times H : (x, v) \longmapsto (\alpha x, a(x) v), \quad a(x) = A(x, 1),$$

- be a linear extension (LSPF) of the dynamical system α. For simplicity we will assume that the set of nonperiodic points of α is dense in X.

Recall [7,2], that LSPF is called hyperbolic (or exponentially dichotomic) on X, if there exist an $\hat{\alpha}$-invariant projection-valued continuous function $P : x \longmapsto P(x) \epsilon L(\mathbf{R}^m)$ and constants $c, \beta > 0$ such that for any $x \epsilon X$ the following estimates hold:

$$\|A(x,k)P(x)A^{-1}(x,n)\| \leq ce^{-\beta(k-n)}, -\infty < n \leq k < \infty$$

$$\|A(x,k)(I-P(x))A^{-1}(x,n)\| \leq ce^{-\beta(n-k)}, -\infty < k \leq n < \infty.$$

Here $\hat{a}$-invariance means that $A(x,k)P(x) = P(\alpha^k x)A(x,k)$ or, and that is the same, the decomposition $X \times H = \mathbf{S} \dotplus \mathbf{U}$ is $\hat{a}$-invariant, and stable $\mathbf{S}_x$ and unstable $\mathbf{U}_x$ fibres are defined by the rule $\mathbf{S}_x = ImP(x), \mathbf{U}_x = Im(I-P(x))$.

Let μ be a quasiinvariant Borel measure on X, which is positive on open sets. Let us consider in the space $L_2(X,\mu;H)$ of H-valued functions the weighted composition operator associated with LSPF $\hat{\alpha}$ and acting by the formula

$$(T_a f)(x) = (\frac{d\mu\alpha^{-1}}{d\mu}(x))^{1/2} a(\alpha^{-1}x) f(\alpha^{-1}x), x\epsilon X.$$

As it was shown in [5] (see also [4]), the hyperbolicity of $\hat{a}$ is equivalent to the hyperbolicity of T_a (that is, to the condition $\sigma(T) \cap \Gamma = \phi$, where $\sigma(T)$ is the spectrum of T and Γ is a unit circle). Moreover, the Riesz projection $\mathbf{P}$ of T, corresponding to $\sigma(T) \cap \{z\epsilon \mathbf{C} : |z| < 1\}$ is related to the function $P(\cdot)$ from the definition of hyperbolicity by the formula $(\mathbf{P}f)(x) = P(x)f(x)$.

The dynamical spectrum Σ of LSPF $\hat{a}$ was defined in [7] as a set of $w\epsilon\mathbf{R}$ for which the extension $\hat{a}_w(x,v) = (\alpha x, e^{-w}a(x)v)$ is nonhyperbolic. It is known [5], that $\Sigma = \ln|\sigma(T)|$ and thus Σ consists (cf. [7]) from closed segments $[r_k^-, r_k^+], k = 1, \ldots, N, N \leq m$. Note finally, that if γ and γ^1 belong to different connected components of $\mathbf{R} \setminus \Sigma$ (or, and that is the same, the circles with radia e^γ and e^{γ^1} separate $\sigma(T_a)$), then Piesz projections $P_\gamma, P_\gamma^{\gamma^1}, P^{\gamma^1}$ corresponding to the parts $\sigma(T) \cap \{|z| < e^\gamma\}, \sigma(T) \cap \{e^\gamma < |z| < e^{\gamma^1}\}, \sigma(T) \cap \{|z| > e^{\gamma^1}\}$ of $\sigma(T)$ are the multiplication operators on continuous projection-valued function. They define [5] continuous spectal subbundles [7]:

$$X \times H = \mathbf{S}^\gamma \dotplus E^{\gamma\gamma^1} \dotplus \mathbf{U}^{\gamma^1}, \mathbf{S}_x^\gamma = ImP_\gamma(x), E_x^{\gamma\gamma^1} = ImP_\gamma^{\gamma^1}(x), \mathbf{U}_x^{\gamma^1} = ImP^{\gamma^1}(x).$$

Recall now the content of the multiplicative ergodic theorem [6]. Let $\nu\epsilon Erg-$ be an α-ergodic measure on X. Then there exists the set $X_\nu \subset X, \nu(X_\nu) = 1$, such that for all $x\epsilon X_\nu$ there exist exact Lyapunov exponents

$$\lambda_\nu(x,v) = \lim_{k\to\infty} k^{-1}\ln\|A(x,k)v\|$$

and measurable decompositions

$$X_\nu \times H = W_\nu^{s_\nu} \dotplus \ldots \dotplus W_\nu^1, \quad s_\nu \leq m.$$

There are s_ν different numbers among $\lambda_\nu(x,v), \lambda_\nu^1 > \lambda_\nu^2 > \ldots > \lambda_\nu^{s_\nu}$, and also $\lambda_\nu^j = \lambda_\nu(x,v)$ iff $v\epsilon W_x^j, j = 1,2,\ldots,s_\nu$.

Note that λ_ν^j and $m_\nu^j = \dim W_\nu^j$ depends on $\nu\epsilon Erg$ and does not depend on $x\epsilon X$. We will call the set $L_{\hat{a}} = \{\lambda_\nu^j : j = 1,\ldots,s_\nu, \nu\epsilon Erg\}$ Lyapunov spectrum of the LSPF $\hat{a}$.

Lemma 2.1 [5,4]. *The spectral radious $R(T_a)$ in L_2 can be calculated by the formula*

$$\ln R(T_a) = \sup\{\lambda_\nu^1 : \nu \epsilon Erg\}.$$

3. Result.

Definition. *The linear extension $\hat{a}$ is called to be exponentially separated if there exists a continuous $\hat{a}$-invariant decomposition $H = \tilde{\mathbf{S}}_x \dot{+} \tilde{\mathbf{U}}_x, x \epsilon X$, and constants $C, \beta > 0$ such that*

$$\|A_u(x,n)\| \geq C e^{\beta n} \|A_s(x,n)\|, n \epsilon \mathbf{N},$$

where A_s and A_u are cocycles constructed by the restrictions $a_s(x) = a(x)|\tilde{\mathbf{S}}_x$ and $a_u(x) = a(x)|\tilde{\mathbf{U}}_x$.

Note, that in accordance with corollary 6.31 from [1], p. 107, one can choose metric in $X \times H$ (the so-called Lyapunov metric) for which $C = 1$ in the definition. From now on we will assume $C = 1$.

Let us define the scalar function

$$f(x) = (\|a_s(x)\| \cdot \|a_u(x)\|^\bullet)^{-1/2}, \quad \text{where} \quad \|a\|^\bullet = \inf\{\|av\| : \|v\|_{\mathbf{R}^m} = 1\},$$

and construct the scalar cocycle $F(x,n) = f \circ \alpha^{k-1}(x) \cdot \ldots \cdot f(x), n \epsilon \mathbf{N}$.

We will use the following lemma, due to I. Bronshtein [1], p. 105, th. 6.30:

Lemma 3.1. *The LSPF $\alpha^n(x,v) = (\alpha^n x, A(x,n)v)$ is exponentially separated if and only if the LSPF $\alpha_f^n(x,v) = (\alpha^n x, F(x,n)A(x,n)v)$, constructed by the cocycle $B = FA$, is hyperbolic.*

For description of exponential separability we need the sets $\{\lambda_\nu^j : j = 1, \ldots s_\nu\}$ and $\{W_\nu^j : j = 1, \ldots, \nu\}$ of Lyapunov exponents and measurable subbundles, defined in the multiplicative ergodic theorem.

Theorem. *For exponential separability of $\hat{a}$ the fulfillment of the following conditions are necessary and sufficient:*

1) For any measure $\nu \epsilon Erg$ there exists number i_ν such that subbundles $W_\nu^1 \dot{+} \ldots \dot{+} W_\nu^{i_\nu}$ and $W_\nu^{i_\nu+1} + \ldots + W_\nu^{s_\nu}$ have $\hat{a}$-invariant continuous extensions $\tilde{\mathbf{U}}$ and $\tilde{\mathbf{S}}$ correspondingy from X_ν up to entire X.

2) The set

$$\Sigma_{\hat{\alpha},f} = \{\lambda_\nu^j + \int_X f d\nu : j = 1, \ldots, s_\nu, \nu \epsilon Erg\}$$

is separated from zero.

Before to prove it, note, that this theorem describes all $\hat{a}$-invariant decompositions $H = \tilde{\mathbf{S}}_x + \tilde{\mathbf{U}}_x$ in the definition of exponentially separated LSPF. If for some measure $\nu_0 \epsilon Erg$ there exists i_{ν_0} such that subbundles $W_{\nu_0}^1 + \ldots + W_{\nu_0}^{i_{\nu_0}}$ and $W_{\nu_0}^{i_{\nu_0}+1} + \ldots + W_{\nu_0}^{s_{\nu_0}}$ admit the $\hat{a}$-invariant continuous extensions $\mathbf{U}$ and $\mathbf{S}$ from X_{ν_0} up to entire X, then for any measure $\nu \epsilon Erg$ one can find i_ν with the same property.

Proof. Let F be a cocycle, defined before lemma 3.1, and $L_{\hat{a}_f} = \{\eta_\nu\}$ - be the Lyapunov spectrum of $\hat{a}_f$. So as F is scalar cocycle, $\eta_\nu^j = \lambda_\nu^j + \int_X f d\nu$.

Suppose $\hat{a}$ is exponentially separated. Then by Lemma 3.1 $\hat{a}_f$ is hyperbolic. Property 2) follows immediately from the definitions of hyperbolicity and Lyapunov exponents. Property 1) follows from the following simple lemma, proved in [5].

Lemma 3.2. *For hyperbolic LSPF $\hat{a}$ the following equalities hold:*

$$W_{\nu,x}^1 + \ldots + W_{\nu,x}^{j_\nu} = Im(I - P(x)), \quad W_{\nu,x}^{j_\nu+1} + \ldots + W_{\nu,x}^{s_\nu} = Im P(x), \quad x \epsilon X_\nu,$$

where j_ν for every $\nu \epsilon Erg$ is defined by the inequality $\lambda_\nu^{j_\nu+1} < 0 < \lambda_\nu^{j_\nu}$ (set $j_\nu = 0$, if $\lambda_\nu^1 < 0$).

Suppose now 1) and 2) are hold. To prove the hyperbolicity of $\hat{a}_f$ (or, and it is the same, the hyperbolicity of the operator $T_b, b = fa$) let us consider, using 1), the decomposition $L_2 = L_2^s \dot{+} L_2^u$, where functions from L_2^s and L_2^u take their values in the fibres $\tilde{S}_x$ and $\tilde{U}_x$ correspondingly, and decomposition $T_b = T_b^s \dot{+} T_b^u, T_b^s = T_b|L_2^s, T_b^u = T_b|L_2^u$. Note, that $T_b^s = T_{b_s}, T_b^u = T_{b_u}$, where $b_s(x) = b(x)|\tilde{S}_x, b_u(x) = b(x)|\tilde{U}_x, x \epsilon X$. Apply lemma 2.1 to the operators T_{b_s} and $(T_{b_u})^{-1} = T_{b_u^{-1}, \alpha^{-1}}$. From condition 2) of the theorem we get the hyperbolicity of T_b.

Let us describe the procedure of inductive "chipping off" the unstable subbundles, which define the exponentially separated LSPF.

1. Calculate $r_1^+ = \ln R(T_a) = \sup\{\lambda_\nu^1 : \nu \epsilon Erg\}$.
2. For $\nu \epsilon Erg$ calculate the sums $n_\nu(k) = \Sigma_{j=1}^k \dim W_\nu^j, k = 1, 2, \ldots, s_\nu$, and choose $k = k(\nu)$ so that $n = n_\nu(k)$ does not depend on $\nu \epsilon Erg$.
3. For $\nu \epsilon Erg$ check the existence of $\hat{a}$-invariant continuous subbundles $\tilde{S}$ and $\tilde{U}$ over X that extend the subbundles $W_\nu^{s_\nu} + \ldots + W_\nu^{k(\nu)+1}$ and $W_\nu^{k(\nu)} \dot{+} \ldots + W_\nu^1$ respectively from X_ν. If there is no such extension we turn back to step 2) and increase n.
4. Calculate

$$r_1^- = \inf\{\lambda_\nu^{k(\nu)} : \nu \epsilon Erg\}, \rho = \sup\{\lambda_\nu^{k(\nu)+1} : \nu \epsilon Erg\}.$$

If $\rho < r_1^-$, we have already found the splitting $X \times H = \mathbf{S} \dot{+} \mathbf{U}$ with respect to which LSPF $\hat{a}_\gamma$ is hyperbolic for $\gamma \epsilon (\rho, r_1^-)$, and thus $\hat{a}$ is exponentially separated. If $\rho \geq r_1^-$ check the condition $0 \notin L_{\hat{a}_f}$. If the condition is fulfilled then $\hat{a}$ is exponentially separated with respect to decomposition $X \times H = \mathbf{S} + \mathbf{U}$; otherwise come back to step 2) and increase n.

5. Build a cocycle $a'(x) = a(x)|\mathbf{S}_x$ and apply to it the procedure descirbed above.

References

[1] Bronshtejn, I.U., Nonautonomous dynamical systems, Kishinev (1984).

[2] Johnson, R.A., K. Palmer and G. Sell, Ergodic properties of linear dynamical systems, SIAM J. Math. Anal. 1 (1987) 1 - 33.

[3] Latushkin, Y.D. and A.M. Stepin, Weighted composition operator on topological Markov chain, Functional Anal. Appl. 22 (1988) no 4, 330 - 331.

[4] Latushkin, Y.D. and A.M. Stepin, Linear skew-product flows and semigroups of weighted composition operators. Lect. Notes Math.(1991), to appear.

[5] Latushkin, Y.D. and A.M. Stepin, Weighted composition operators, spectral theory of linear skew-product flows and the multiplicative ergodic theorem, Matem. Sbornik 181 (1990) no. 6, 723 - 742.

[6] Oseledec V.I., A multiplicative ergodic theorem. Lyapunov charactersitic numbers for dynamical systems, Trans. Moscow Math. Soc. 19 (1968) 197 - 231.

[7] Sacker, R.T. and G.R. Sell, A spectral theory for linear differential systems, J. Diff. Eqns. 27 (1978) 320 - 338.

[8] Sacker, R.T. and G. Sell, Existence of dichotomies and invariant splitting for linear differential systems, J.Diff. Eqns. I - 15 (1974), 429 - 455; II, III - 22 (1976) 473 - 522; IV - 27 (1978) 106 - 137.

U LEDZEWICZ-KOWALEWSKA
Optimality conditions for abnormal optimal control problems with additional equality constraints

In this paper an optimal control problem with additional equality constraints as in [7] is considered in the, so called, abnormal case. The abnormal case occurs when a process satisfies the classical Pontryagin - type Maximum Principle in the trivial way, i.e. $\lambda_0=0$ and the Maximum Principle does not depend on the minimized functional. Such a situation is usually caused by the fact that the classical Lusternik theorem about the tangent space to the equality constraint is satisfied only in the case of regular operators.

Avakov [1] proposed a generalization of the Lusternik theorem to the case of nonregular operators at the cost of the additional assumption of twice Fréchet differentiability. Applying this result, he obtained the nondegenerate conditions of optimality in the nonregular case for the smooth problem (see [1]), the smooth-convex problem (see [2]) and proved the Maximum Principle for certain class of optimal control problems (see [2],[3]). In [8] Avakov's generalization of the Lusternik theorem from [1] was applied to the optimization problems with inequality and equality constraints, as in the Dubovitski - Milyutin formalism, in the case when the operator that determines the equality constraints is twice Fréchet differentiable but not necessarily regular. Results from [8] were then applied to the classical optimal control problem with terminal data, and the extended local Maximum Principles was obtained in [10] generalizing results of [4] to the abnormal case.

In this paper the result of [8] is applied to the optimal control problem with an additional equality constraint of [7] and an extended version of the local Maximum Principle is presented generalizing results from [7] to the abnormal case. This extended version reduces to the one of [7] if the problem becomes regular in the sense of the optimization theory. The extended local Maximum Principle for the optimal control problem with terminal data of [10] can be considered as a special case of the result obtained here when the additional equality constraint is not present.

&1. The Optimal Control Problem

Consider an optimal control problem with equality constraints as the one discussed in [7]. Let $\overline{W}_{11}^n(0,1)$ be the subspace of the space $W_{11}^n(0,1)$ containing those functions which satisfy the condition $x(0) = 0$.

Consider the problem of minimizing the functional

$$I(x,u) = \int_0^1 f^0(x(t),u(t),t)dt \tag{1}$$

under the constraints

$$\dot{x}(t) = f(x(t),u(t),t) \quad \text{with} \quad x(0) = 0; \tag{2}$$

$$q(x(1)) = 0; \tag{3}$$

$$g(x(t),u(t),t) = 0; \tag{4}$$

$$u(.) \in U; \tag{5}$$

where $U = \{u(.) \in L_\infty^r(0,1): u(t) \in M \text{ for } t \in [0,1]\}$, $x(.) \in \overline{W}_{11}^n(0,1)$, $u(.) \in L_\infty^r(0,1)$, functions $f^0: R^n \times R^r \times R \to R$, $f: R^n \times R^r \times R \to R^n$, $q: R^n \to R^k$ and $g: R^n \times R^r \times R \to R^m$.

We will consider the above problem (1)–(5) under the following assumptions:

(A1) the functions f, f^0, g are continuous with respect to (x,u) with any $t \in [0,1]$ and measurable with respect to t for every (x,u);

(A2) the function g is continuously Fréchet differentiable and g_x and g_u are bounded for any bounded (x,u);

(A3) there exist derivatives f_x^0, f_u^0, f_x, f_u, f_{xx}, f_{xu}, f_{uu}, q_x, q_{xx}, g_{xx}, g_{xu} and g_{uu} which are bounded for any bounded (x,u), measurable with respect to t for every (x,u) and continuous with respect to (x,u) for any $t \in [0,1]$;

(A4) the set M is closed and convex and possesses a nonempty interior in R^r.

The above problem given by (1)–(5) under assumptions (A1)–(A4) will be called Problem I. We say that the pair (x,u) is an admissible process in Problem I if and only if (x,u) satisfies constraints (2)–(5).

&2. An extension of the local Maximum Principle

In this section we are going to apply results in [8] to the optimal control problem with equality constraints, Problem 1, introduced in Section I. We will formulate the extension of the local Maximum Principle for this problem which will generalize the Maximum Principle from [7] to the case of abnormal optimal controls.

We then introduce an operator $F: \overline{W}_{11}^n(0,1) \times L_\infty^r(0,1) \to \overline{W}_{11}^n(0,1) \times L_\infty^m(0,1) \times R^k$ given by

$$F(x,u) = \left(x(t) - \int_0^t f(x(t),u(t),t)dt, \ g(x(t),u(t),t), \ q(x(1))\right). \tag{6}$$

Let (x_0, u_0) be an admissible process in Problem I. In order to simplify our notation, we assume that f_x^0, f_u^0, f_x, f_u, f_{xx}, f_{xu}, f_{uu}, q_x, q_{xx}, g_{xx}, g_{ux} and g_{uu} denote partial derivatives at the point $(x_0(t), u_0(t),t)$. Following the construction in [8], for every

$(h,v) \in \overline{W}_{11}^n(0,1) \times L_\infty^r(0,1)$ we can introduce the operator

$$G(x_0,u_0,h,v): \overline{W}_{11}^n(0,1) \times L_\infty^r(0,1) \to \overline{W}_{11}^n(0,1) \times L_\infty^m(0,1) \times R^k$$
$$\times (\overline{W}_{11}^n(0,1) \times L_\infty^m(0,1) \times R^k)/\text{Im } F'(x_0,u_0),$$

given in the form

$$G(x_0,u_0,h,v)(\overline{h},\overline{v})(t) = \left(F'(x_0,u_0)(\overline{h},\overline{v})(t), \, \pi(F''(x_0,u_0,h,v)(\overline{h},\overline{v})(t))\right) \tag{7}$$

for every $(\overline{h},\overline{v}) \in \overline{W}_{11}^n(0,1) \times L_\infty^r(0,1)$, where π is a quotient map from the space $\overline{W}_{11}^n(0,1) \times R^k$ onto the space $\overline{W}_{11}^n(0,1) \times R^k/\text{Im } F'(x_0,u_0)$.

Let us define the set of parameters (h,v) which satisfy conditions (8a)-(8e) listed below:

(a) $dh/dt - f_x h - f_u v = 0,$ \hfill (8a)

(b) $g_x h + g_u v = 0,$ \hfill (8b)

(c) $q_x(x_0(1)) h(1) = 0,$ \hfill (8c)

(d) $\left\{ -\int_0^t \left(f_{xx} h^2 + 2 f_{xu} hv + f_{uu} v^2\right) dt, \, g_{xx} h^2 + 2 g_{xu} h v + g_{uu} v^2, \right.$

$\left. q_{xx}(x_0(1)) h^2 \right\} \in \text{Im } F'(x_0,u_0);$ \hfill (8d)

(e) $G(x_0, u_0, h, v)$ given by (7) has a closed image in

$$\overline{W}_{11}^n(0,1) \times L_\infty^m(0,1) \times R^k \times (\overline{W}_{11}^n(0,1) \times L_\infty^m(0,1) \times R^k/\text{Im } F'(x_0,u_0); \tag{8e}$$

For given admissible process (x_0,u_0) the set of the parameters (h,v) satisfying (8a-e) forms the set $P(x_0,u_0)$. Now, by using Theorem 2.2 in [8], the extended version of the local Maximum Principle for this optimal control Problem I can be obtained. Complete proof of the below theorem for the problem which includes also inequality constraints is given in [9].

Theorem 1.(Local Maximum Principle) Let

1. an admissible process (x_0,u_0) be optimal in Problem I;
2. the operator $A: \overline{W}_{11}^n \times L_\infty^r \to \overline{W}_{11}^n \times L_\infty^m$ given by the formula

$$A(\overline{x},\overline{u})(t) = \left(\overline{x} - \int_0^t (f_x \overline{x} + f_u \overline{u}) \, dt, \, g_x \overline{x} + g_u \overline{u}\right)$$

has closed image in $\overline{W}_{11}^n \times L_\infty^m$,
then for every $(h,v) \in P(x_0,u_0)$ there exist $\lambda_0 = \lambda_0(h,v) \geq 0$, $a = a(h,v) \in R^k$, $b = b(h,v) \in R^k$, functions $\rho(.) = \rho(h,v)(.)$, $\psi(.) = \psi(h,v)(.): [0,1] \to R^n$, $\delta(.) = \delta(h,v)(.)$ and

$\omega(.)=\omega(h,v)(.) \in b.^m a(0,1)$, *not vanishing simultaneously, such that*

$$\dot{\psi} = -\lambda_0 f_x^0 - f_x^* \psi + g_x^* \hat{\omega} - f_{xx}^*(h)\rho - f_{ux}^*(v)\rho + g_{xx}^*(h)\hat{\delta} + g_{ux}^*(v)\hat{\delta} \qquad (9)$$

with terminal condition

$$\psi(1) = -q_x^*(x_0(1))\, a - (q_{xx}(x_0(1))h(1))^* \, b \qquad (10)$$

where functions $\rho(.), \delta(.)$ *and vector* b *satisfy*

$$\dot{\rho} = -f_x^* \rho + g_x^* \hat{\delta} \qquad (11)$$

with terminal condition

$$\rho(1) = -q_x^*(x_0(1))b \qquad (12)$$

and such that

$$\left(f_u^* \rho - g_u^* \hat{\delta}, u_0 \right) = \left(f_u^* \rho - g_u^* \hat{\delta}, u \right) \qquad (13)$$

for every $u \in M$ *and* $t \in [0,1]$, *where,* $\hat{\omega}(t), \hat{\delta}(t)$ *are integrable functions corresponding to functions* $\omega(t)$ *and* $\delta(t)$ *in the sense that for every* $x(.) \in L_\infty^m$

$$\int_0^1 x(t)\, d\omega = \int_0^1 x(t)\, \hat{\omega}(t)\, dt + f_1^s(x(t)), \quad \int_0^1 x(t)\, d\delta = \int_0^1 x(t)\, \hat{\delta}(t)\, dt + f_2^s(x(t)), \qquad (14)$$

where f_1^s, f_2^s *are certain singular functionals and the following Maximum condition holds*

$$\lambda_0 \int_0^1 f_x^0 u_0\, dt + \int_0^1 f_u^* \psi u_0\, dt - \int_0^1 g_u u_0\, d\omega + \int_0^1 \left(f_{xu}^*(h)\, \rho\, u_0 + f_{uu}^*(v)\, \rho\, u_0 \right) dt$$

$$- \int_0^1 \left(g_{xu}(h) u_0 + g_{uu}(v) u_0 \right) d\delta \le \lambda_0 \int_0^1 f_x^0 u\, dt + \int_0^1 f_u^* \psi u\, dt - \int_0^1 g_u u\, d\omega$$

$$+ \int_0^1 \left(f_{xu}^*(h) u\, \rho + f_{uu}^*(v) u\, \rho \right) dt - \int_0^1 \left(g_{xu}(h) u + g_{uu}(v) u \right) d\delta \qquad (15)$$

for all $u(.) \in U$.

In the case when $\lambda_0 = 0$, $b = 0$ *and* $\rho(t) \equiv 0$, $\delta(t) \equiv 0$, *there exists* $\hat{u}(.) \in U$ *such that*

$$\int_0^1 f_u(\hat{u} - u_0)\psi\, dt - \int_0^1 \left(g_u(\hat{u} - u_0) \right) d\omega < 0. \qquad (16)$$

Remark 1. *Since the multipliers in the above theorem depend on parameters* (h,v) *let us notice that equation* (9) *represents a family of "extended" adjoint equations and condition* (15) *represents a family of "extended" Maximum conditions which depend on* parameters (h,v) from the set $P(x_0,u_0)$.

If the operator F given by (6) is regular, i.e. $\text{Im } F'(x_0,u_0) = \overline{W}_{11}^n(0,1) \times L_\infty^m(0,1) \times R^k$, conditions (11),(12) and (13) implies that $\rho=0$ and $b=0$. Then we have the following changes in the formulation of Theorem 1:

— the family of "extended" adjoint equations (9) reduces to the form of the adjoint equation of [7];

— equation (11) with the terminal condition (12) and Maximum condition (13) disappear;

— the family of "extended" Maximum conditions (15) reduces to the Maximum condition of [7],

i.e. the extended local Maximum Principle from this paper reduces to the classical Pontryagin type local Maximum Principle from [7].

REFERENCES

[1] Avakov, E.R.,(1985) Extremum conditions for smooth problems with equality-type constraints. U.S.S.R. Computational Mathematics and Mathematical Physics, Vol. 25, No.5, pp.680-693.

[2] Avakov, E.R.,(1988) Necessary conditions for the minimum for the nonregular problems in Banach spaces. Maximum principle for abnormal problems of optimal control. Trudy Matematiceskovo Instituta, A.N. SSSR, Vol. 185, pp. 3-29, (in Russian).

[3] Avakov, E.R.,(1988) The maximum principle for abnormal optimal control problems, Soviet Math.Dokl., Vol.37, pp. 231-134.

[4] Girsanov, I.V.,(1972) Lectures on Mathematical Theory of Extremum Problems, Lecture Notes in Economics and Mathematical Systems, 67, Springer Verlag.

[5] Ioffe, A.D. and Tikhomirov,W.M.,(1979) Theory of Extremal Problems, North Holland.

[6] Ledzewicz-Kowalewska, U., (1986) On some specification of the Dubovitskii-Milytin method, Vol.10, No.12, pp.1367-71.

[7] Ledzewicz-Kowalewska, U., (1988) Application of some specification of the Dubovitskii-Milyutin method to problems of optimal control , Nonlinear Analysis, Vol. 12, No. 2, pp. 101-108.

[8] Ledzewicz-Kowalewska, U., (1991) On the Euler-Lagrange equation in the case of nonregular operator equality constraints, Journal of Optimization Theory and Applications , Vol. 71, No.3.

[9] Ledzewicz-Kowalewska, U., On abnormal optimal control problems with mixed equality and inequality constraints, to appear in Journal of Mathematical Analysis and Applications.

[10] Ledzewicz-Kowalewska, U., An extension of the local Maximum Principle to abnormal optimal control problems, to appear.

Urszula Ledzewicz-Kowalewska
Department of Mathematics and Statistics,
Southern Illinois University at Edwardsville, Edwardsville, ILL 62026, USA

Research partially supported by SIUE Research Scholar Award and by N.S.F. Grant D.M.S. – 9109324.

X B LIN
Exponential dichotomy and stability of long periodic solutions in predator-prey models with diffusion

1. Predator-prey equations can have stable periodic solutions that bifurcate from homoclinic or heteroclinic cycles, [3,6]. Such solutions also satisfy partial differential equations obtained by adding diffusions and Neumann or periodic boundary conditions. In this paper we study the stability of these solutions in the intermediate spaces $D_A(\theta)$, $0<\theta<1$. Our main tool is the extension of the notion of exponential dichotomies to $D_A(\theta)$. General results related to the stability of diffusively perturbed solutions can be found in [2,7]. Except for Theorem 5, all the resuts in this paper are proved in [8].

2. Let A be a densely defined closed linear operator in a Banach space X. Let $\rho(A)$ contain the sector

$$S_{\omega,\phi} = \{ z \in \mathbb{C},\ z \neq \omega,\ |\arg(z-\omega)| \leq \phi \}. \tag{1}$$

where $\omega \in \mathbb{R}$ and $\pi/2 < \phi < \pi$. Assume that for $\lambda \in S_{\omega,\phi}$,

$$\|R(\lambda, A)\|_{L(X)} \leq M|\lambda - \omega|^{-1}. \tag{2}$$

Then A is the infinitesimal generator of a C_0 analytic semigroup e^{At} in X. For each $0<\theta<1$, define a Banach space

$$D_A(\theta) = \{ x \in X |\ \lim_{t \to \infty} t^{1-\theta} A e^{At} x = 0 \}.$$

Also define $D_A(\theta+1) = \{ x \in D_A | Ax \in D_A(\theta) \}$. Their norms are $\|x\|_\theta = \sup_{0 \leq t \leq 1} \|t^{1-\theta} A e^{At} x\|_X + \|x\|_X$, and $\|x\|_{\theta+1} = \|Ax\|_\theta + \|x\|_X$. These intermediate spaces are used to obtain maximal space regularity for abstract parabolic equations, [1,5,9].

Let $A(t): D \to X$, $a \leq t \leq b$, possibly $a = -\infty$ or $b = +\infty$, be linear operators such that the graph norm of $A(t)$ is equivalent to the norm in D. Assume that
(a) for each $t \in [a, b]$, $A(t): D \to X$ satisfies (1) and (2);
(b) $D_{A(t)}(\theta+k)$ is independent of t and is denoted by $D_A(\theta+k)$, $k=0, 1$;

Partially supported by NSF grant DMS9002803

(c) the function $t \to A(t)$ is in $C([a,b]; L(D,X)) \cap C([a,b]; L(D_A(\theta+1), D_A(\theta)))$.

Consider the initial value problem

$$u'(t)=A(t)u(t)+f(t), \quad u(s)=x, \quad a \leq s \leq t \leq b. \qquad (3)$$

It is well known that for each $x \in D_A(\theta+1)$ and $f \in C([a,b]; D_A(\theta))$, (3) has a unique solution $u \in C^1([s,b]; D_A(\theta)) \cap C([s,b]; D_A(\theta+1))$. We denote the solution by $u(t)=T(t,s)x$ when $f=0$. $T(t,s)$ can be extended to $D_A(\theta)$ by continuation. See [1,8,9]. The solution of (3) can be written by the variation of constants formula

$$u(t)=T(t,s)x+\int_s^t T(t,\tau)f(\tau)d\tau, \quad a \leq s \leq t \leq b. \qquad (4)$$

Solutions for nonlinear parabolic equations can be obtained by (4) and the contraction principle in suitable function spaces.

3. Exponential dichotomies for ODEs are discussed in detail in [4]. They are extended to parabolic periodic systems in [10].

Definition 1. The evolution operator $T(t,s) \in L(D_A(\theta+1))$ is said to have a (pseudo) exponential dichotomy on $I=[a,b]$, possibly $a=-\infty$ or $b=\infty$, if there exist projections $P_u(t)+P_s(t)=I$, $t \in I$, on $D_A(\theta+1)$, and constants K, C_0, $h >0$ and $\beta < \alpha$ with the following properties for s and t in I:

(i) $\|\int_\tau^t T(t,s)f(s)ds\|_{\theta+1} \leq C_0 \sup_{\tau \leq s \leq t} \|f(s)\|_\theta$ for $[\tau,t] \subset I$ and $t-\tau \leq h$.

(ii) $P_\nu(t)T(t,s)x=T(t,s)P_\nu(s)x$, $\nu=u,s$, for each $x \in D_A(\theta+1)$ and $s \leq t$.

(iii) $T(t,s): P_u(s)D_A(\theta+1) \to P_u(t)D_A(\theta+1)$, $s \leq t$ is a homeomorphism with the inverse denoted by $T(s,t)$.

(iv) $\|T(t,s)P_s(s)x\|_{\theta+1} \leq Ke^{\beta(t-s)}\|x\|_{\theta+1}$ and $\|T(s,t)P_u(t)x\|_{\theta+1} \leq Ke^{-\alpha(t-s)}\|x\|_{\theta+1}$ for $s \leq t$.

It is useful to extend the exponential dichotomy to $D_A(\theta)$.

Theorem 1. For each $t<b$ and $x \in D_A(\theta)$, $P_\nu(t)x$, $\nu=u,s$, can be defined by continuation. Moreover, if $\|T(t,s)x\|_\theta \leq C_1 \|x\|_\theta$ for $s. t \in I$ and $t-s \leq h$, we have

$$P_\nu(t)T(t,s)x=T(t,s)P_\nu(s)x, \quad s \leq t \text{ and } \nu=u,s.$$

$$\|T(t,s)P_s(s)x\|_\theta \leq \bar{K}_1 e^{\beta(t-s)}\|x\|_\theta, \quad s \leq t < b.$$

$$\|T(s,t)P_u(t)x\|_\theta \leq \bar{K}_2 e^{-\alpha(t-s)}\|v\|_\theta \quad s \leq t \leq b.$$

where $\bar{K}_1 = C(1+(b-s)^{-1})$ and $\bar{K}_2 = C(1+(b-t)^{-1})$.

Let $f \in C(I; D_A(\theta))$ and $\beta < \gamma < \alpha$. Define $\|f\|_{\gamma,\theta,\tau_0} = \sup_{t \in I} \|e^{-\gamma(t-\tau_0)} f(t)\|_\theta$. Such functions f with $\|f\|_{\gamma,\theta,\tau_0} < \infty$ form a Banach space, denoted by $C_{\gamma,\theta,\tau_0}(I)$.

Theorem 2. a) Let $I = \mathbb{R}$ and $f \in C_{\gamma,\theta,\tau_0}(I)$. Then there exists a unique solution $u(t) \in C_{\gamma,\theta,\tau_0}(I) \cap C^1(I; D_A(\theta)) \cap C(I; D_A(\theta+1))$ for $u'(t) = A(t)u(t) + f(t)$. Also

$$u(t) = \int_{-\infty}^{t} P_s(t) T(t,s) f(s) ds + \int_{\infty}^{t} T(t,s) P_u(s) f(s) ds,$$

$$\|u(t)\|_{\theta+1} + \|u'(t)\|_\theta \leq C e^{\gamma(t-\tau_0)} \|f\|_{\gamma,\theta,\tau_0}.$$

b) Let $I = [a,b]$ be a finite interval. For $\phi_a \in P_s(a) D_A(\theta+1)$, $\phi_b \in P_u(b) D_A(\theta+1)$ and $f \in C(I; D_A(\theta))$, there exists a unique solution $u \in C(I; D_A(\theta+1)) \cap C^1(I; D_A(\theta))$ for the boundary value problem

$$u'(t) = A(t) u(t) + f(t),$$

$$P_s(a) u(a) = \phi_a, \text{ and } P_u(b) u(b) = \phi_b.$$

We also have

$$u(t) = T(t,a)\phi_a + T(t,b)\phi_b + \int_a^t P_s(t) T(t,s) f(s) ds + \int_b^t T(t,s) P_u(s) f(s) ds,$$

$$\|u(t)\|_{\theta+1} + \|u'(t)\|_\theta \leq K e^{\beta(t-s)} \|\phi_a\|_{\theta+1} + K e^{\alpha(t-b)} \|\phi_b\|_{\theta+1} + C e^{\gamma(t-\tau_0)} \|f\|_{\gamma,\theta,\tau_0}.$$

Here the constant C only depend on K, C_0, α, β and γ.

c) Similar results also hold for $I = (-\infty, b]$ or $I = [a, \infty)$.

Let $B(t) \in L(D_A, X)$ and $t \to B(t)$ be continuous for $t \in I$. Assume that $\sup_{t \in I} \|B(t)\|_{L(D_A;X)} + \sup_{t \in I} \|B(t)\|_{L(D_A(\theta+1); D_A(\theta))} \leq \delta$ where δ is small. Then we can show that $u'(t) = (A(t) + B(t)) u(t)$ define an evolution operator $T_B(t,s)$. Also $T_B(t,s)$ has an exponential dichotomy if $T(t,s)$ has one on I.

Theorem 3. (roughness of the exponential dichotomies) Let $\tilde\alpha$ and $\tilde\beta$ be two constants with $\beta < \tilde\beta < \tilde\alpha < \alpha$. If δ is sufficiently small then $T_B(t,s)$ has an exponential dichotomy on I. The exponents, for $T_B(t,s)$, are $\tilde\alpha$ and $\tilde\beta$ and the constants $\tilde K$ and $\tilde C_0$ depend on K, C_0 and h only. The new projection $\tilde P_s(t)$ for $T_B(t,s)$ is $O(\delta)$ close to $P_s(t)$.

4. We will study the diffusively perturbed predator-prey equations

$$u_t = d_1 u_{xx} + uM(u,v),$$
$$v_t = d_2 v_{xx} + vN(u,v), \quad 0<x<1, \tag{5}$$

with Neumann or periodic boundary conditions at $x=0$ and $x=1$, and the related ODE model with $d_1=d_2=0$.

Conway and Smoller [3] studied an ODE model that describes the asocial behavior of the prey. Let $M=f(u,v)-v$ and $N=u-\gamma$. Assume that

(A) $f(b)=f(k)=0$, where $0<b<k$; $f(u)>0$ for $b<u<k$: and $f(u)<0$ otherwise;

(B) $f'>0$ on $[b,m)$, $f'<0$ on $(m,k]$, $f''<0$ for $u\in\mathbb{R}$;

It was shown that $B=(b,0)$ and $K=(k,0)$ are hyperbolic equilibria. Let $-\lambda_B^- < 0 < \lambda_B^+$ and $-\lambda_K^- < 0 < \lambda_K^+$ be the eigenvalues at the equilibria. There is a unique $\gamma=\gamma_0 \in (b,k)$ such that the ODE has a heteroclinic cycle connecting $(b,0)$ and $(k,0)$. When $\gamma_0 < \gamma < \gamma_0 + \varepsilon$, the heteroclinic cycle breaks to form a stable periodic solution $p(t,\gamma)$, if condition

(C) $\lambda_K^+ \lambda_B^+ - \lambda_K^- \lambda_B^- < 0$

is satisfied. Cf [8]. Using the tool of exponential dichotomies we can prove the folloing stability result after diffusions are added.

Theorem 4. (i) Assume that the Neumann boundary conditions are imposed. Then for each positive (d_1,d_2) satisfying

$$(k-\gamma)(bf'(b)-\pi^2 d_1)+(k-\gamma-\pi^2 d_2)(\gamma-b)<0, \tag{6}$$

there exists $\varepsilon_0>0$, depending on d_1 and d_2 such that the periodic solution $p(t,\gamma)$, $\gamma_0<\gamma<\gamma_0+\varepsilon_0$, is asymptotically stable in $D_A(\theta+1)$.

(ii) If the periodic boundary condition is imposed, then the same result is valid with (6) replaced by

$$(k-\gamma)(bf'(b)-4\pi^2 d_1)+(k-\gamma-4\pi^2 d_2)(\gamma-b)<0.$$

Our second example is the predator-prey model studied by Freedman and Wolkowicz [6]. Let u be the prey and v be the predator in equations (6). Their system models the group defense of prey against the predator so that $N_u<0$. In their model $N=N(u)$ is independent of v. When a parameter $k=k_0$, the system has a hyperbolic equilibrium E and a homoclinic solution asymptotic to E. The homoclinic solution bifurcates to a long period solution when $k_0-\varepsilon<k<k_0$. Assuming that the Jacobian matrix at the equilibrium E is

$$\begin{pmatrix} A & B \\ C & D \end{pmatrix}$$

where $A=uM_u$, $B=uM_v<0$, $C=vN_v<0$ and $D=0$, We can prove the following result when diffusions are added to the ODE model.

Theorem 5. If the Neumann boundary condition is imposed, then for each positive (d_1,d_2), satisfying

$$(\pi^2 d_1 - A)\pi^2 d_2 > BC,$$

there exists $\varepsilon_0 > 0$, depending on (d_1, d_2), such that the periodic solution is asymptotically stable in $D_A(\theta+1)$ if $k_0 - \varepsilon_0 < k < k_0$.

(ii) The same result is valid if the periodic boundary condition is imposed provided that

$$(4\pi^2 d_1 - A)4\pi^2 d_2 > BC.$$

References

1. Buttu, A. [1990], On the evolution operator for a class of non autonomous abstract parabolic equations, preprint.

2. Conway, E., Hoff, D. and Smoller, J. [1978], Large time behavior of solutions of systems of nonlinear reaction-diffusion equations, SIAM J. Appl. Math., 35, 1-16.

3. Conway, E. and Smoller, J. [1986], Global analysis of a system of predator- prey equations, SIAM J. Appl. Math., 46, 630-642.

4. Coppel, W.A. [1978], Dichotomies is stability theory, Lecture Notes in Mathematics, 629, Springer, Berlin.

5. Da Prato, G. and Grisvard, P. [1979], Equations d'evolution abstraites non linéaires de type parabolique, Ann. Mat. Pura. Appl., 120, 329-396.

6. Freedman, H.I. and Wolkoowicz, G.S. [1986], Predator-prey systems with group defense: the paradox of enrichment revisited, Bull Math. Bio., 48, 493-508.

7. Hale, J.K. [1986], Large diffusivity and asymptotic behavior in parabolic systems, J. Math. Anal. Appl. 118, 455-466.

8. Lin, X.B. [1991], Exponential dichotomies in intermediate spaces with applications to a diffusively perturbed predator-prey model, pre-print.

9. Lunardi, A. [1987], On the evolution operator for abstract parabolic equations, Israel J. Math., 60, 281-314.

10. Lunardi, A [1988], Stability of the periodic solutions to fully nonlinear equations in Banach space, Diff. and Int. Equations, 1, 253-279.

Xiao-Biao Lin   Department of Mathematics

North Carolina State University   Raleigh, N.C. 27695-8205

A LUNARDI
Stability of travelling waves in free boundary parabolic problems arising as combustion models

We present here some results - obtained in collaboration with C.-M. Brauner, C. Schmidt-Lainé, and F. Alabau - about the behavior near travelling wave solutions of some free boundary parabolic problems with interface conditions. Such problems arise as specific models in one-dimensional combustion phenomena.

Our method can be applied to a number of equations and systems. However, we focus here on a Deflagration-Detonation transition model, introduced in [6], [13], [14]:

$$\begin{cases} u_t = u_\eta + u_{\eta\eta}, \ t \geq 0, \ \eta \in \mathbf{R}, \ \eta \neq \xi(t) \\ u(\xi(t), t) = u_*, \ [u_\eta]_{\eta=\xi(t)} = -1, \ t \geq 0, \\ u(-\infty, t) = 0, \ u(+\infty, t) = u_\infty, \ t \geq 0, \\ u(\eta, 0) = u_0(\eta), \ \eta \in \mathbf{R}, \end{cases} \quad (1)$$

and on a system in Near Equidiffusional Flame Theory, introduced in [9]:

$$\begin{cases} \chi_t(\eta, t) = \chi_{\eta\eta}(\eta, t), \ t \geq 0, \ \eta < \xi(t), \\ \chi(\eta, t) = 1, \ t \geq 0, \ \eta \geq \xi(t), \\ S_t(\eta, t) = S_{\eta\eta}(\eta, t) + \ell \chi_{\eta\eta}(\eta, t) - \gamma(\chi_\eta(\eta, t))^2, \\ \qquad t \geq 0, \ \eta \neq \xi(t), \end{cases} \quad (2)$$

$$\begin{cases} [\chi]_{\eta=\xi(t)} = [S]_{\eta=\xi(t)} = 0, \ t \geq 0, \\ [\chi_\eta]_{\eta=\xi(t)} = -exp(S(\xi(t), t)/2), \ t \geq 0, \\ [S_\eta]_{\eta=\xi(t)} = (\ell - \gamma/2)exp(S(\xi(t), t)/2), \ t \geq 0, \\ \chi(-\infty, t) = S(-\infty, t) = 0, \ t \geq 0, \\ |S(+\infty, t)| < +\infty, \ t \geq 0, \end{cases} \quad (3)$$

$$\xi(0) = 0, \ \chi(\eta, 0) = \chi_I(\eta), \ S(\eta, 0) = S_I(\eta), \ \eta \in \mathbf{R}. \quad (4)$$

Here $[f]_{\eta=\xi(t)} = f(\xi(t)^+, t) - f(\xi(t)^-, t)$ denotes the jump of the function f at $\eta = \xi(t)$, and the data u_*, u_∞, γ are positive numbers. For the physical meaning of such problems, see [6], [13], [14], [2], [9]. Travelling waves are solutions such that ξ is linear, say $\xi(t) = \xi_0 - ct$, and u (respectively, ξ, S) are of the form $u(t, \eta) = U_0(\eta + ct)$ (resp. $\chi(t, \eta) = \chi_0(\eta + ct)$, $S(t, \eta) = S_0(\eta + ct)$), with $U_0, \chi_0, S_0 : \mathbf{R} \to \mathbf{R}$. It is not difficult to see that if $u_\infty > \sqrt{2}$, $2/u_\infty < u_* < 2/u_\infty + u_\infty$, then problem (1) has a (unique up to translations) travelling wave solution, with speed $c = 1/u_\infty + u_\infty/2$, and $U_0(y) \sim const.e^{2\alpha_1 y}$ as $y \to -\infty$, $U_0(y) \sim u_\infty + const.e^{-2\alpha_2 y}$ as $y \to +\infty$, with $\alpha_1, \alpha_2 > 0$. Similarly, problem (2)-(3)-(4) has a (unique up to translations) travelling wave solution for every value of the parameters ℓ and γ, with speed $c = 1$.

After several transformations, which we describe briefly below, we reduce problems (1) and (4) (2)-(3) to evolution equations in suitable Banach spaces, in such a way that orbital stability/instability of the TW solution is equivalent to stability/instability of the null solution of the abstract equation. Due to the interface conditions, the final equation is fully nonlinear, i.e. of the form

$$\begin{cases} u'(t) = Lu(t) + F(u(t)), & t \geq 0 \\ u(0) = u_0, \end{cases} \quad (5)$$

where L is a sectorial operator in a Banach space X, and the nonlinear smooth perturbation F is defined on the domain $D(L)$ of L (more precisely, on a neighborhood of 0 in $D(L)$) and has values in X. In addition, $D(L)$ is not dense in X. Therefore, we cannot apply the well known results about abstract semilinear parabolic equations (contained, for instance, in the book of Henry [5]), but we need more specific tools, which can be found in the papers [7], [8], [4], [2]. Geometric theory of equations such as (5) includes results of local existence, uniqueness, regularity, stability, Hopf bifurcation. In particular, the principle of linearized stability holds, so that the problem of determining the stability of the TW solution of problems (1) and (2)-(3)-(4) is reduced to the problem of finding out the spectrum of L. In our case, L is nothing but the realization of a second order elliptic differential operator with suitable boundary conditions in the space X, and it is possible to characterize its spectrum. As a consequence, we are able to give a rigorous mathematical justification to the statements of [10] about problem (2) and of [9] about problem (2)-(3)-(4).

Let us describe how to transform our problems to abstract parabolic equations. To be definite, let us consider problem (1), which has been studied in [2]. After the obvious change of coordinates $x = \eta - \xi(t)$, setting $z(x,t) = u(\eta,t) = u(x + \xi(t), t)$, $z_0(x) = u_0(x + \xi(t))$, the free boundary $\eta = \xi(t)$ is transformed into the fixed boundary $x = 0$. Moreover we introduce the differences $s(t) = \xi(t) + ct$, $v(x,t) = z(x,t) - U_0(x)$, $v_0(x) = z_0(x) - U_0(x)$, which should satisfy the system

$$\begin{cases} v_t = v_{xx} - cv_x + \dot{s}U' + \dot{s}v_x + (Uv)_x + vv_x, & x \in \mathbf{R}, \ x \neq 0, \\ v(0,t) = [v_x(t)] = 0, \\ v(-\infty, t) = v(+\infty, t) = 0, \\ v(x,0) = v_0(x), \ s(0) = 0, \end{cases} \quad (6)$$

where we denote by $[f]$ the jump of f at $x = 0$. Now we define the Banach spaces

$$\mathcal{X} = \{v : \mathbf{R} \to \mathbf{R} \mid x \mapsto e^{-\alpha_1 x} v(x) (\text{resp.} x \mapsto e^{\alpha_2 x} v(x))$$
$$\text{is uniformly continuous and bounded on } \mathbf{R}_-(\text{resp. } \mathbf{R}_+)\},$$

$$D(\mathcal{L}) = \{v \in \mathcal{X}/v', v'' \in \mathcal{X}, [v'] = (c - u_*)[v], v(0^-)U'(0^+) = v(0^+)U'(0^-)\}$$

and the differential operator $\mathcal{L} : D(\mathcal{L}) \to \mathcal{X}$, $\mathcal{L}v = v'' - cv'$. $\mathcal{X}$ is endowed with its natural norm, $D(\mathcal{L})$ is endowed with the graph norm.

Problem (6) may be rewritten as an evolution equation in $\mathcal{X}$:

$$\begin{cases} \dot{v} = \mathcal{L}v + \dot{s}U' + \dot{s}v' + vv', & t \geq 0 \\ v(0) = v_0. \end{cases} \quad (7)$$

Some remarks are now due about the choice of $\mathcal{X}$ and $D(\mathcal{L})$. First, we consider weighted spaces because of the nice spectral properties of the realizations of elliptic operators in such spaces (see next Lemma 1). Second, the boundary conditions in the definition of $D(\mathcal{L})$ are those which arise in linearizing (6) arond 0 and decoupling it. Indeed, the linearized system is

$$\begin{cases} \hat{v}_t = \hat{v}_{xx} - c\hat{v}_x + \dot{s}U', & t \geq 0, \ x \neq 0, \\ \hat{v}(0,t) = 0, \ [\hat{v}_x] = 0, & t \geq 0, \\ \hat{v}(x,0) = v_0(x), & x \in \mathbf{R}, \end{cases} \quad (8)$$

so that the difference $\hat{w} = \hat{v} - s(t)U'$ satisfies

$$\begin{cases} [\hat{w}]_{x=0} = s(t), & t \geq 0, \\ \hat{w}_t = \hat{w}_{xx} - c\hat{w}_x, & t \geq 0, \ x \neq 0, \\ [\hat{w}_x] = (c - u_*)[\hat{w}], \ \hat{w}(0^-)U'(0^+) = \hat{w}(0^+)U'(0^-), & t \geq 0, \\ \hat{w}(x,0) = v_0(x), & x \in \mathbf{R}, \end{cases} \quad (9)$$

and (8) is decoupled.

The spectral properties of $\mathcal{L}$ are the following. The proof can be found in [10].

(i) $\mathcal{L}$ is a sectorial operator in $\mathcal{X}$.

(ii) $\sigma(\mathcal{L}) = \sigma_1 \cup \{0\} \cup \sigma_2$, where $\sigma_1 = (-\infty, -c^2/4 + 1/2]$, and σ_2 consists of a real negative eigenvalue if $2/u_\infty < u_* \leq u_0 = c - \sqrt{c^2 - 1} - \sqrt{2c^2 - 3}$, it is empty if $u_0 < u_* \leq u_c = c + \sqrt{c^2 - 1}$, and it consists of a real positive eigenvalue if $u_c < u_* < u_\infty + 2/u_\infty$.

(iii) 0 is a simple eigenvalue, and the projection on the kernel of $\mathcal{L}$ which commutes with $\mathcal{L}$ is given by

$$Pv = \frac{1}{u_\infty}(\int_{-\infty}^{+\infty} v dx)U'.$$

We remark that property (ii) is due to the fact that $\mathcal{X}$ is endowed with a suitably weighted norm. If we replace $\mathcal{X}$ by a nonweighted space, the continuous spectrum can reach the imaginary axis, which is a very bad property in the study of the stability (an interesting discussion on the choice of the weights in the stability analysis of nonsingular travelling wave solutions of parabolic equations can be found in [11], [12]). However, if one is interested only in local existence and regularity, it is possible to work also in nonweighted spaces, but we will not develop this approach here.

Now we are able to decouple system (7), by applying P and $I - P$ and setting

$$v = Pv + (I - P)v = w + p(t)U'.$$

We find

$$\begin{cases} (i) \ \dot{s} = [\dot{w}] = [Lw] + [\dot{s}v' + vv'] = [Lw], \\ (ii) \ \begin{cases} \dot{w} = Lw + [Lw]([w]U' + w') + \frac{d}{dx}\frac{1}{2}([w]U' + w')^2, \\ w(0) = (I - P)v_0 = w_0. \end{cases} \end{cases} \quad (10)$$

Problem (10)(ii) is an evolution equation in the space $X = (I - P)(\mathcal{X})$. The linear operator L is the part of $\mathcal{L}$ in X, that is $D(L) = (I - P)(D(\mathcal{L}))$, $Lw = \mathcal{L}w$ for $w \in D(L)$. Since $\mathcal{L}$ is sectorial in $\mathcal{X}$, then L is sectorial in X. Moreover, problem (10)(ii) can be written in the form (5), by setting

$$F(w) = [Lw]([w]U' + w') + \frac{d}{dx}\frac{1}{2}([w]U' + w')^2, \quad \forall w \in D(L).$$

As easily seen, F is analytic, and $F(0) = 0$, $F'(0) = 0$. Therefore, we are able to apply the results of local existence, regularity, and stability mentioned above. In particular, we find that for every initial datum $w_0 \in D$ satisfying the compatibility condition $Lw_0 + F(w_0) \in \overline{D}$, problem (10)(ii) has a unique local regular solution, which is analytic for $t > 0$. Concerning stability, we have $\sigma(L) = \sigma_1 \cup \sigma_2$, because we got rid of the eigenvalue 0 by the projection step. Consequently, the null solution of (10) is exponentially asymptotically stable if $u_* \leq u_c$, and it is unstable if $u_c < u_* < u_\infty + 2/u_\infty$.

Coming back to the original problem (1) we find easily

$$z(\cdot, t) = [w(t)]U_0' + w(t) + U, \quad s(t) = \int_0^t p'(\sigma)d\sigma = [w(t)] - \frac{1}{u_\infty}\int_{-\infty}^{+\infty} v_0(x)dx.$$

Therefore, the free boundary is analytic for $t > 0$. Moreover, if $u_* \leq u_c$, the wave U_0 is exponentially asymptotically orbitally stable, and there exists a final shift s_∞ such that

$$\begin{cases} \sup_{\eta \in \mathbf{R}} |u(\eta, t) - U_0(\eta - s_\infty + ct)| \to 0, & \text{as } t \to +\infty \\ \zeta(t) + ct \to s_\infty, & \text{as } t \to +\infty. \end{cases}$$

If $u_c < u_* < u_\infty + 2/u_\infty$, the wave is unstable.

Problem (2)-(3)-(4) was studied in [1], following the same procedure. From the technical point of wiew it is a bit more complicated than problem (1), because of the nonlinear boundary condition. Such a difficulty can be overcome by using a suitable relevament operator, as it is common in parabolic problems with nonlinearities at the boundary. We arrive however at a final problem of the type (5), where the spectrum of L is given by $\sigma(L) = \sigma_c(L) \cup \sigma_p(L)$, and

$$\sigma_c(L) \subset (-\infty, -1/4]$$

$$\sigma_p(L) = \{\frac{\ell^2 - 8\ell - 32 - \sqrt{\ell^4 - 16\ell^3}}{128}, \frac{\ell^2 - 8\ell - 32 + \sqrt{\ell^4 - 16\ell^3}}{128}\}.$$

In particular, for $-2 < \ell < \ell_c = 4(1 + \sqrt{3})$, all the elements of the spectrum of L have negative real part, so that the travelling wave is orbitally stable, whereas if $\ell < -2$ or $\ell > \ell_c$ there are eigenvalues with positive real part, so that the travelling wave is unstable. Moreover, at $\ell = \ell_c$, two eigenvalues cross transversally the imaginary axis. So it is possible to apply the result of [4] about Hopf bifurcation, obtaining the existence of a pulsating front around the travelling wave for suitable values of ℓ near ℓ_c.

References

[1] F. ALABAU, A. LUNARDI, *Behavior near the travelling wave solution of a free boundary problem in Combustion Theory*, submitted.

[2] C.-M. BRAUNER, A. LUNARDI, CL. SCHMIDT-LAINÉ, *Stability of travelling waves with interface conditions*, to appear in Nonlinear Analysis TMA.

[3] C.-M. BRAUNER, S. NOOR EBAD, CL. SCHMIDT-LAINÉ, *Nonlinear stability analysis of singular travelling waves in combustion : a one-phase problem*, Nonlinear Analysis TMA **16**(1991), 881-892.

[4] G. DA PRATO, A. LUNARDI, *Hopf bifurcation for fully nonlinear equations in Banach space*, Ann. Inst. H. Poincaré, Analyse non linéaire, **3** (1986), 315-329.

[5] D. HENRY, *Geometric theory of semilinear parabolic equations*, Lect. Notes in Math. **840**, Springer-Verlag, New-York (1981).

[6] G. S. S. LUDFORD, D. S. STEWART, *Fast deflagration waves*, J. Méc. Th. Appl., **2** (1983), 463-487.

[7] A. LUNARDI, *On the local dynamical system associated to a fully nonlinear parabolic equation*, in : "Nonlinear Analysis and Applications", V. Lakshmikantham Ed., Marcel Dekker Publ. (1987), 319-326.

[8] A. LUNARDI, *Time analyticity for solutions of nonlinear abstract parabolic evolution equations*, J. Funct. Anal. **71** (1987), 294-308.

[9] B.J. MATKOVSKI, A. VAN HARTEN, *A new model in flame theory*, SIAM J. Appl. Math. **42** (1982), 850-867

[10] S. NOOR EBAD, CL. SCHMIDT-LAINÉ, *Further numerical results on a deflagration to detonation transition model* (submitted).

[11] D. H. SATTINGER, *On the stability of waves of nonlinear parabolic systems*, Adv. in Math., **22**(1976), 312-355.

[12] D. H. SATTINGER, *Weighted norms for the stability of traveling waves*, J. Diff. Eqns. **25** (1977), 130-144.

[13] D. S. STEWART, *Transition to detonation on a model problem*, J. Méc. Th. Appl., **4** (1985), 103-137.

[14] D. S. STEWART, G. S. S. LUDFORD, *The acceleration of fast deflagration waves*, Z. A. M. M., **63** (1983), 291-302.

A N LYBEROPOULOS
On the ω-limit set of solutions of scalar balance laws on S^1

Consider the initial value problem

$$\partial_t u(x,t) + \partial_x f(u(x,t)) = g(u(x,t)), \quad -\infty < x < \infty, \quad 0 < t < \infty, \tag{1}$$

$$u(x,0) = u_0(x), \quad -\infty < x < \infty, \tag{2}$$

where $f(\cdot)$ and $g(\cdot)$ are given smooth functions defined on $(-\infty, \infty)$ and taking values, along with the unknown $u(x,t)$, in $(-\infty, \infty)$. The importance in studying the above problem stems from the fact that equation (1), which is commonly called a *balance law* (or *conservation law*, in the absence of the source term g), serves as the prototypical example for systems of equations that govern the evolution of one-dimensional continuous media with "elastic" response.

Despite its apparent simplicity, (1), (2) is well-posed in the classical sense only locally in time. As a matter of fact, even if the initial data $u_0(\cdot)$ are analytic, solutions generally stay smooth only up to a critical time beyond which the nonlinear structure of $f(\cdot)$ leads to the development of discontinuities (shocks) that tend to dominate the wave pattern thereafter. However (1), (2) is globally well-posed in a weak sense. Indeed, when $g(u)$ is either dissipative or it grows no faster than linear in u, then, for any bounded $u_0(\cdot)$ of locally bounded variation on $(-\infty, \infty)$, there exists a unique global solution of (1), (2) in the class BV of functions of locally bounded variation in the sense of Tonelli and Cesari [19], [12], which satisfies, in the sense of distributions, the so-called *entropy admissibility criterion*

$$\partial_t \eta(u(x,t)) + \partial_x q(u(x,t)) \geq \eta'(u(x,t)) g(u(x,t)), \tag{3}$$

for every pair of functions $\eta(u), q(u)$, where $\eta(u)$ is concave and

$$q(u) := \int_0^u f'(v) \eta'(v) dv. \tag{4}$$

For a thorough discussion on the motivation of the admissibility condition (3), as well as, other analogous conditions see [8].

Our objective in this note is to discuss the large time behavior of admissible BV-solutions of (1) under initial data that are L–periodic, i.e.

$$u_0(x+L) = u_0(x), \quad L > 0, \quad -\infty < x < \infty. \tag{5}$$

Since for each fixed $t \in [0,\infty)$, $u(\cdot,t)$ will also be L–periodic, it is clear that we can equivalently consider the problem on the circle, i.e. with $x \in S^1 = \mathbf{R}/L\mathbf{Z}$. Moreover, we assume throughout that:

(H1) $f(\cdot) \in C^2(-\infty,\infty)$, normalized so that $f(0) = f'(0) = 0$ and $f''(\cdot)$ does not vanish identically on any interval.

(H2) $g(\cdot) \in C^1(-\infty,\infty)$ and there is $K > 0$ such that $|g(u)| \le K|u|$ for all u in $(-\infty,\infty)$.

Let $X = BV(S^1)$ denote the Banach space of functions $w : S^1 \mapsto \mathbf{R}$ which have bounded variation on S^1. Then, Eq. (1) generates a global *semiflow* on X by associating to any initial datum $u_0 \in X$ a solution curve $\Phi_t(u_0) := u(\cdot,t)$, $t \in [0,\infty)$. For any *trajectory* $\gamma(u_0) := \{\Phi_t(u_0) : t \ge 0\}$ we define its ω-*limit* set by

$$\omega(u_0) := \{w \in X : \exists t_k \uparrow \infty \text{ such that } \lim_{k \to \infty} \Phi_{t_k}(u_0) = w\}. \tag{6}$$

Using standard a priori estimates and dynamical systems arguments, it can be shown that if $\omega(u_0) \ne \emptyset$ then it is a closed, compact, invariant set and has no proper subset possessing these three properties; see e.g. [3] and the references there.

The problem of obtaining more refined information on the structure of $\omega(u_0)$ when $g(\cdot)$ vanishes has been addressed by several authors, who showed that in this case $\omega(u_0)$ is consisting of exactly one constant state. More precisely, if $f(\cdot)$ is strictly convex on $(-\infty,\infty)$ then, confinement of waves and their strong interaction induces $O(t^{-1})$ decay of the solution to its (constant) mean value on S^1 [13], [4], while, if $f(\cdot)$ has a single inflection point at the origin and $u_0(\cdot)$ has zero mean, then $f'(u(x,t))$ decays as $O(t^{-1})$ at a uniform rate, independent of the size of the data [2], [7], [11]. Furthermore, using techniques from topological dynamics it has been shown in [3] that, if $f''(\cdot)$ vanishes on a set with no finite accumulation points then, $u(x,t)$ decays in $L^1_{loc}(-\infty,\infty)$ to its mean value on S^1 but, unfortunately, no information on the rate of decay can be extracted.

By contrast, very little seems to be known about the large time behavior of solutions of the general balance law (1). In particular, it can be shown that if $f(\cdot)$ is uniformly convex then the total variation of $u(\cdot,t)$ on S^1 decays to zero provided that $g(\cdot)$ is appropriately dissipative [6], but no similar result has been established so far in the case where no convexity restrictions on $f(\cdot)$ are imposed. It is worth noting here that if $g(\cdot)$ is dissipative then, as is expected, it collaborates with all the dissipative mechanisms that are present in the homogeneous case ($g \equiv 0$) and, as a result, induces stronger decay, but in the opposite case it induces competition the outcome of which affects crucially the time growth of the solution. At the same time and in contrast to the convex case, the dissipative mechanisms that affect the asymptotic behavior of the solution when $f(\cdot)$ has inflection points, become much more complicated to unravel. The reason is that now solutions contain, in general, not only *genuine shocks* but *contact shocks* as well. Genuine shocks have the tendency to simplify the wave pattern by absorbing all incoming signals and never producing outgoing ones. Contact shocks, on the other hand, can generate wave fans which contain *rarefaction waves* that radiate out of contact discontinuities and make the geometric structure of the wave pattern subtantially more intricate.

A good example for exploring the interplay between the above features is provided by the following equation

$$\partial_t u(x,t) + \partial_x u^m(x,t) = u(x,t), \qquad m \in \mathbb{N}, \quad m \geq 2, \tag{7}$$

in which all the dissipative mechanisms (most pronounced in the formation of strong shocks and their subsequent interaction thereof) coexist and compete with the excitation induced by a linear source field. Of course, in order to ensure that $w(u_0) \neq \emptyset$, it is necessary to assume here that the initial data have zero average on S^1. In the convex case (m even), solutions of (7) no longer decay as $t \to \infty$ but, instead, approach asymptotically a *standing wave* that contains an, at most, countable number of shocks [15]. However, in the nonconvex case (m odd) the situation turns out to be considerably more complicated. Indeed, pursuing detailed investigation and after a rather long line of arguments, it is shown in [16] that, as $t \to \infty$, the solution stays pointwise bounded and the bound is independent of the size of the initial data! Furthermore, when the data are of class C^0 and t is sufficiently large then, generically, the number of shocks attains a finite value which remains constant thereafter. In particular, all shocks become contact discontinuities while their evolution is governed by a nonlinear system of functional and delay-differential equations whose dynamical behavior (not yet understood) determines that of the solution $u(x,t)$. Nevertheless, some interesting properties incorporated by that system have been derived.

In the sequel we announce, without proof, a new result concerning the large time behavior of solutions of the general balance law (1) on S^1 with strictly convex $f(\cdot)$ and generic $g(\cdot)$; for details see the author's forthcoming article [17].

Theorem. Let $u(x,t)$, with $(x,t) \in S^1 \times [0,\infty)$, be the admissible BV–solution of (1), (2), (5), where $f(\cdot)$ is convex and satisfies (H1) while $g(\cdot)$ vanishes only at a finite number of points and satisfies (H2). Then, if $u(\cdot,t)$ stays bounded in $L^\infty(S^1)$ as $t \to \infty$, exactly one of the following two alternatives happens : Either $u(\cdot,t)$ converges to a *constant state* or approaches asymptotically a *rotating wave*, i.e. a nonconstant admissible weak solution of (1) of the form $\tilde{u}(x,t) = \varphi(x - ct)$, $c \in \mathbb{R}$.

The proof is based on two fundamental tools. The first one is a "set–valued" Liapounov functional similar to that introduced in [15]. The second is the concept of a generalized characteristic as introduced in [5]. Recall that a *classical characteristic* of (1), associated with a C^1–solution $u(x,t)$ is a trajectory $\chi(\cdot)$ of the ordinary differential equation

$$\dot\chi(t) = f'(u(\chi(t),t)) . \tag{8}$$

For admissible BV–solutions the same definition is adopted, but now (8) is interpreted as a contingent equation in the sense of Filippov [10]. As a consequence of the Filippov theory, through any point $(\bar{x},\bar{t})$ of the upper half-plane passes at least one *forward* characteristic, defined on a right maximal interval $[\bar{t},\tau)$, $\tau \leq \infty$, and at least one *backward* characteristic defined on a left maximal interval $(\sigma,\bar{t}]$, $\sigma \geq 0$. The set of forward (or backward)

113

characteristics through ($\bar{x},\bar{t}$) spans the funnel confined between a *minimal* and a *maximal* forward (or backward) characteristic through ($\bar{x},\bar{t}$). Of course, these extremal forward (or backward) characteristics need not be distinct. It turns out that all generalized characteristics propagate with either classical characteristic speed or with shock speed, but it is the special properties of the extremal backward ones, that play the key role in the analysis.

We should point out here that our results obtained in [15], [16] and [17] are closely related with those of [1], [9] and [18]. In fact, the authors there examine the large time behavior of the scalar reaction diffusion equation

$$u_t = \varepsilon u_{xx} + h(x, u, u_x), \quad x \in S^1, \quad t \geq 0, \quad \varepsilon > 0, \tag{9}$$

and show that, under some assumptions on the structure of the function h, the ω-limit set of the solution either contains one periodic solution, or consists of solutions tending to equilibrium as $t \to \pm\infty$. At the same time, they address as an open question the issue on whether or not a similar result holds for (9) in the viscosity limit as $\varepsilon \downarrow 0$. Clearly, comparison between our results and theirs becomes even more interesting if one takes into account that when $h(x, u, u_x) \equiv -f(u)_x + g(u)$, admissible solutions of (1) can be viewed as viscosity limits of solutions of (9) [12].

Finally, we would like to mention that an exhaustive study of the asymptotic properties of solutions of the model equation (1) could provide the necessary technical background for exploring similar questions for systems of hyperbolic balance laws [14].

References

[1] S.B. ANGENENT & B. FIEDLER, The dynamics of rotating waves in scalar reaction diffusion equations, *Trans. Amer. Math. Soc.* **307** (1988), 545–568.

[2] J.G. CONLON, Asymptotic behavior for a hyperbolic conservation law with periodic initial data, *Comm. Pure Appl. Math.* **32** (1979), 99–112.

[3] C.M. DAFERMOS, Applications of the invariance principle for compact processes II. Asymptotic behavior of solutions of a hyperbolic conservation law, *J. Differential Equations* **11** (1972), 416-424.

[4] C.M. DAFERMOS, Characteristics in hyperbolic conservation laws, *in* "Nonlinear Analysis and Mechanics" (R.J. Knops, Ed.), pp. 1–58, Research Notes in Math. No. 17, Pitman, London, 1977.

[5] C.M. DAFERMOS, Generalized characteristics and the structure of solutions of hyperbolic conservation laws, *Indiana Univ. Math. J.* **26** (1977), 1097–1119.

[6] C.M. DAFERMOS, Asymptotic behavior of solutions of hyperbolic balance laws, *in* "Bifurcation Phenomena in Mathematical Physics" (C. Bardos and D. Bessis, Eds.), pp. 521–533, D. Reidel, Dordrecht, 1979.

[7] C.M. DAFERMOS, Regularity and large time behavior of solutions of a conservation law without convexity, *Proc. Royal Soc. Edinburgh Sect. A* **99** (1985), 201–239.

[8] C.M. DAFERMOS, Hyperbolic systems of conservation laws, *in* "Systems of Nonlinear Partial Differential Equations" (J.M. Ball, Ed.), pp. 25–70, NATO ASI Series C, No. 111, D. Reidel, Dordrecht, 1983.

[9] B. FIEDLER & J. MALLET-PARET, A Poincaré–Bendixson theorem for scalar reaction diffusion equations, *Arch. Rat. Mech. Anal.* **107** (1989), 325–345.

[10] A.F. FILIPPOV, Differential equations with discontinuous right-hand side, *Mat. Sbornik* (N.S.) **51** (1960), 99–128; (English translation: *Amer. Math. Soc. Transl. Ser. 2* **42** (1964), 199-231).

[11] J.M. GREENBERG & D.D. TONG, Decay of periodic solutions of $\partial u/\partial t + \partial f(u)/\partial x = 0$, *J. Math. Anal. Appl.* **43** (1973), 56–71.

[12] S.N. KRUŽKOV, First order quasilinear equations in several independent variables, *Mat. Sbornik* (N.S.) **81** (1970), 228–255; (English translation: *Math. USSR-Sbornik* **10** (1970), 217–243).

[13] P.D. LAX, Hyperbolic systems of conservation laws II, *Comm. Pure Appl. Math.* **10** (1957), 537–566.

[14] T.-P. LIU, Quasilinear hyperbolic systems, *Comm. Math. Phys.* **68** (1979), 141–172.

[15] A.N. LYBEROPOULOS, Asymptotic oscillations of solutions of scalar conservation laws with convexity under the action of a linear excitation, *Quart. Appl. Math.* **48** (1990), 755–765.

[16] A.N. LYBEROPOULOS, Large time structure of solutions of scalar conservation laws without convexity in the presence of a linear source field, *J. Differential Equations* (to appear in 1992).

[17] A.N. LYBEROPOULOS, (in preparation).

[18] H. MATANO, Asymptotic behavior of solutions of semilinear heat equations on S^1, *in* "Nonlinear Diffusion Equations and their Equilibrium States II" (W.-M. Ni, L.A. Peletier and J. Serrin, Eds.), pp. 139–162, Springer Verlag, New York, 1988.

[19] A.I. VOL'PERT, The space BV and quasilinear equations, *Mat. Sbornik* **73** (1967), 255–302; (English translation: *Math. USSR-Sbornik* **2** (1967), 225–267).

1980 *Mathematics subject classifications*: Primary 35B40; Secondary 35B10, 35L65

Department of Mathematics, University of Oklahoma, Norman, OK 73019, USA

M MAESUMI
Triple-shock elementary waves for two-phase incompressible flow in porous media

1. Introduction

Elementary waves are scale invariant traveling wave solutions of hyperbolic conservation laws [1, 2]. In this paper we study the configurations of triple-shock elementary waves (TSEWs) in two-phase incompressible flow in homogeneous porous media. These "Y"-shaped formations may occur in the collision of two shock fronts. Given fixed physical parameters there are only a finite number of such waves. In general, exactly one of the three opening angles between the shocks is less than 90°. Moreover the magnitude of this angle is bounded by the second derivative of the fractional flow function. For a miscible flow the TSEWs reduce to essentially plane wave solutions.

2. Elementary Waves

We consider an idealized two-phase (water and oil) two-dimensional flow, governed by

$$\phi s_t + \mathbf{v} \cdot \nabla f(s) = 0, \qquad \mathbf{v} = -\lambda(s) \nabla P, \qquad \nabla \cdot \mathbf{v} = 0. \qquad (1)$$

Here ϕ is porosity and $\mathbf{v}$ is the total flux. The saturation of one of the phases (e.g. water) is denoted s. The fractional flow function f is defined so that the volumetric velocity of the water phase is $\mathbf{v}f$. P is the pressure and $\lambda > 0$ is the mobility of the flow. We assume the porous medium is homogeneous and ϕ, f and λ do not depend on position.

Hyperbolic equations tend to form discontinuous solutions called shock fronts. Here a pair of jump conditions are satisfied across the shock. They express the continuity of the pressure and the normal flux. Assume $(\mathbf{n}, \mathbf{t})$ are the local normal and tangential unit vectors to a shock front at $(\mathbf{x}_0, t)$. For any quantity $q(\mathbf{x}, t)$, define q_- and q_+ at $(\mathbf{x}_0, t)$ by $q_\pm = \lim_{\varepsilon \to 0\pm} q(\mathbf{x}_0 + \mathbf{n}\varepsilon, t)$. Then the jump conditions can be written as $\mathbf{v}_- \cdot \mathbf{n} = \mathbf{v}_+ \cdot \mathbf{n} \equiv N$, $\mathbf{v}_+ \cdot \mathbf{t}/\lambda_+ = \mathbf{v}_- \cdot \mathbf{t}/\lambda_- \equiv T$, where N is the normal flux and T is the negative of tangential pressure gradient. The "primary" shock speed is defined as $\sigma = (f(s_+) - f(s_-))/(s_+ - s_-)$ while the propagation speed of the shock, in the direction of $\mathbf{n}$, is $W = \sigma N/\phi$. Denote the distance between f and its secant by $\Delta(s; r, l) = f(s) - ((s-r)f(l) + (l-s)f(r))/(l-r)$. Then the entropy condition requires $N(s_+ - s_-)\Delta(s; s_+, s_-) \geq 0$ for all s between s_+ and s_-. For miscible flow $f(s) = s$, $\sigma = 1$, and the discontinuous solutions are called contact discontinuities. For uniformity we use the word "shock" for these solutions as well.

The simplest case of the interaction of shocks is described by elementary waves. An elementary wave $E(\mathbf{x},t) = (s,\mathbf{v})$ is a scale invariant traveling wave solution of (1). Thus

$$E(\mathbf{x},t) = E(\alpha\mathbf{x}, \alpha t), \quad \text{for all } \alpha > 0, \qquad E(\mathbf{x} + \tau\mathbf{U}, t + \tau) = E(\mathbf{x},t), \quad \text{for all } \tau. \qquad (2)$$

Here $\mathbf{U}$ is the propagation velocity of the wave. At time t the node of the wave is at the point $t\mathbf{U}$. The following lemma follows from (1) and (2). (See [2].)

Lemma 1. *E is constant on rays which start at the node. Moreover $\mathbf{v}$ is constant where it is smooth. If s is continuous at a point $\mathbf{x}$ and $\mathbf{n}$ is a normal vector to the ray $\mathbf{x} - t\mathbf{U}$ then $f'(s)\mathbf{v}\cdot\mathbf{n} = \phi\mathbf{U}\cdot\mathbf{n}$. If s is discontinuous, i.e. across a shock, $\sigma\mathbf{v}\cdot\mathbf{n} = \phi\mathbf{U}\cdot\mathbf{n}$.*

In this paper we assume s is piecewise constant, in other words the variation in s is only due to jump discontinuity. Also we define *essentially plane waves* as elementary waves in which $\mathbf{v}$ is discontinuous across at most one line.

3. Triple-Shock Elementary Waves

Assume the three sectors of a TSEW are labeled (in the counter-clockwise direction) by i or $j = 1, 2, 3$. In these sectors the saturations are denoted by s_i, the mobilities by $\lambda_i = \lambda(s_i)$, the opening angles by α_i and the velocities by $\mathbf{v}_i = -\lambda_i(a_i, b_i)$. The subscript ij on a quantity indicates its value on the interface between the ith and (for $\mathbf{n}_{ij}$, into) the jth sector.

Assume at time $t = 0$ the three shocks pass through origin with the 1-2 interface coinciding with the positive x-axis. Then the equations of the three shocks at $t = 1$ are

$$W_{12} - y = 0, \quad W_{23} + x\sin\alpha_2 - y\cos\alpha_2 = 0, \quad W_{31} - x\sin\alpha_1 - y\cos\alpha_1 = 0. \qquad (3)$$

Since the TSEW is propagating by translation these lines should intersect at a common point. Therefore the determinant of their coefficients is zero. Writing for W in terms of σN we get

$$N_{12}\sigma_{12}\sin\alpha_3 + N_{23}\sigma_{23}\sin\alpha_1 + N_{31}\sigma_{31}\sin\alpha_2 = 0, \quad \text{where} \quad \alpha_1 + \alpha_2 + \alpha_3 = 2\pi. \qquad (4)$$

The three pairs of jump conditions across the three interfaces, $ij = 12, 23, 31$, are

$$\lambda_i(a_i, b_i)\cdot\mathbf{n}_{ij} = \lambda_j(a_j, b_j)\cdot\mathbf{n}_{ij} = -N_{ij}, \quad (a_i, b_i)\cdot\mathbf{t}_{ij} = (a_j, b_j)\cdot\mathbf{t}_{ij} = -T_{ij}. \qquad (5)$$

(4) and (5) constitute 14 equations where λ_i and σ_{ij} are known and the remaining 15 quantities $(a_i, b_i, \alpha_i, T_{ij}, N_{ij})$ are to be determined. For this we need to determine the signs of the trigonometric terms that appear in the equations. Indicating an angle that is in quadrant m by q_m, $(m-1)\pi/2 < q_m < m\pi/2$, $m = 1, 2, 3, 4$, we can distinguish six *formations* for the α_i's (up to a rotation). These are listed in Table 1.

Proposition 1. *The solution of (4,5) is either essentially a plane wave or*

$$\frac{b_1^2}{a_1^2} = \frac{-\lambda_2}{\lambda_1\lambda_3}\frac{\sigma_{12}\lambda_1\lambda_2(\lambda_1 - \lambda_2) + \sigma_{23}\lambda_2\lambda_3(\lambda_2 - \lambda_3) + \sigma_{31}\lambda_3\lambda_1(\lambda_3 - \lambda_1)}{\sigma_{12}\lambda_3(\lambda_1 - \lambda_2) + \sigma_{23}\lambda_1(\lambda_2 - \lambda_3) + \sigma_{31}\lambda_2(\lambda_3 - \lambda_1)}. \quad (6)$$

Moreover Formations 5 and 6 do not occur in TSEWs.

Formation	1	2	3	4	5	6
α_1	q_2	q_1	q_2	q_1	q_1	q_2
α_2	q_2	q_2	q_1	q_1	q_1	q_2
α_3	q_1	q_3	q_3	q_4	q_3	q_2

Table 1. Six possible *formations* for the angles of TSEWs.

Proof. The proof consist of computing all variables in terms of a_1 and b_1 from (5) and then using (4) to find b_1/a_1. Here we highlight important intermediate results. The trivial solutions of (4,5) are essentially plane waves, where for some i and j, $s_i = s_j$ or $s_i \neq s_j$ but $\lambda_i = \lambda_j$. In these solutions there is at most one line across which velocity is discontinuous. If for some i, $\alpha_i = 0$ or π or 2π then the corresponding solution is essentially a plane wave. To obtain the nontrivial solutions we use (5) to get

$$\tan\alpha_1 = \frac{(\lambda_3 - \lambda_2)a_1 b_1}{\lambda_2 a_1^2 + \lambda_3 b_1^2}, \quad \tan\alpha_2 = \frac{(\lambda_1 - \lambda_3)\lambda_2 a_1 b_1}{\lambda_2^2 a_1^2 + \lambda_1\lambda_3 b_1^2}, \quad \tan\alpha_3 = \frac{(\lambda_2 - \lambda_1)a_1 b_1}{\lambda_2 a_1^2 + \lambda_1 b_1^2}. \quad (7)$$

Let $g_i = \text{sgn}(\cos\alpha_i)$, $c = (\lambda_2^2 a_1^2 + \lambda_3^2 b_1^2)^{1/2}$ and $e_i = g_i/c$ then from (5,7) we derive

$$N_{12} = -\lambda_1 b_1, \quad N_{23} = -e_2\lambda_3 b_1(\lambda_2^2 a_1^2 + \lambda_1^2 b_1^2)^{1/2}, \quad N_{31} = -e_1\lambda_1\lambda_3 b_1(a_1^2 + b_1^2)^{1/2}. \quad (8)$$

For each of the six formations in Table 1 we determine the g_i's. For the first four formations we arrive at $g_1 g_2 g_3 = 1$ and for the last two formations we get $g_1 g_2 g_3 = -1$. For distinct λ_i's, however, Formations 5 and 6 are impossible. To see this note that in both formations the $\tan\alpha_i$'s have the same sign while the presence of the cyclic differences of the λ_i's in (7) indicates that the signs are not identical. Therefore $g_1 g_2 g_3 = 1$ and using (8) in (4) we get (6).

4. Triple-Shock Elementary Waves for a Miscible Flow

The degree of nonlinearity of the fractional flow function f determines the extent to which the resulting TSEWs deviate from a plane wave. If f is nearly linear, i.e. $f'' \approx 0$, then $\sigma_{12} \approx \sigma_{23} \approx \sigma_{31}$ and the denominator of (6) is close to zero. Since the numerator is generally nonzero, we must have $a_1 \approx 0$. Then (7) implies that all angles are nearly a multiple of π. In particular for miscible flow $a_1 = 0$ and it easily follows from (4,5) that the resulting configuration is essentially a plane wave. We summarize this result in the following proposition.

Proposition 2. *If the λ_i's are distinct and the flow is miscible then the angles between the shock fronts in a triple-shock elementary wave are multiples of π (i.e. it is a plane wave).*

5. The Entropy Satisfying Solutions

In this section we first explore the admissibility criteria, for solutions of (4,5), imposed by the entropy condition and the convexity properties of f and then those imposed by the convexity of λ. Let $\Delta_{ij}(s) = \Delta(s; s_i, s_j)$. Then the entropy condition requires

$$N_{ij}(s_j - s_i)\Delta_{ij}(s) \geq 0 \qquad \text{for } ij = 12,\ 23,\ 31. \tag{9}$$

Let $\xi_1 < \xi_2 < \xi_3$ be three saturation values and assume $\Delta(s; \xi_i, \xi_j)$ has a fixed sign for s between ξ_i and ξ_j. Then there are $2^3 = 8$ possible *combinations* for these signs. These combinations are shown in Table 2. Notice that in combination 7 and 8 f has a cusp at ξ_2.

Combination	1	2	3	4	5	6	7	8
$\Delta(s; \xi_1, \xi_2)$	−	−	−	+	+	+	−	+
$\Delta(s; \xi_2, \xi_3)$	−	+	+	+	−	−	−	+
$\Delta(s; \xi_1, \xi_3)$	−	−	+	+	+	−	+	−

Table 2. Eight *combinations* for the relative position of f with respect to three secants.

Proposition 3. *If f is a C^1 function then Formations 4 and 5 of Table 1 are impossible.*

Proof. From (8) it follows that in Formation 4 (and Formation 5 which was ruled out in Proposition 2) the sign of all the N_{ij}'s are the same (they are equal to the sign of $-b_1$). Therefore to satisfy the entropy condition the signs of $(s_j - s_i)\Delta_{ij}(s)$'s should be the same. This is only possible when f is as shown in Combinations 7 or 8 in Table 2, where f has a cusp and is not C^1, or when f is linear on an interval containing s_1, s_2 and s_3. In the latter event, as we saw for the miscible flow, α_i's should be a multiple of π and the TSEW is a plane wave. QED.

We are left with Formations 1, 2 and 3. Note that exactly one of the angles in these three remaining formations is less than 90°. To classify the possible solutions for each formation we have to satisfy three requirements *simultaneously*:

(a) The right hand side of (6) is positive. This results in an algebraic inequality which restricts the free parameters of the problem to a certain subset.

(b) The angles in Table 1 are compatible with (7). The sign of right hand sides of these formulas should match the sign of the $\tan \alpha_i$'s of Table 1. Therefore the values of λ_i's and the sign of $a_1 b_1$ should satisfy the inequalities in Table 3.

(c) The entropy condition is satisfied across each shock.

We analyze the sign of each term in (9) *independently*. The saturations, s_1, s_2 and s_3, can be ordered in one of 3! = 6 different ways and the sign of $s_j - s_i$ can be determined from these orderings. The distance functions Δ_{ij}'s can have one of 8 different sign combinations. And we have already analyzed the sign of N_{ij}'s in terms of the sign of b_1 in (8). By combining all of these results we arrive at Table 4. This table correlates the ordering of saturations, Formation (as labeled in Table 1), the sign of b_1 and Combination (as labeled in Table 2). The combinations are denoted in the columns with the heading "C".

In practice the mobility function $\lambda(s)$ is convex. This gives additional information about the sign of $a_1 b_1$ in Table 3. For example if λ is convex and $s_1 < s_2 < s_3$ then it is impossible to have $\lambda_1 < \lambda_3 < \lambda_2$, hence Formation 1 can occur only if $a_1 b_1 < 0$. In Table 4 the allowed signs of a_1 are denoted in the columns with the heading "a_1".

Formation	1	2	3
$a_1 b_1 > 0$	$\lambda_1 < \lambda_3 < \lambda_2$	$\lambda_1 < \lambda_2 < \lambda_3$	$\lambda_3 < \lambda_1 < \lambda_2$
$a_1 b_1 < 0$	$\lambda_2 < \lambda_3 < \lambda_1$	$\lambda_3 < \lambda_2 < \lambda_1$	$\lambda_2 < \lambda_1 < \lambda_3$

Table 3. Necessary conditions for compatibility of (7) and the formations in Table 1.

Formation	1				2				3			
Sign of b_1	+		−		+		−		+		−	
	C	a_1	C	a_1	C	a_1	C	a_1	C	a_1	C	a_1
$s_1 < s_2 < s_3$	2	−	5	+	3	±	6	±	1	−	4	+
$s_2 < s_3 < s_1$	4	±	1	±	5	−	2	+	3	+	6	−
$s_3 < s_1 < s_2$	6	+	3	−	1	+	4	−	5	±	2	±
$s_1 < s_3 < s_2$	1	±	4	±	6	−	3	+	2	+	5	−
$s_3 < s_2 < s_1$	3	−	6	+	2	±	5	±	4	−	1	+
$s_2 < s_1 < s_3$	5	+	2	−	4	+	1	−	6	±	3	±

Table 4. Necessary compatibility conditions for the triple-shock elementary waves. Formations 1, 2, 3 are from Table 1. $C = 1, \cdots, 6$ is a Combination from Table 2. If $\lambda(s)$ is convex the sign of a_1 is restricted to indicated values.

6. Acknowledgement.
The author thanks Professors Glimm and Lindquist for guidance.

References

1. J. Glimm and D. H. Sharp, "Elementary Waves for Hyperbolic Equations in Higher Dimensions: An Example from Petroleum Reservoir Modeling," *Contemporary Mathematics*, vol. 60, pp. 35-41, 1987.

2. M. Maesumi, "Symmetry Breaking and the Interaction of Hyperbolic Waves," *IMPACT of Computing in Science and Engineering*, to appear.

Mathematics Department, Lamar University, Beaumont, Texas 77710, USA

G N HILE AND C MATAWA
Existence and uniqueness of solutions to quasilinear elliptic equations

1 Introduction

In this paper we will present new results concerning the existence and uniqueness of solutions $u \in W^{2,p}_{loc}(R^n)$ to the nonlinear second order elliptic equation

$$a(x) \cdot D^2 u(x) + H(x, u(x), Du(x)) = 0 \tag{1}$$

where $H = H(x, u, p)$ with $x, p \in R^n$, $u \in R$, maps R^{2n+1} into R, and is assumed measurable in all variables, Lipschitz continuous in u and p. Decay conditions on H and on the associated Lipschitz constants at infinity are also assumed. The real symmetric $n \times n$ matrix a is strictly positive definite in R^n, approaches the Laplace operator at infinity, and is Lipschitz continuous with again the Lipschitz constant decaying appropriately at infinity. The results we obtain extend those of H. Begehr and G. N. Hile in [3] for classical solutions of the related linear equation

$$Lu := a(x) \cdot D^2 u(x) + b(x) \cdot Du(x) + c(x)u(x) = f(x). \tag{2}$$

Following the prevailing terminology in the literature we shall refer to solutions of these equations in all of R^n as *entire solutions*. The details of the proofs will be published in [11]. For a vector $x \in R^n$ we let $r(x) := |x|$ denote the Euclidean norm of x. For $n \times n$ matrices $A = (a_{i,j})$, $B = (b_{i,j})$, in $R^{n \times n}$ we define $|A| = \sqrt{A \cdot A}$. For $\phi = \phi(x)$ a real valued function defined on some subset Ω of R^n with Sobolev derivatives up to order 2, we let $D\phi$ represent the $1 \times n$ vector of first partial derivatives of ϕ and $D^2 \phi$ the $n \times n$ matrix of second partial derivatives of ϕ. Let L be the differential operator defined as in equation (2) on real valued functions $u = u(x)$, $x \in R^n$. The matrix $a(x)$ will always be assumed to be $n \times n$, real, symmetric, with measurable entries; $b(x)$ an $n \times 1$ real vector with measurable entries, and $c(x)$ a measurable real scalar function. The functions $a(x), b(x)$, and $c(x)$ are defined for all $x \in R^n$. We shall always assume that there is a positive constant λ such that for all $x, y \in R^n$, $a(x)\xi \cdot \xi \geq \lambda |\xi|^2$, and that there exist constants $0 < \alpha \leq 1$ and $K \geq 0$, such that for all $x, y \in R^n$,

$$|a(x) - a(y)| \leq K \frac{|x-y|^\alpha}{(1 + |x| + |y|)^\alpha}. \tag{3}$$

By a solution of $Lu = f$ for a given real valued function f, we shall mean a function $u \in W^{2,p}_{loc}(R^n)$ for which the equation $Lu = f$ holds almost everywhere in R^n. For $1 \leq p \leq \infty$, and functions u defined on R^n with values in R, R^n, or $R^{n \times n}$, we define

$$\|u\|_p := \begin{cases} [\int_{R^n} |u(y)|^p \, dy]^{1/p}, & p < \infty \\ \text{ess sup } |u|, & p = \infty \end{cases} \tag{4}$$

Let I denote the $n \times n$ identity matrix. We shall assume the following condition on the coefficients of the linear operator L:

(L) There exist nonnegative constants δ and Λ such that

$$\left\|(1+r)^\delta(a-I)\right\|_\infty, \left\|(1+r)^{\delta+1} b\right\|_\infty, \left\|(1+r)^{\delta+2} c\right\|_\infty \leq \Lambda. \tag{5}$$

2 The Linear Equation

Before studying solutions to the general linear equation (2), We use estimates for bounded domains found in the book of Gilbarg and Trudinger [8] to obtain an existence and uniqueness theorem for entire solutions to the Poisson Equation and also *a priori* bounds for these solutions. We then use the results for the Poisson Equation to obtain the following *a priori* bound for entire solutions of equation (2) with prescribed behaviour at infinity. We shall denote by Γ the fundamental solution

$$\Gamma(x) := \begin{cases} \frac{|x|^{2-n}}{(2-n)\omega_n}, & n > 2 \\ \frac{1}{2\pi} \log |x|, & n = 2 \end{cases} \tag{6}$$

where ω_n is the surface area of the unit ball in R^n.

Theorem 1 *Assume condition (L) with $0 < \delta \leq 1$, that $\sigma \in R$, $n < p < \infty$, and that $u \in W^{2,p}_{loc}(R^n)$.*
a) If $n \geq 3$, $-\frac{n}{p} < \sigma < n - 2 - \frac{n}{p}$, $c \leq 0$ in R^n, and if $u(\infty) = 0$, then

$$\|(1+r)^\sigma u\|_p + \left\|(1+r)^{\sigma+1} Du\right\|_p + \left\|(1+r)^{\sigma+2} D^2 u\right\|_p$$
$$\leq C(n, p, \alpha, \lambda, \Lambda, K, \sigma) \left\|(1+r)^{\sigma+2} Lu\right\|_p. \tag{7}$$

b) If $n = 2$, $-\frac{2}{p} < \sigma$, $c \equiv 0$, and if $u - \gamma \Gamma$ vanishes at infinity for some real constant γ, then for $\tau < -\frac{2}{p}$,

$$\|(1+r)^\tau u\|_p + \left\|(1+r)^{\tau+1} Du\right\|_p + \left\|(1+r)^{\tau+2} D^2 u\right\|_p$$
$$\leq C(p, \alpha, \lambda, \Lambda, K, \sigma, \tau) \left\|(1+r)^{\sigma+2} Lu\right\|_p. \tag{8}$$

Moreover, if $\sigma < -\frac{2}{p} + \delta$, then

$$\left\|(1+r)^{\sigma+2} \Delta u\right\|_p \leq C(p, \alpha, \lambda, \Lambda, K, \sigma, \delta) \left\|(1+r)^{\sigma+2} Lu\right\|_p, \tag{9}$$

$$|\gamma| \leq C(p, \alpha, \lambda, \Lambda, K, \sigma, \delta) \left\|(1+r)^{\sigma+2} Lu\right\|_p, \tag{10}$$

and for $|x| \geq 1$,

$$|u(x) - \gamma \Gamma(x)| \leq C(p, \alpha, \lambda, \Lambda, K, \sigma, \delta) \left\|(1+r)^{\sigma+2} Lu\right\|_p |x|^{-\sigma-2/p}. \tag{11}$$

3 The Nonlinear Equation

We now consider the equation (1). We write $H = H(x, u, p)$, where $x \in R^n$, $u \in R$, and $p \in R^n$. We assume the following:

(N1) There exist $\Lambda \geq 0$ and $\delta > 0$, such that for $x, p, q \in R^n$ and $u, v \in R$,

$$|a(x) - I| \leq \Lambda(1 + |x|)^{-\delta},$$
$$|H(x, u, p) - H(x, v, p)| \leq \Lambda |u - v| (1 + |x|)^{-2-\delta},$$
$$|H(x, u, p) - H(x, v, q)| \leq \Lambda |p - q| (1 + |x|)^{-1-\delta},$$

(N2) For all x, u, v and p with $u \neq v$,

$$\frac{H(x, u, p) - H(x, v, p)}{u - v} \begin{cases} \leq 0, & n \geq 3 \\ = 0, & n = 2 \end{cases}, \tag{12}$$

(N3) For some p and σ, with $n < p < \infty$ and $-\frac{n}{p} < \sigma$,

$$\left\| (1 + r)^{\sigma+2} H(\cdot, 0, 0) \right\|_p < \infty. \tag{13}$$

Theorem 2 *a) Assume conditions (N1) to (N3) on a and H and that $n \geq 3$. Then there exists a unique entire solution u in $W_{loc}^{2,p}(R^n)$ of equation (1) such that $u(\infty) = 0$, and for this solution if $-\frac{n}{p} < \sigma < n - 2 - \frac{n}{p}$ and $0 < \delta \leq 1$, we have the bound*

$$\left\| (1+r)^\sigma u \right\|_p + \left\| (1+r)^{\sigma+1} Du \right\|_p + \left\| (1+r)^{\sigma+2} D^2 u \right\|_p$$
$$\leq C(n, p, \alpha, \lambda, \Lambda, K, \sigma) \left\| (1+r)^{\sigma+2} H(\cdot, 0, 0) \right\|_p. \tag{14}$$

b) Assume conditions (N1) to (N3) on a and H and that $n = 2$. Then there exists a unique entire solution u in $W_{loc}^{2,p}(R^2)$ of equation (1) such that $u - \gamma\Gamma$ vanishes at infinity for some real constant γ, and for this solution if $\tau < -\frac{2}{p} < \sigma$ and $0 < \delta \leq 1$, we have the bound

$$\left\| (1+r)^\tau u \right\|_p + \left\| (1+r)^{\tau+1} Du \right\|_p + \left\| (1+r)^{\tau+2} D^2 u \right\|_p$$
$$\leq C(p, \alpha, \lambda, \Lambda, K, \sigma, \tau) \left\| (1+r)^{\sigma+2} H(\cdot, 0, 0) \right\|_p, \tag{15}$$

and further if $\sigma < -\frac{2}{p} + \delta$, then

$$\left\| (1+r)^{\sigma+2} \Delta u \right\|_p \leq C(p, \alpha, \lambda, \Lambda, K, \sigma, \delta) \left\| (1+r)^{\sigma+2} H(\cdot, 0, 0) \right\|_p, \tag{16}$$

and

$$|\gamma| \leq C(p, \alpha, \lambda, \Lambda, K, \sigma, \delta) \left\| (1+r)^{\sigma+2} H(\cdot, 0, 0) \right\|_p, \tag{17}$$

with, for $|x| \geq 1$,

$$|u(x) - \gamma\Gamma(x)| \leq \left\| (1+r)^{\sigma+2} H(\cdot, 0, 0) \right\|_p |x|^{-\sigma-2/p}. \tag{18}$$

To prove uniqueness we let u and v be two solutions and show by linearization that $w = u-v$ solves an equation of the form

$$a \cdot D^2 w + B(\cdot, v, Du, Dv) \cdot Dw + C(\cdot, u, v, Du) \cdot w = 0, \tag{19}$$

where B and C satisfy the conditions of Theorem 1 We then apply Theorem 1 to show that $w \equiv 0$. To establish the bounds we show that any entire solution $u \in W^{2,p}_{loc}(R^n)$ solves

$$a \cdot D^2 w + B(\cdot, 0, Du, 0) \cdot Du + C(\cdot, u, 0, Du) \cdot u + H(\cdot, 0, 0) = 0. \tag{20}$$

and again apply Theorem 1. To prove existence we define for $0 \leq t \leq 1$

$$a_t = ta + (1-t)I, \tag{21}$$

and consider the equation

$$a \cdot D^2 u + t[H(\cdot, u, Du) - H(\cdot, 0, 0)] = F \tag{22}$$

where $\|(1+r)^{\sigma+2} F\|_p < \infty$. We show by arguments similar to those in [3] that we can solve that equation with the prescribed conditions on u at infinity for any such F and any t in $[0,1]$; then the theorem is established by taking $t=1$ and $F = -H(\cdot, 0, 0)$.

We use Theorem 2 to prove a more general result:

Theorem 3 *Assume conditions (N1) to (N3) on a and H.*

i) For any harmonic polynomial P of degree less than δ, there exists a unique entire solution u in $W^{2,p}_{loc}(R^n)$ of equation (1) such that

    a) if $n \geq 3$, then $u - P$ vanishes at infinity

    b) if $n = 2$, then $u - P - \gamma\Gamma$ vanishes at infinity for some $\gamma \in R$

ii) If u is an entire solution in $W^{2,p}_{loc}(R^n)$ of equation (1) such that

$$|u(x)| = O(|x|)^\tau \tag{23}$$

as $x \to \infty$ for some τ, $0 \leq \tau < \delta$, then there exists a unique harmonic polynomial P of degree no larger than τ such that (a) and (b) of i) hold.

References

[1] R. A. Adams "Sobolev Spaces", New York : Academic Press, 1975 .

[2] A. D. Aleksandrov *Uniqueness conditions and estimates for the solution of the Dirichlet problem*, Vestnik Leningrad Univ **18** 3 (1963), 5–29. (Russian). English transl in Amer. Math. Soc. Transl. (2) **68** (1968), 89–119.

[3] H. Begehr and G. N. Hile *Schauder estimates and existence theory for entire solutions of linear elliptic equations*, Proc. Roy. Soc. Edinburgh **110A** (1988), 101–123.

[4] E. Bohn and L. K. Jackson *The Louisville theorem for a quasilinear elliptic partial differential equation*, Trans. Amer. Math. Soc. **104** (1962), 392–397.

[5] J. M. Bony *Principe du maximum dans les spaces de Sobolev*, C. R. Acad. Sci. Paris **265** (1967), 333–336.

[6] A. Friedman *Bounded entire solutions of elliptic equations*, Pacific J. Math. **44** (1973), 497–507.

[7] D. Gilbarg and J. Serrin *On isolated singularities of solutions of second order elliptic differential equations*, J. Analyse Math. **4** (1954-6), 309–340.

[8] D. Gilbarg and N. S. Trudinger, "Elliptic Partial Differential Equations of Second Order," Second Edition, Berlin : Springer, 1983.

[9] S. Hildebrandt and K. Wildman *Sätze vom Louivilleeschen Typ für quasilineare elliptische Gleichungen und Systemme* Nachr. Akad. Wiss. Göttingen Math. Phys. Kl. **II** 4 (1979), 41–59.

[10] G. N. Hile *Entire solutions of linear elliptic equations with Laplacian principal part*, Pacific J. Math. **62** (1976), 127–140.

[11] G. N. Hile and C. P. Mawata *Louiville theorems for nonlinear elliptic equations of second order*, in preperation.

[12] A. V. Ivanov *Local estimates for the first derivatives of solutions of quasilinear second order elliptic equations and their application to Louiville type theorems*, Sem. Steklov Math. Inst. Leningrad **30** (1972), 40–50. Translated in J. Sov. Math. **4** (1975), 335–344.

[13] C. P. Mawata *Schauder estimates and existence theory for entire solutions of linear parabolic equations*, Differential and Integral Equations *2* 3 (1989), 251–274.

[14] L. Peletier and J. Serrin *Gradient bounds and Liouville theorems for quasilinear elliptic equations*, Ann Scuola Norm. Sup. Pisa Cl. Sci. Ser. 4 **5** (1978), 65–104.

[15] R. Redheffer *On the inequality $\Delta u \geq f(u, |grad(u)|)$*, J. Math. Anal. **1** (1960) 277–299.

[16] J. Serrin *Entire solutions of nonlinear Poisson equations* Proc. London Math. Soc. **24** (1972), 348–366.

[17] J. Serrin *Liouville theorems and gradient bounds for quasilinear elliptic systems*, Arch. Rational Mech. Anal. **66** (1977), 295–310.

[18] N. Weck *Liouville theorems for linear elliptic systems*, Proc. Roy. Soc. Edinburgh **94A** (1983), 309–322.

A FONDA AND J MAWHIN

An iterative method for the solvability of semilinear equations in Hilbert spaces and applications

1 An existence and uniqueness theorem

Let H be a Hilbert space. We are interested in the unique solvability of abstract equations of the form

$$Lu = Nu + h \qquad (1)$$

where $L : D(L) \subset H \to H$ is a selfadjoint linear operator, $N : H \to H$ is a continuous gradient operator and $h \in H$ is arbitrary. We will provide some consequences and applications of the following abstract result, which we proved in [3].

Theorem 1. *Let $A, B : H \to H$ be two linear, continuous selfadjoint operators such that the following conditions hold.*

(i) *$N - A$ and $B - N$ are monotone.*
(ii) *$L - (1 - \mu)A - \mu B$ has a bounded inverse for every $\mu \in [0, 1]$.*

Then equation (1) has for every $h \in H$ a unique solution $u \in D(L)$ which can be obtained through the iterative process defined by

$$Lu_{k+1} - (1/2)(A + B)u_{k+1} = Nu_k - (1/2)(A + B)u_k + h,$$

with u_0 arbitrary in H.

As pointed out in [3], there is a quite vast literature on this problem. Our approach seems to be rather natural and easy, since a change of variable simply reduces equation (1) to a fixed point problem for a contraction mapping. For the reader's convenience, we reproduce here the proof of Theorem 1.

Proof. By condition (i), $B - A$ is nonnegative. As the set of operators with a bounded inverse is open, we can find, by condition (ii), an $\epsilon > 0$ such that $L-(1-\mu)(A-\epsilon I)-\mu(B+\epsilon I)$ has a bounded inverse for every $\mu \in [0, 1]$. The operator $S = B - A + 2\epsilon I$ is selfadjoint, positive and invertible. Let $S^{1/2}$ and $S^{-1/2}$ be the square roots of S and S^{-1} respectively. We now operate the change of variable $v = S^{1/2}u$. Setting

$$\tilde{L} = S^{-1/2}(L - A + \epsilon I)S^{-1/2}, \quad \tilde{N} = S^{-1/2}(N - A + \epsilon I)S^{-1/2},$$

equation (1) becomes equivalent to the equation

$$\tilde{L}v = \tilde{N}v + S^{-1/2}h, \tag{2}$$

where $v \in D(\tilde{L}) = S^{1/2}(D(L))$. Conditions (i) and (ii) become respectively

(i)' $\tilde{N}$ and $I - \tilde{N}$ are monotone;
(ii)' $\sigma(\tilde{L}) \cap [0,1] = \emptyset$,

where $\sigma(\tilde{L})$ denotes the spectrum of $\tilde{L}$. Furthermore, equation (2) is equivalent to the fixed point problem

$$v = (\tilde{L} - (1/2)I)^{-1}[\tilde{N}v - (1/2)v + S^{-1/2}h] := Tv.$$

By (i)' and Lemma 1 in [7], one has that $\tilde{N} - (1/2)I$ is Lipschitz continuous with Lipschitz constant $1/2$. By (ii)' and the selfadjointness of $\tilde{L}$, one has

$$\|(\tilde{L} - (1/2)I)^{-1}\| < 2.$$

Consequently, T is a contraction, and the conclusion is a straightforward consequence of the fixed point theorem for contractive mappings. ∎

2 More explicit assumptions

Condition (ii) in Theorem 1 can be written more explicitly in case L commutes with A and B. Since this happens in many applications, we analyze this case below. This is based upon the following observation.

Proposition 1. *Let $D : H \to H$ be a continuous linear selfadjoint operator which commutes with L. If $\sigma(L) \cap \sigma(D) = \emptyset$, then $L - D$ has a bounded inverse.*

Proof. Since D and L commute, their spectral families commute ([9], ch. 8), and it is possible to find a spectral family $E(\lambda, \lambda')$ on the product space so that

$$L - D = \int_{-\infty}^{+\infty} \int_{-\infty}^{+\infty} (\lambda - \lambda') \, dE(\lambda, \lambda'),$$

(see [8], ch. 9). Since the spectra $\sigma(L)$ and $\sigma(D)$ are a positive distance apart, 0 cannot be in the spectrum of $L - D$. ∎

Remark that, in general, the two propositions "$\sigma(L) \cap \sigma(D) = \emptyset$" and "$L - D$ has a bounded inverse" do not imply each other ([3]). Moreover, even when L and D commute, the propositions are not equivalent, as shown by the following example in $H = \mathbb{R}^2$:

$$L(x_1, x_2) = (2x_1, x_2), \quad D(x_1, x_2) = (x_1, 0).$$

Corollary 1. *Let A, B be two continuous linear selfadjoint operators which commute with L and are such that*

(i) $N - A$ and $B - N$ are monotone;
(ii) $\sigma(L) \cap \sigma[(1-\mu)A + \mu B] = \emptyset$ for every $\mu \in [0,1]$.

Then the conclusion of Theorem 1 holds.

Having in mind applications to systems of differential equations, we consider now the case where $H = E^m$, with E a Hilbert space.

Corollary 2. *Let* A *and* B *be two symmetric* $m \times m$ *matrices with respective eigenvalues* $\alpha_1 \leq \ldots \leq \alpha_m$ *and* $\beta_1 \leq \ldots \leq \beta_m$. *Denote by* $A, B : E^m \to E^m$ *the induced "constant" operators, and assume that they commute with* L. *If the following conditions are satisfied.*

(i) $N - A$ and $B - N$ are monotone.
(ii) $\sigma(L) \cap \bigcup_{i=1}^{m} [\alpha_i, \beta_i] = \emptyset$.

Then the conclusion of Theorem 1 holds.

The above result includes the ones of Amann [1], Theorem 3.2, and Dancer [2], Theorem 3. It is a straightforward consequence of Corollary 1.

3 Applications to differential equations

We want to give now applications to some boundary value problems for ordinary and partial differential equations. To this aim, let Ω be a measurable subset of $\mathbb{R}^n$, and $H = [L^2(\Omega)]^m$. Our first applications will deal with "diagonal" operators of a particular form, which will commute with every "constant" operator.

Corollary 3. *Assume that L is of the form* $L(u_1, \ldots, u_m) = (\Lambda u_1, \ldots, \Lambda u_m)$, *where* $\Lambda : D(\Lambda) \subset L^2(\Omega) \to L^2(\Omega)$ *is a selfadjoint operator. Let* $G : \mathbb{R}^m \to \mathbb{R}$ *be a differentiable function with gradient* ∇G, *and* A, B *be two symmetric* $m \times m$ *matrices with eigenvalues* $\alpha_1 \leq \ldots \leq \alpha_m$ *and* $\beta_1 \leq \ldots \leq \beta_m$, *respectively. Assume that the following conditions hold.*

(j) $\langle A(v_1 - v_2), v_1 - v_2 \rangle \leq \langle \nabla G(v_1) - \nabla G(v_2), v_1 - v_2 \rangle \leq \langle B(v_1 - v_2), v_1 - v_2 \rangle$, for every $v_1, v_2 \in \mathbb{R}^m$.
(jj) $\sigma(\Lambda) \cap \bigcup_{i=1}^{m} [\alpha_i, \beta_i] = \emptyset$.

Then the equation
$$(Lu)(t) = \nabla G(u(t)) + h(t)$$
has, for every $h \in [L^2(\Omega)]^m$, *a unique solution* $u \in D(L)$ *which can be obtained, from any* $u_0 \in [L^2(\Omega)]^m$, *through the iterative process defined by*

$$(Lu_{k+1})(t) - (1/2)(A + B)u_{k+1}(t)$$
$$= \nabla G(u_k(t)) - (1/2)(A + B)u_k(t) + h(t).$$

When dealing with differential equations, the above operators L and Λ are generally both denoted by Λ. Notice that for G twice differentiable, condition (j) of Corollary 3 is equivalent to the condition

(j)' $\mathbf{A} \leq G''(v) \leq \mathbf{B}$, for every $v \in \mathbf{R}^m$,

which was first introduced by Lazer [6].

Here are some examples of application for Corollary 3.

a) *Ordinary differential equations*

$$-u''(t) = \nabla G(u(t)) + h(t), \text{ in }]0, T[,$$
$$u(0) = u(T), \ u'(0) = u'(T).$$

In this case, $\sigma(\Lambda) = \{(kT/2\pi)^2 : k \in \mathbf{N}\}$.

b) *Elliptic problems*

$$-\Delta u(x) = \nabla G(u(x)) + h(x), \text{ in } \Omega,$$
$$u = 0 \text{ or } \partial u/\partial \nu = 0 \text{ on } \partial \Omega.$$

If Ω is a bounded domain, $\sigma(\Lambda)$ is made of eigenvalues with do not accumulate at any finite point. More general elliptic operators can be considered as well.

c) *Hyperbolic problems*

$$\Box u(t,x) = \nabla G(u(t,x)) + h(t,x) \text{ in }]0, T[\times \mathcal{D},$$
$$u(0,x) = u(T,x), \ u_t(0,x) = u_t(T,x) \text{ for } x \in \mathcal{D},$$
$$u(t,x) = 0 \text{ in }]0, T[\times \partial \mathcal{D}.$$

When $\mathcal{D}$ is a bounded domain, $\sigma(\Lambda)$ is the closure of the set obtained by making all the possible differences between the eigenvalues $(-kT/2\pi)^2$ of the corresponding periodic problem and those of the corresponding Dirichlet problem on $\mathcal{D}$. When $\mathcal{D}$ is the cube $]0, \pi[^n$, one has

$$\sigma(\Lambda) = \{m_1^2 + \ldots + m_n^2 - (kT/2\pi)^2 : m_i \in \mathbf{N}^*, k \in \mathbf{N}, i = 1, \ldots, n\}.$$

Other types of boundary conditions can be treated similarly, as e.g. periodic-Neumann, Dirichlet-Dirichlet, Dirichlet-Neumann.

d) *Schrödinger equations*

$$-\Delta u(x) + V(x)u(x) = \nabla G(u(x)) + h(x) \text{ in } \mathbf{R}^n, \ u \in [L^2(\mathbf{R}^n)]^m.$$

Here $V : \mathbf{R}^n \to \mathbf{R}$ is the potential. Corollary 3 applies once sufficient conditions can be given on V in order to guarantee the existence of gaps in the spectrum of $\Lambda = -\Delta + V$. We don't need extra technical conditions as in [1].

We now turn our attention toward operators which do not satisfy the commutativity properties of Corollaries 1 to 3. In those cases, we will apply directly Theorem 1.

e) Hamiltonian systems

Consider the Hamiltonian system
$$Ju'(t) = \nabla G(u(t)) + h(t), \text{ in }]0, T[,$$
$$u(0) = u(T),$$

where $m = 2l$, $J = \begin{pmatrix} 0 & I_l \\ -I_l & 0 \end{pmatrix}$ is the symplectic matrix. Let A and B be two symmetric $m \times m$ matrices, and let $A, B : [L^2(0, T)]^m \to [L^2(0, T)]^m$ be the induced 'constant' operators. Notice that the operator $L = J\frac{d}{dt}$ does not commute in general with A and B. Assume for instance that condition

(h) $\quad \mathsf{A} \le G''(v) \le \mathsf{B}, \ (v \in \mathbf{R}^n)$

holds. In order to apply Theorem 1, we need to check its assumption (ii), i.e., as L has compact resolvent, that the problem
$$Ju' + (1 - \mu)\mathsf{A}u + \mu\mathsf{B}u = 0, \ u(0) = u(T),$$

has only the trivial solution for every $\mu \in [0, 1]$. It is easy to check that this is equivalent to the condition

(hh) $\quad \bigcup_{\mu \in [0,1]} \sigma[(1-\mu)J\mathsf{A} + \mu J\mathsf{B}] \cap \frac{2\pi}{T}i\mathbf{Z} = \emptyset.$

We can conclude that, if conditions (h) and (hh) are satisfied, the Hamiltonian system above has a unique T-periodic solution for every $h \in [L^2(0, T)]^m$, and the iterative process of Theorem 1 converges to the solution. We can remark that condition (hh) generalizes the one found when dealing with second order systems. Moreover, condition (hh) can be shown to be equivalent to the following one, introduced by Amann [1], via a finite dimensional reduction.

Condition (hh)'. *Suppose that*
$$\sigma(J\mathsf{A}) \cap \frac{2\pi}{T}i\mathbf{Z} = \emptyset = \sigma(J\mathsf{B}) \cap \frac{2\pi}{T}i\mathbf{Z},$$

and choose $\beta \in \mathbf{R}_+ \setminus \frac{2\pi}{T}\mathbf{Z}$ such that $-\beta I \le \mathsf{A} \le \mathsf{B} \le \beta I$. Denote by E the sum of the eigenspaces of L belonging to the eigenvalues in $]-\beta, \beta[$, and assume, with m the Morse index, that $m((L - A)|_E) = m((L - B)|_E)$.

f) Dirac equations

Consider the problem of finding $u \in L^2(\mathbb{R}^3, \mathbb{C}^4)$ such that

$$\sum_{j=1}^{3}[\alpha_j \frac{\partial u}{\partial x_j}(x) - Q_j(x)\alpha_j u(x)] + m\alpha_4 u(x) = \nabla G(u(x)) + h(x),$$

where $Q = (Q_1, Q_2, Q_3) : \mathbb{R}^3 \to \mathbb{R}^3$ is the magnetic potential, $m > 0$, and the α_j are the 4×4 matrices defined by

$$\alpha_j = \begin{pmatrix} 0 & \sigma_j \\ \sigma_j & 0 \end{pmatrix}, \ (1 \leq j \leq 3), \ \alpha_4 = \begin{pmatrix} I_2 & 0 \\ 0 & -I_2 \end{pmatrix},$$

where the σ_j are the Pauli matrices satisfying the relations

$$\sigma_j \sigma_k + \sigma_k \sigma_j = 2\delta_{jk}, \ \sigma_1 \sigma_2 = i\sigma_3, \ ;\sigma_2 \sigma_3 = i\sigma_1, \ \sigma_3 \sigma_1 = i\sigma_2.$$

One can take for instance,

$$\sigma_1 = \begin{pmatrix} 0 & 1 \\ 1 & 0 \end{pmatrix}, \ \sigma_2 = \begin{pmatrix} 0 & -i \\ i & 0 \end{pmatrix}, \ \sigma_3 = \begin{pmatrix} 1 & 0 \\ 0 & -1 \end{pmatrix}.$$

It is known that the spectrum $\sigma(L)$ of the Dirac operator defined above is such that

$$\sigma(L) \cap]-m, m[= \emptyset,$$

(see e.g. [4] and [5]). Assume that, for some $\rho \in [0, m[$ and $a_j, b_j \in \mathbb{R}$, $(j = 1, 2, 3)$, we have

$$-\rho I_4 - \sum_{j=1}^{3} a_j \alpha_j \leq G''(v) \leq \rho I_4 + \sum_{j=1}^{3} b_j \alpha_j,$$

for every $v \in \mathbb{C}^4$. The, by Theorem 1, the problem has a unique solution for every $h \in L^2(\mathbb{R}^3, \mathbb{C}^4)$. In fact, condition (ii) of Theorem 1 is verified since, for every $\mu \in [0, 1]$, the kernel of the operator

$$\sum_{j=1}^{3}[\alpha_j \frac{\partial}{\partial x_j} - Q_j \alpha_j] + m\alpha_4 - (1-\mu)[-\rho I_4 - \sum_{j=1}^{3} a_j \alpha_j] - \mu[\rho I_4 + \sum_{j=1}^{3} b_j \alpha_j] =$$

$$\sum_{j=1}^{3}[\alpha_j \frac{\partial}{\partial x_j} - (Q_j - (1-\mu)a_j + \mu b_j)\alpha_j] + m\alpha_4 + (1 - 2\mu)\rho I_4$$

is trivial, as $(1 - 2\mu)\rho \in]-m, m[$.

References

[1] H. Amann, *On the unique solvability of semilinear operator equations in Hilbert spaces*, J. Math. Pures Appl. 61 (1982), 149-175

[2] E.N. Dancer, Order intervals of self-adjoint linear operators and nonlinear homeomorphisms, *Pacific J. Math.* 115 (1984), 57-72

[3] A. Fonda and J. Mawhin, Iterative and variational methods for the solvability of some semilinear equations in Hilbert spaces, *J. Differential Equations*, to appear

[4] B. Helffer, J. Nourrigat and X.P. Wang, Sur le spectre de l'équation de Dirac (dans $\mathbb{R}^2$ ou $\mathbb{R}^3$) avec champ magnétique, *Ann. Scient. Ecole Norm. Sup.* (4) 22 (1989), 515-533

[5] V.Y. Ivrii, Sharp spectral asymptotics for the Dirac operator with a strong magnetic field, *Soviet Math. Dokl.* 41 (1990), 122-125

[6] A.C. Lazer, Applications of a lemma on bilinear forms to a problem in nonlinear oscillations, *Proc. Amer. Math. Soc.* 33 (1972), 89-94

[7] J. Mawhin, Semilinear equations of gradient type in Hilbert spaces and applications to differential equations, in *Nonlinear Differential Equations, Invariance, Stability and Bifurcation*, Academic Press, New York, 1981, 269-282

[8] F. Riesz and B.S. Nagy, *Leçons d'analyse fonctionnelle*, Akadémiai Kiado, Budapest, 1952

[9] M.H. Stone, *Linear Transformations in Hilbert space*, Amer. Math. Soc., Providence R.I., 1932

P S MILOJEVIĆ
Nonlinear Fredholm theory and applications

In this paper, we shall present some extensions of the Fredholm theory to nonlinear and semilinear maps of (pseudo) A-proper type and their applications to BVP's for nonlinear ODE's and elliptic PDE's in (non)divergence form and to periodic-boundary value problems for semilinear hyperbolic equations and Hamiltonian systems.

1. Nonlinear Fredholm Theory

In the first section, we begin with some extensions of the first Fredholm theorem to nonlinear A-proper and condensing maps dealing with the unique approximation solvability of

$$Tx = f \quad (x \in D(T) \subset X, \ f \in Y) \tag{1.1}$$

and error estimates of its approximate solutions, where X and Y are Banach spaces and $T : D(T) \subset X \to Y$ a suitable nonlinear map. We complete this section by discussing these questions for semilinear operator equations

$$Ax + Nx = f, \quad (x \in D(A), \ f \in Y) \tag{1.2}$$

where A is a linear and N a nonlinear map. Many extensions of the first Fredholm theorem (but without uniqueness) to Eq.'s (1.1)–(1.2) with T asymptotically close to a suitable map, and of the Fredholm alternative to Eq.(1.2) with A a Fredholm map of index zero are given in Sections 1.2-1.4. Regarding N, we assume either that it is asymptotically close to A or that it is asymptotically $\{B_1, B_2\}$-quasilinear.

1.1 Unique (approximation) solvability. In this section we shall discuss various conditions such that the existence of at most one solution of an (associated) homogeneous equation implies the unique (approximation) solvability of $Tx = f$ for each $f \in X$. Error estimates are also discussed when the equations are approximation solvable.

Let $\{X_n\}$ and $\{Y_n\}$ be finite dimensional subspaces of Banach spaces X and Y respectively such that $\dim X_n = \dim Y_n$ for each n and $\text{dist}(x, X_n) \to 0$ as $n \to \infty$ for each $x \in X$. Let $P_n : X \to Y_n$ and $Q_n : Y \to Y_n$ be linear projections onto X_n and Y_n respectively such that $P_n x \to x$ for each $x \in X$ and $\delta = \sup \|Q_n\| < \infty$. Then $\Gamma = \{X_n, P_n; Y_n, Q_n\}$ is a projection scheme for (X, Y).

Definition 1.1. A map $T : D \subset X \to Y$ is said to be *approximation-proper* (A-proper for short) with respect to Γ if (i) $Q_n T : D \cap X_n \to Y_n$ is continuous for each n and (ii) whenever $x_{n_k} \in D \cap X_{n_k}$ is bounded and $\|Q_{n_k} T x_{n_k} - Q_{n_k} f\| \to 0$ for some $f \in Y$, then a subsequence $x_{n_{k(i)}} \to x$ and $Tx = f$. T is said to be pseudo A-proper w.r.t. Γ if in (ii) above we do not require that a subsequence of x_{n_k} converges to x for which $Tx = f$.

For many examples of (pseudo) A-proper maps we refer to [Pe, Mi-1-6]. For example, ball-condensing and, in particular, compact and k-contractive, perturbations of Fredholm maps of index zero, maps of type $(S+)$, sums of ball-condensing and strongly monotone maps are all A-proper maps. Monotone like maps and such perturbations of closed linear maps with $\dim \ker(A) \le \infty$ are pseudo A-proper maps.

For A-proper maps, we have ([Mi-2,7])

Theorem 1.1 *(a) Let $T : X \to Y$ be A-proper w.r.t. Γ and satisfy condition $(+)$, i.e., $\{x_n\}$ is bounded whenever $Tx_n \to f$. Suppose that $U \subset X$ is a neighborhood of 0 and $A : U \to 2^Y$ is such that for some $c > 0$ and $n_0 \ge 1$*

$$\|Q_n u\| \ge c\|x\| \quad \text{for all } x \in U \cap X_n, \, u \in Ax, \, n \ge n_0 \tag{1.3}$$

$$Tx - Ty \in A(x-y) \quad \text{whenever } x - y \in U. \tag{1.4}$$

Then $T(X) = Y$. If $U = X$, Eq.(1.1) is uniquely approximation solvable for each $f \in Y$.

(b) If, in addition, $U = X$ and $c_1 \|x\| \le \|u\| \le c_2 \|x\|$ for all $x \in X$, $u \in Ax$ and some $c_1, c_2 > 0$, then T is a homeomorphism and, for each $f \in Y$ and $\epsilon \in (0,c)$, the approximate solutions $x_n \in X_n$ of Eq.(1.1), i.e., $Q_n T x_n = Q_n f$, satisfy

$$\|x_n - x_0\| \le (c - \epsilon)^{-1} \|Tx_n - f\| \quad \text{for } n \ge n_1 \ge n_0 \tag{1.5}$$

$$\|x_n - x_0\| \le k\|P_n x_0 - x_0\| \le k_1 \, dist(x_0, X_n) \tag{1.6}$$

where the constant $k = k(c, c_2, \epsilon, \delta)$, $k_1 = 2k\delta_1$, $\delta_1 = \sup \|P_n\|$.

Theorem 1-(a) has been proven in [Mi-2] by showing that the pair (X_n, Q_n) is a covering space for Y_n and using the simple connectedness of Y_n and condition $(+)$. Part (b) follows from an approximation solvability and error estimate result of the author [Mi-6] and

Theorem 1.2 *([Mi-6]) Let $T : X \to Y$ be continuous, A-proper w.r.t. Γ, satisfy condition $(+)$ and be locally invertible on X. Then T is a homeomorphism onto Y and Eq.(1.1) is uniquely approximation solvable for each $f \in Y$.*

To state a related result for ϕ-condensing maps, we recall that the *set measure of noncompactness* of a bounded set $D \subset X$ is defined as $\gamma(D) = \inf\{d > 0 : D$ has a finite covering by sets of diameter less than $d\}$. The *ball-measure of noncompactness* of D is defined as $\chi(D) = \inf\{r > 0 | D \subset \cup_{i=1}^n B(x_i, r), x \in X, n \in N\}$. Let ϕ denote either the set or the ball-measure of noncompactness. Then a map $T : D \subset X \to 2^Y$ is said to be ϕ-*condensing* if $\phi(T(Q)) < \phi(Q)$ whenever $Q \subset D$ and $\phi(Q) \ne 0$. Let $CK(X)$ be the set of all bounded closed and convex subsets of X.

Theorem 1.3 *([Mi-1,7]) (a) Let $U \subset X$ be a neighborhood of 0, $A : U \to CK(X)$ be u.s.c. and ϕ-condensing and $x = 0$ if $x \in Ax$. Let $N : X \to X$ be continuous and $Nx - Ny \in A(x - y)$ whenever $x - y \in U$. Then $T = I - N : X \to X$ is bijective.*

(b) If, in addition, A and N are ball-condensing on X, then the equation $Tx = f$ is uniquely approximation solvable w.r.t. $\Gamma = \{X_n, P_n\}$ for each $f \in X$. If A is also positively homogeneous, the solutions $x_n \in X_n$ of $P_n T x = P_n f$ satisfy (1.5)–(1.6).

Proof. The proof of part (a) consists in showing that the pair $(X, I - T)$ is a covering space for X and using the simple connectedness of X (cf. [Mi-1] for details). If A and N are ball-condensing, we claim that for each $f \in X$ there are an $r > 0$ and $n_0 \ge 1$ such that $\deg(I - P_n N, B(0, r) \cap H_n, P_n f) = \deg(I - N, B(0, r), f) \ne 0$ for each $n \ge n_0$. Indeed, for a given $f \in X$, select $r > 0$ such that $f \in (I - N)(B(0, r))$. Since $I - N$ is a homeomorphism, $\deg(I - N, B(0, r), f) \ne 0$. Then the homotopy $H(t, x) = tP_n Nx + (1-t)Nx$

on $[0,1] \times \overline{B(0,r)}$ is admissible and $\deg(I - N, B(0,r), f) = \deg(I - P_nN, B(0,r), P_nf) = \deg(I - P_nN, B(0,r) \cap H_n, P_nf)$. Hence, the claim is valid and the approximate equations $x - P_nNx = P_nf$ are solvable. The rest of part (b) follows from Theorem 1-(b)(cf. [Mi-6] for details). □

Theorem 1.3-(a) is due to Lasota-Opial [L-O] when A is compact.

Now, we turn to the unique approximation solvability of Eq.(1.2) with $\dim \ker(A) \leq \infty$.

Theorem 1.4 ([Mi-6]) *Let $A : D(A) \subset H \to H$ and $C : H \to H$ be selfadjoint maps such that $\|(A - C)x\| \geq c\|x\|$ for all $x \in H$ and some $c > 0$, and $N : H \to H$ be a continuous map such that for some $a < c$*

$$\|Nx - Ny - C(x - y)\| \leq a\|x - y\| \text{ for all } x, y \in H.$$

Then Eq.(1.2) is uniquely approximation solvable w.r.t. $\Gamma = \{(A - C)^{-1}(H_n), P_n\}$ for each $f \in H$ and the approximate solutions satisfy (1.5). If the spectrum of A, $\sigma(A)$, consists only of eigenvalues and the corresponding eigenvectors form a basis in H, then the error estimate (1.6) is also valid.

Corollary 1.1 (cf.[Mi-6,7]) *Let $A : D(A) \subset H \to H$ be selfadjoint, $N : H \to H$ be nonlinear. Then the conclusions of Theorem 1.4 are valid if either one of the following conditions holds:*

(a) N is α-strongly monotone and β-contractive with $\beta^2 < k\alpha$ and $\lambda = \lambda_1/2 > 0$, where $k = \inf\{|\lambda_i| \, | \, \lambda_i \in \sigma(A) \setminus \{0\}\}$.

(b) N is Gateaux differentiable and for some $\alpha, \beta \in R$ with $[\alpha, \beta] \subset (\lambda_k, \lambda_{k+1})$:

$$\alpha\|x - y\|^2 \leq (Nx - Ny, x - y) \leq \beta\|x - y\|^2 \text{ for } x, y \in H.$$

(c) Let $N : H \to H$ be a continuous gradient map and $B^{\pm} : H \to H$ be linear continuous and selfadjoint maps such that

(i) $N - B^-$ and $B^+ - N$ are monotone, and either

(ii) $H = H^- \oplus H^+$ for some closed subspaces $H^{\pm}$ and the projections $P^{\pm} : H \to H^{\pm}$ are such that $P^{\pm}(D(A)) \subset D(A)$ and for some $\gamma > 0$

$$((A - B^-)x, x) \leq -\gamma\|x\|^2, \quad x \in D(A) \cap H^- \tag{1.7}$$

$$((A - B^+)x, x) \geq \gamma\|x\|^2, \quad x \in D(A) \cap H^+ \tag{1.8}$$

or

(iii) $A - (1-t)B^- - tB^+$ has a bounded inverse for each $t \in [0,1]$.

(d) N has a symmetric Gateaux derivative on H and $B^- \leq N'(x) \leq B^+$ for each $x \in H$, where B^- and B^+ are as in (c) and (iii) holds.

Part (a) is a variant of Smiley [S], and the unique solvability in (c) is due to Tersian [T] with (i) -(ii) and to Fonda-Mawhin [F-M] with (i) and (iii). Their results extend the corresponding ones in Amann [A], Dancer [D] and many earlier ones (see [F-M] and [Mi-6]). In a different context, it has been shown in [Mi-6] that (ii) implies (iii) (cf. Lemma 1.5 in Section 1.4).

1.2 Fredholm alternative for semilinear equations. We begin with the following generalized first Fredholm theorem for nonlinear maps that are asymptotically close to a suitable map.

Theorem 1.5. *([Mi-1,2]) Let $A, T : X \to Y$ be nonlinear maps and for some function $c : R \to R$ with $c(r) \to \infty$ as $r \to \infty$*

$$\|Q_n Ax\| \geq c(\|x\|) \text{ for } x \in X_n, n \geq n_0 \geq 1$$

$$|T - A| = \limsup_{\|x\| \to \infty} \|Tx - Ax\|/c(\|x\|) < 1/\delta.$$

Let T be pseudo A-proper w.r.t. Γ and either A is odd or $T = A + N$ and $\deg(Q_n A, B(0, r) \cap X_n, 0) \neq 0$ for each r large and each $n \geq n_0(r)$. Then $T(X) = Y$.

Proof. Let $f \in Y$ be fixed. If A is odd, then there are $r > 0$ and $n_0 \geq 1$ such that $Q_n(Tx - f) \neq tQ_n(T(-x) - f)$ for all $x \in \partial B(0, r) \cap X_n$, $t \in [0, 1]$, $n \geq n_0$. By Borsuk's antipodes theorem, $Q_n Tx = Q_n f$ is solvable in $\overline{B(0, r)} \cap X_n$, and $Tx = f$ for some x by the pseudo A-properness of T (cf. [Mi-1,2] for more details). In the second case, define the homotopy $H_n(t, x) = Q_n Ax + tQ_n Nx$. Then, it is easy to see that $H_n(t, x) \neq tQ_n f$ on some $\partial B(0, r) \cap X_n$ for $n \geq n_0$ (cf. [Mi-6]). Hence, $\deg(Q_n T, B(0, r) \cap X_n, Q_n f) \neq 0$ for $n \geq n_0$, and the conclusion follows as above. □

In particular, if T is asymptotically positively k-homogeneous, i.e., $T = A + N$ with the quasinorm $|N|$ of N, defined as above with $c(r) = r^k$, sufficiently small and A is positively k-homogeneous, $k > 0$, A-proper w. r.t. Γ and $x = 0$ if $Ax = 0$, then Theorem 1.5 is applicable (cf. [Mi-1,2]). When T is asymptotically linear, Theorem 1.5 includes many earlier results when $T = I - N$, N compact, or T is of type (S) or A-proper ([K], [N-1,2], [H], [Pe], etc.).

The following special case is suitable for studying semilinear equations without resonance at infinity (cf. [Mi-5] and Sections 2.2-2.5.)

Theorem 1.6 *([Mi-5]) Let $A : D(A) \subset X \to Y$ and $N_\infty : X \to Y$ be linear maps, $V = (D(A), \|\cdot\|_0)$ be a Banach space densely and continuously embedded in X and $N : X \to Y$ be a nonlinear map such that $A + N : V \to Y$ is pseudo A-proper w.r.t. $\Gamma = \{X_n, Y_n, Q_n\}$ and $Q_n(A + N_\infty)x = (A + N_\infty)x$ on X_n. Suppose that for some positive constants a, b, c and r, with $a < c$,*

$$\|Nx - N_\infty x\| \leq a\|x\|_0 + b \text{ for } \|x\|_0 \geq r \quad (1.9)$$

and either $\|(A - N_\infty)^{-1} y\|_0 \leq c\|y\|$ on Y, or $X = Y = H$ is a Hilbert space, A and N_∞ are selfadjoint and

$$0 < a < \min\{|\mu| \,|\, \mu \in \sigma(A - N_\infty)\}. \quad (1.10)$$

Then Eq.(1.2) is solvable for each $f \in Y$.

Next, we state a full extension of the Fredholm Alternative for Eq.(1.2).

Theorem 1.7 *([Mi-5]) (Fredholm Alternative) Let $A : X \to Y$ be a linear continuous Fredholm map of index zero and $N : Y \to Y$ be continuous and have a sufficiently small quasinorm $|N|$. Let A and $T = A + N$ be A-proper w.r.t. Γ with $\ker(A) \subset X_n$. Then*

(i) either $\ker(A) = 0$ and Eq.(1.2) is approximation solvable for each $f \in Y$, or

(ii) $\ker(A) \neq 0$ and if $\mathrm{codim} R(A) = m > 0$ and $R(N) \subset R(A)$, then for each $f \in R(A)$ $(= N(A^*)^\perp)$, and only such ones, there is a connected closed subset K of $T^{-1}(f)$ whose dimension at each point is at least m.

The proof of (i) follows from Theorem 1.5, while (ii) is based on the covering dimension result in [F-M-P]. Theorem 1.7, but without the dimension assertion, is due to Kachurovsky

[K] when $T = I - N$ with N compact and $|N| = 0$, to Hess [H] and Petryshyn [Pe] when T is asymptotically linear and of type (S) or A-proper, respectively. Fredholm alternatives for (multivalued) pseudo A-proper and condensing maps can be found in [Mi-3,5]. Applications to ODE's and Hammerstein integral equations can be found in [K, Mi-3,5].

1.3 Semilinear equations with positive homogeneous nonlinearities. Let H be a Hilbert space, V and W be its closed subspaces such that $\dim V < \infty$ and $H = V \oplus W$. Let $A : D(A) \subset H \to H$ be a linear densely defined map with $A(V) \subset V$ and $A(D(A) \cap W) \subset W$, $N : H \to H$ be a nonlinear continuous and bounded map and $\Gamma = \{H_n, P_n\}$ be a projection scheme for H with $V \subset H_n$ and $P_n Ax = Ax$ on H_n for all large n. Let $P : H \to V$ be the orthogonal projection.

Since N_∞ is bounded and symmetric, $A - N_\infty$ is selfadjoint. Hence $c_0 = \min\{|\mu| \,|\, \mu \in \sigma(A - N_\infty)\} = \|(A - N_\infty)^{-1}\|^{-1}$ and condition (1.10) is equivalent to $a < c_0$. Set $A_1 = A|D(A) \cap W$ and assume that $N_\infty(W) \subset W$. Then $c = c(W) = \|(A_1 - N_\infty)|_W\|^{-1} \geq c_0$. Thus, if we replace (1.10) by $a < c$, then condition (1.9) allows a bigger nonlinearity N. In this case, we prove the first Fredholm theorem for Eq.(1.2) for positive homogeneous N.

Let X be a Banach space compactly embedded in H, with the norm $\|\cdot\|_1$. Set $X_n = X \cap H_n$. Let $G : X \to H$ be a linear map having a densely defined adjoint and $P_n Gx = Gx$ on X. We say that $A + N$ satisfies condition (*) if whenever $x_n \in H_n$ is bounded, $0 < \epsilon_n \to 0$, $\epsilon_n G x_n \to 0$ and $Ax_n + P_n N x_n + \epsilon_n G x_n = P_n f$, then there is an $x \in H$ such that $Ax + Nx = f$. Throughout this section we assume that $N_\infty(W) \subset W$ and $P_n N_\infty x = N_\infty x$ on H_n.

Theorem 1.8 *Let $A + N : D(A) \subset H \to H$ be pseudo A-proper w.r.t. Γ, N be bounded, positively homogeneous and $(Nx, x) > 0$ for $x \neq 0$, and $A + tN + \epsilon G + (1 - t)\epsilon I : D(A) \cap X \subset X \to H$ be A-proper w.r.t. $\Gamma = \{X_n, P_n\}$ for (X, H) for each $\epsilon > 0$ small and $t \in [0, 1]$. Suppose that (1.9)–(1.10) for $A - N_\infty|W$ hold, $Gx = 0$ implies that $(Ax, x) \geq 0$ and*

$$(Ax, Gx) = 0, \ (Nx, Gx) = 0, \ (Gx, x) = 0 \ \text{for all } x \in X_n. \tag{1.11}$$

Let $x = 0$ be the only solution of the equation $Ax + Nx = 0$. Then

(a) Eq.(1.2) is solvable for each $f \in D(G^*)$.

(b) If, in addition, condition (*) holds, Eq.(1.2) is solvable for each $f \in H$.

The proof will be preceded by a couple of lemmas (cf. [Mi-8]).

Lemma 1.1 *Let $A + N : D(A) \subset H \to H$ be pseudo A-proper w.r.t. Γ, N be bounded and conditions (1.9)–(1.10) for $A - N_\infty|W$ hold. Then, for each $f \in H$, each $v \in V$ and each large n, the equation*

$$Aw_n + (I - P)P_n N(v + w_n) = (I - P)P_n f \ \text{for} \ w_n \in W \cap H_n \tag{1.12}$$

is solvable, and there are constants $c_1 > 0$ and $c_2 > 0$, independent of n, v and w_n, such that

$$\|w_n\| \leq c_1 \|v\| + c_2(\|f\| + b). \tag{1.13}$$

Lemma 1.2 *Let the conditions of Theorem 1.8 hold. Then the equations $Ax + P_n Nx + \epsilon Gx = P_n f$ and $Ax + Nx + \epsilon Gx = f$ are solvable for each $f \in H$, $n \geq n_0$, and $\epsilon > 0$ small.*

Proof of Theorem 1.8 (a) Let $f \in H$ be fixed. By Lemma 1.2, for each $\epsilon_n > 0$ with $\epsilon_n \to 0$ there is an $x_n \in H_n$, with $n \geq n_0$, such that $Ax_n + P_n N x_n + \epsilon_n G x_n = P_n f$. We claim that the set $\{x_n | n \geq n_0\}$ is bounded in H. If not, we may assume that $\|x_n\| \to \infty$ as $n \to \infty$. Hence,

$$Ax_n + P_n N x_n = p_n, \; p_n = P_n f - \epsilon_n G x_n.$$

Taking the scalar product of it with $G x_n$, we get $\epsilon_n \|G x_n\| \leq \|f\|$. Hence, $\|p_n\| \leq C$ for all $n \geq n_0$ and some $C > 0$. Since $H_n = V \oplus W_n$, we can write $x_n = v_n + w_n$ with $v_n \in V$ and $w_n \in W_n = (I - P)(H) \cap H_n$. Moreover, since $A : D(A) \cap W \to W$, we have that

$$Aw_n + (I - P) P_n N x_n = (I - P) p_n.$$

By Lemma 1.1, there are $c_1 > 0$ and $c_2 > 0$, independent of n, such that

$$\|w_n\| \leq c_1 \|v_n\| + c_2 \|p_n\|.$$

We claim that $\{v_n\}$ is bounded in H. If not, we may assume that $\|v_n\| \to \infty$ as $n \to \infty$. Set $u_n = x_n / \|v_n\|$. Then $\|P u_n\| = 1$ for each n and, since $\dim V < \infty$, we may assume that $P u_n \to v \in V$ and $\|v\| = 1$. But,

$$\|(I - P) u_n\| = \|w_n\|/\|v_n\| \leq c_1 + c_2 \|p_n\|/\|v_n\| \to c_1 \; as \; n \to \infty.$$

Hence, $\{u_n\}$ is bounded and

$$A u_n + P_n N u_n = p_n / \|P x_n\| \to 0 \; as \; n \to \infty.$$

Since $A + N$ is pseudo A-proper, a subsequence $u_{n_k} \rightharpoonup u_0 \in H$ and $A u_0 + N u_0 = 0$ with $\|P u_0\| = \|v\| = 1$. This contradicts one of the assumptions of the theorem and therefore, $\{x_n\}$ is bounded in H.

Now, we get as above that for $f \in D(G^*)$

$$\epsilon_n^2 \|G x_n\|^2 \leq \epsilon_n \|G^* f\| \|x_n\| \to 0 \; as \; n \to \infty.$$

Hence, $\epsilon_n \|G x_n\| \to 0$ and

$$A x_n + P_n N x_n = P_n f - \epsilon_n G x_n \to f \; as \; n \to \infty.$$

Since $A + N$ is pseudo A-proper, there is an $x \in D(A)$ such that $Ax + Nx = f$.

(b) We may assume that $x_n \rightharpoonup x$ and $\epsilon_n G x_n \rightharpoonup u$ in H as $n \to \infty$. Since $(\epsilon_n G x_n, h) = \epsilon_n (x_n, G^* h)$ for each $h \in D(G^*)$, it follows that $(u.h) = 0$. Hence, $\epsilon_n G x_n \rightharpoonup 0$ by the density of $D(G^*)$ in H, and $A x_n + P_n N x_n - \epsilon_n G x_n = P_n f$. Therefore, by condition (*), there is an $x \in D(A)$ such that $Ax + Nx = f$ for each $f \in H$. □

Proposition 1.1 Let $A : D(A) \subset H \to H$ be a linear selfadjoint map with closed range $R(A)$ and compact partial inverse $A^{-1} : R(A) \to H$. Let $N : H \to H$ be a continuous bounded monotone map. Then $A + N : D(A) \subset H \to H$ is pseudo A-proper w.r.t. $\Gamma = \{H_n, P_n\}$ with $P_n A x = A x$ on H_n and satisfies condition (*) with $G : X \to H$ such that $(Gx, x) = 0$ for $x \in X$.

Proof. The pseudo A-properness of $A + N$ is known ([Mi-6]). Suppose that $x_n \in H_n$ is bounded, $0 < \epsilon_n \to 0$, $\epsilon_n G x_n \rightharpoonup 0$ and $A x_n + P_n N x_n + \epsilon_n G x_n = P_n f$. Let

138

$P : H \to R(A)$ be the orthogonal projection. Since the graph of A is weakly closed, we may assume that $x_n \rightharpoonup x$ and $Ax_n \rightharpoonup Ax$. Since A^{-1} is compact, we may assume that $Px_n = A^{-1}(P_n f - P_n N x_n - \epsilon_n G x_n) \to Px$ and therefore, $(Ax_n, x_n) = (Ax_n, Px_n) \to (Ax, Px) = (Ax, x)$. For each $h \in \cup_{n \geq 1} H_n$, $h \in H_k$ for some large k and for $n \geq k$,

$$(P_n f - Ax_n - Nh, x_n - h) = (P_n N x_n - Nh, x_n - h) - \epsilon_n(Gx_n, h)$$
$$= (Nx_n - Nh, x_n - h) - \epsilon_n(Gx_n, h).$$

Passing to the limit as $n \to \infty$, we get

$$(f - Ax - Nh, x - h) \geq 0 \; for \; all \; h \in \cup_{n \geq 1} H_n.$$

Since $\cup_{n \geq 1} H_n$ is dense in H, this inequality holds for all $h \in H$. Now, we shall use Minty's trick. For $u \in H$ and $\tau > 0$, set $h = x - \tau u$, to get, after dividing by τ, that $(f - Ax - N(x - \tau u), u) \geq 0$. As $\tau \to 0$, we get $(f - Ax - Nx, u) \geq 0$ for all $u \in H$. Hence, $Ax + Nx = f$. □

Propositon 1.2 *Let* $N : V_1 \to H$ *be compact and* $\Gamma = \{H_n, H_n, P_n\}$ *be a scheme for* (V_2, H) *with* $P_n Ax = Ax$ *and* $P_n Gx = Gx$ *for* $x \in H_n$. *If* $A + \epsilon G + (1-t)\epsilon I : D(A) \cap V_2 \to H$ *is continuously invertible for* $t \in [0,1]$, *then* $H_{t\epsilon} = A + tN + \epsilon G + (1-t)\epsilon I : D(A) \cap V_2 \to H$ *is A-proper w.r.t.* Γ *for* $\epsilon > 0$ *and* $t \in [0,1]$.

Theorems 1.6 and 1.8 apply also to the following new class of A-proper maps ([Mi-8]).

Proposition 1.3 *Let* $A : D(A) \subset H \to H$ *be a linear selfadjoint map,* $N : H \to H$ *be a continuous gradient map and* $\alpha \leq \beta$ *be real numbers such that* $[\alpha, \beta] \cap \sigma(A)$ *consists of a (positive) finite number of eigenvalues of* A *having finite multiplicity and*

$$\alpha \|x - y\|^2 \leq (Nx - Ny, x - y) \leq \beta \|x - y\|^2 \; for \; x, y \in H.$$

Then $A + N : D(A) \subset H \to H$ *is A-proper w.r.t.* $\Gamma = \{H_n, P_n\}$ *with* $P_n Ax = Ax$ *on* H_n.

1.4 Semilinear equations with $\{B_1, B_2\}$-quasilinear perturbations. We continue our study of Eq.(1.2) with asymptotically $\{B_1, B_2\}$-quasilinear nonlinearities N, where $B_1, B_2 : H \to H$ are selfadjoint maps with $B_1 \leq B_2$, i.e. $(B_1 x, x) \leq (B_2 x, x)$ for $x \in H$. A fixed point theory for such maps has been developed by Perov [P] and Krasnoselskii-Zabreiko [K-Z] assuming that $\{B_1, B_2\}$ is a regular pair. These maps have been studied extensively in the context of semilinear equations by the author [Mi-5,6].

Definition 1.2 a) A nonlinear map $K : H \to H$ is $\{B_1, B_2\}$-quasilinear on a set $S \subset H$ if for each $x \in S$ there exists a linear selfadjoint map $B : H \to H$ such that $B_1 \leq B \leq B_2$ and $Bx = Kx$;

b) A map $N : H \to H$ is said to be asymptotically $\{B_1, B_2\}$-quasilinear if there is a $\{B_1, B_2\}$-quasilinear outside some ball map K such that

$$|N - K| = limsup_{\|x\| \to \infty} \frac{\|Nx - Kx\|}{\|x\|} < \infty.$$

This class of maps is rather large. For example, let $N : H \to H$ have a selfadjoint weak Gateaux derivative $N'(x)$ on H. Assume that $B_1 \leq N'(x) \leq B_2$ for each x and some selfadjoint maps B_1 and B_2. Then N is asymptotically $\{B_1, B_2\}$-quasilinear with $|N - K| = 0$ (cf. [Mi-6]).

In the nondifferentiable case, if $Nx = B(x)x + Mx$ for some nonlinear map M with the quasinorm $|M| < \infty$ and selfadjoint maps $B(x) : H \to H$ with $B_1 \leq B(x) \leq B_2$ for each $x \in H$, then N is asymptotically $\{B_1, B_2\}$-quasilinear.

Our first result for these maps is an extension of Theorem 1.6.

Theorem 1.9 *Let $A : D(A) \subset H \to H$ be a linear densely defined map and $N : H \to H$ be bounded and asymptotically $\{B_1, B_2\}$-quasilinear. Assume that there is a $c > |N - K|$ such that for each selfadjoint map C with $B_1 \leq C \leq B_2$ we have that*

$$\|Ax - Cx\| \geq c\|x\|, \quad x \in D(A) \setminus B(0, R). \quad (1.14)$$

(a) If $A - tN$ is A-proper w.r.t. $\Gamma = \{H_n, P_n\}$ for $t \in [0, 1]$, then Eq.(1.2) is approximation solvable for each $f \in H$.

(b) If $A - K$ is A-proper, $\deg(P_n(A - K), B(0, r) \cap H_n, 0) \neq 0$ for each $r > R$ and each $n \geq n_0(r) \geq 1$, $|N - K|$ sufficiently small and $A - N$ is pseudo A-proper w.r.t. Γ, then Eq.(1.2) is solvable for each $f \in H$.

Proof. (a) Let $f \in H$ be fixed and define $H(t, x) = Ax - tNx$ on $[0, 1] \times D(A)$. Then there is an $r = r(f) > R$ such that

$$H(t, x) \neq f \text{ for } x \in \partial B(0, r) \cap D(A), t \in [0, 1]. \quad (1.15)$$

If not, then there are $x_n \in H$ and $t_n \in [0, 1]$ such that $\|x_n\| \to \infty$ and the equality holds in (1.15). Let $\epsilon_0 > 0$ be small such that $\|Nx - Kx\| \leq (|N - K| + \epsilon_0)\|x\|$ for $\|x\| \geq R_1$ and $|N - K| + \epsilon_0 < c$. For each x_n with $\|x_n\| \geq R_1$ there is a C_n such that $B_1 \leq C_n \leq B_2$, $Kx_n = C_n x_n$ and $\|Ax_n - C_n x_n\| \geq c\|x_n\|$. Hence,

$$c\|x_n\| \leq \|Ax_n - C_n x_n\| \leq (|N - K| + \epsilon_0)\|x_n\| + \|f\|.$$

Dividing by $\|x_n\|$, this leads to a contradiction and (1.15) holds. Using the A-properness of $A - tN$, we get that for all large n

$$P_n H(t, x) \neq P_n f \text{ for } x \in \partial B(0, r) \cap H_n, t \in [0, 1].$$

The conclusion now follows form the homotopy theorem and the A-properness of $A - N$.

(b) Define $H(t, x) = Ax - Kx - t(N - K)x$. By (1.14), $\|Ax - Kx\| \geq c\|x\|$ for each $x \in H \setminus B(0, R)$. By the A-properness of $A - K$, for each $r \geq R$ there is a $c_1 > 0$ and an n_0 such that

$$\|P_n(A - K)x\| \geq c_1\|x\| \text{ for } x \in \partial B(0, r) \cap H_n, n \geq n_0.$$

Then, arguing as in (a), we see that for each $f \in H$ there is an $r > R$ such that $P_n H(t, x) \neq tP_n f$ for each $x \in \partial B(0, r) \cap H_n$, $t \in [0, 1]$, $n \geq n_0$. Hence, by the Brouwer homotopy theorem and the pseudo A-properness of $A - N$, Eq.(1.2) is solvable for each f. □

Analyzing the above proof, we see that the following result is valid.

Theorem 1.10 *([Mi-5]) Let $A : D(A) \subset X \to Y$ be a linear densely defined map and $N : X \to Y$ be bounded and of the form $Nx = B(x)x + Mx$ for some linear maps $B(x) : X \to Y$. Assume that there is a $c > |M|$ such that*

$$\|Ax - B(x)x\| \geq c\|x\|, \quad x \in D(A) \setminus B(0, R). \quad (1.16)$$

(a) If $A - tN$ is A-proper w.r.t. $\Gamma = \{X_n, Y_n, Q_n\}$ for $t \in [0,1]$, then Eq.(1.2) is approximation solvable for each $f \in Y$.

(b) If, for $Kx = B(x)x$ on X, $A - K$ is A-proper, $\deg(Q_n(A - K), B_r \cap X_n, 0) \neq 0$ for each $r > R$ and each $n \geq n_0(r) \geq 1$, $|N - K|$ is sufficiently small and $A - N$ is pseudo A-proper w.r.t. Γ, then Eq.(1.2) is solvable for each $f \in Y$.

The A-properness of A in Theorems 1.9-10-(a) requires that $\dim \ker(A) < \infty$. The following basic solvability result for Eq.(1.2) is a variant of Theorem 1.9-(b).

Theorem 1.11 (cf. [Mi-6]) *Let $A : D(A) \subset H \to H$ and $N : H \to H$ be bounded asymptotically $\{B_1, B_2\}$-quasilinear. Let there exist $\epsilon > 0$ and $c > |N - K|$ such that for some selfadjoint maps $B^-, B^+ \in L(H)$ with $B_1 \leq B^-$ and $B^+ \leq B_2$ and any selfadjoint map $C \in L(H)$ with $B_1 - \epsilon I \leq C \leq B_2 + \epsilon I$ we have that (1.14) holds. Suppose that for a selfadjoint map $C_0 \in L(H)$ with $B_1 - \epsilon I \leq C_0 \leq B_2 + \epsilon I$ the map $H(t, \cdot) = A - (1-t)C_0 - tN$ is A-proper w.r.t. $\Gamma = \{H_n, P_n\}$ for each $t \in [0, 1)$ and H_1 is pseudo A-proper w.r.t. Γ. Then Eq.(1.2) is solvable for each $f \in H$.*

Proof. Since $N_f x = Nx - f$, $f \in H$, has the same properties as N, it suffices to solve the equation $Ax - Nx = 0$. Let $\epsilon_0 > 0$ be such that $|N - K| + \epsilon_0 < c$. Then there is an $r > 0$ such that

$$\|Nx - Kx\| \leq (|N - K| + \epsilon_0)\|x\| \text{ for each } \|x\| \geq r.$$

Moreover,

$$H(t, x) = Ax - (1-t)C_0 x - tNx \neq 0 \text{ for } x \in \partial B_r \cap D(A), t \in [0, 1]. \quad (1.17)$$

If not, then $H(t, x) = 0$ for some $\|x\| = r$ and $t \in [0, 1]$. Hence, subtracting tKx from both sides, we get

$$\|Ax - tKx - (1-t)C_0 x\| = t\|Nx - Kx\| < c\|x\|$$

Since K is $\{B_1, B_2\}$-quasilinear, there is a selfadjoint map $C^* \in L(H)$ such that $Kx = C^* x$, $B_1 \leq C^* \leq B_2$ and therefore

$$\|Ax - tC^* x - (1-t)C_0 x\| < c\|x\|. \quad (1.18)$$

But, $C = tC^* + (1-t)C_0$ is selfadjoint, $B_1 - \epsilon I \leq C \leq B_2 + \epsilon I$ and therefore (1.14) holds. This contradicts (1.18) and so (1.17) is valid.

Next, since C_0 and N are bounded maps, $H(t, x)$ is an A-proper homotopy on $[0, \epsilon] \times (\overline{B_r} \cap D(A))$ w.r.t. Γ for each ϵ in $(0, 1)$ and is continuous at 1 uniformly for $x \in \overline{B_r} \cap D(A)$. Hence, the solvability of $Ax - Nx = 0$ follows from Theorem 3.1 in [Mi-4]. □

Remark 1.1 If we know that $Nx = B(x)x + Mx$ for some linear maps $B(x)$ and $|M|$ small, then Theorem 1.11 is valid if (1.14) holds for each $C = B(x)$, $x \in H$.

Let us now discuss some conditions that imply (1.14) and (1.16).

Lemma 1.3 ([Mi-5]) *Let $X \subset Y$, $A : D(A) \subset X \to Y$ and $N : X \to Y$ be such that $Nx = B(x)x + Mx$ with $B(x) : X \to Y$ continuous and linear for each $x \in X$ and $|M|$ sufficiently small. For some $\lambda \in R$, let $A_\lambda = A - \lambda I : D(A) \subset X \to Y$ be continuously invertible and $B_\lambda(x) = B(x) - \lambda I$. Suppose that either*

(i) $\limsup_{\|x\|\to\infty}\|B_\lambda(x)\| < 1/\|A_\lambda^{-1}\|$, or

(ii) Let either $Q_n A_\lambda x = A_\lambda x$ on X_n, or $A_\lambda : D(A) \subset X \to Y$ be A-proper w.r.t. $\Gamma = \{X_n, Y_n, Q_n\}$. For some $a, b, c_0 > 0$ with $\delta a < c_0$, let $\|A_\lambda x\| \geq c_0 \|x\|$ for $x \in D(A)$ and $\|B_\lambda(x)x\| \leq a\|x\| + b$ for all $\|x\|$ large.

Then there are $c, R > 0$ such that (1.16) holds and, for $Kx = B(x)x$ on X, $\deg(Q_n(A-K), B_r \cap X_n, 0) \neq 0$ for each $r > R$ and each $n \geq n_0(r)$.

Remark 1.2 If (ii) holds, the solvability of Eq.(1.2) follows also by Theorem 1.6.

Lemma 1.4 *Let $A : D(A) \subset H \to H$ be selfadjoint. Then (1.14) holds for each selfadjoint map C with $B_1 \leq C \leq B_2$ if there is an $a > 0$ such that either*

(i) $0 < a < \min\{|\lambda| \,|\, \lambda \in \sigma(A-C)\}$, or

(ii) *each C commutes with A and $\text{dist}(\sigma(A), \sigma(C)) \geq a$.*

Lemma 1.5 ([Mi-6]) *Let $A : D(A) \subset H \to H$ and $B^\pm \in L(H)$ be selfadjoint with $B^- \leq B^+$ and $H^\pm$ be subspaces with $H = H^- \oplus H^+$ and such that (1.7) and (1.8) hold for some $\gamma_1 > 0$ and $\gamma_2 > 0$, respectively. Then there are $\epsilon > 0$ and $c > 0$ such that for any selfadjoint maps $B_1, B_2, C \in L(H)$ with $B_1 \leq B^-$ and $B^+ \leq B_2$ and $B_1 - \epsilon I \leq C \leq B_2 + \epsilon I$ we have that (1.14) holds on H.*

Corollary 1.2 ([Mi-6]) *Let $A : D(A) \subset H \to H$ satisfy (1.7)–(1.8) and $N : H \to H$ be bounded asymptotically $\{B_1, B_2\}$-quasilinear with $|N-K|$ sufficiently small. Suppose that a selfadjoint map $C_0 \in L(H)$ satisfies $B_1 - \epsilon I \leq C_0 \leq B_2 + \epsilon I$ with sufficiently small ϵ and that $H(t,.) = A - (1-t)C_0 - tN$ is A-proper w.r.t. $\Gamma = \{H_n, P_n\}$ with $P_n A x = A x$ on H_n for each $t \in [0,1]$ and H_1 is pseudo A-proper w.r.t. Γ. Then Eq.(1.2) is solvable for each $f \in H$.*

Let us now discuss some conditions on $B^\pm$ which imply (1.7)–(1.8). Assume, as in Amann [A],

(1.19) a) $A : D(A) \subset H \to H$ is selfadjoint.

b) $B^\pm = \sum_{i=1}^m \lambda_i^\pm P_i^\pm$ commute with A, where $P_i^\pm : H \to \ker(B^\pm - \lambda_i)$ are orthogonal projections, $\lambda_1^\pm \leq \cdots \leq \lambda_m^\pm$ and $\lambda_i^\pm$ are pairwise distinct.

c) $\bigcup_{i=1}^m [\lambda_i^-, \lambda_i^+] \subset \rho(A)$-the resolvent set of A.

Being selfadjoint, A possesses a spectral resolution

$$A = \int_{-\infty}^{\infty} \lambda\, dE_\lambda$$

where $\{E_\lambda \,|\, \lambda \in R\}$ is a right continuous spectral family. Since $B^\pm$ commute with A, it is known that $P_i^\pm$ commute with the resolution of the identity $\{E_\lambda \,|\, \lambda \in R\}$. Hence, the selfadjoint maps $A - B^\pm$ have the spectral resolution

$$A - B^\pm = \sum_{i=1}^m \int_{-\infty}^{\infty} (\lambda - \lambda_i^\pm) dE_\lambda P_i^\pm. \tag{1.20}$$

Define the orthogonal projections $P^\pm$ by

$$P^- = \sum_{i=1}^m E(-\infty, \lambda_i^-) P_i^- \quad \text{and} \quad P^+ = \sum_{i=1}^m E(\lambda_i^+, \infty) P_i^+,$$

where

$$E(\alpha, \beta) = \int_\alpha^\beta dE_\lambda$$

for all $\alpha, \beta \in \rho(A) \cup \{\pm\infty\}$ with $\alpha < \beta$. Define $H^\pm = P^\pm(H)$ and note that by (1.19)-c),

$$P^+ = \sum_{i=1}^m E(\lambda_i^-, \infty) P_i^+,$$

and

$$\gamma = dist\left(\cup_{i=1}^m [\lambda_i^-, \lambda_i^+], \sigma(A)\right) > 0.$$

Moreover, by (1.20), we have that (1.7) and (1.8) hold.

Lemma 1.6 *If (1.19) holds and $P_i^- = P_i^+$ for $1 \le i \le m$, then there are orthogonal subspaces $H^\pm$ such that $H = H^- \oplus H^+$ and conditions (1.7)–(1.8) hold.*

Proof. It remains only to show that H^+ and H^- are orthogonal. Since $P_i^- = P_i^+$ for $i = 1, \ldots, m$, and $E(-\infty, \lambda_i^-) = I - E(\lambda_i^-, \infty)$, we get that $P^+ = I - P^-$ and therefore, $H^+ = (H^-)^\perp$. □

When $B^\pm$ are not of the form (1.19)-b), we need to assume more on the linear part A.

(1.21) Suppose A is selfadjoint possessing a countable spectrum $\sigma(A)$ consisting of eigenvalues and whose eigenvectors form a complete orthonormal system in H.

(1.22) There are selfadjoint maps $C_1, C_2 \in L(H)$ and two consecutive finite multiplicity eigenvalues $\lambda_k < \lambda_{k+1}$ of A such that

$$\lambda_k \|x\|^2 < (C_1 x, x) \le (C_2 x, x) < \lambda_{k+1} \|x\|^2 \ for \ x \in H \setminus \{0\}.$$

Let H^- (resp. H^+) be the subspaces of H spanned by the eigenvectors of A corresponding to the eigenvalues $\lambda_i \le \lambda_k$ (resp. $\lambda_i \ge \lambda_{k+1}$).

Lemma 1.7. *(cf. [Mi-6]) Let (1.21)–(1.22) hold. Then there are $\gamma_1 > 0$ and $\gamma_2 > 0$ such that for any selfadjoint maps $B^\pm \in L(H)$ satisfying $C_1 \le B^-$ and $B^+ \le C_2$ on H, we have that (1.7) and (1.8) hold, respectively.*

Remark 1.3 If λ_k (resp., λ_{k+1}) is of infinite multiplicity, then Lemma (1.7) is still valid if we assume in (1.22)

$$(\lambda_k + \epsilon)\|x\|^2 \le (C_1 x, x) \ \left(resp., (C_2 x, x) \le (\lambda_{k+1} - \epsilon)\|x\|^2\right) \ for \ 0 \ne x \in H.$$

For a discussion of the case when $H^- \oplus H^+ \ne H$ we refer to [A , Mi-6].

2. Applications

In this section, we shall give applications of the above abstract results to BVP's for nonlinear ODE's and elliptic PDE's in (non)divergence form and to periodic-boundary value problems for semilinear hyperbolic equations and Hamiltonian systems.

2.1 Linear BVP's for nonlinear ODE's. We shall consider the unique approximation solvability of linear BVP's for nonlinear ODE's of the form

$$\begin{aligned} x'(t) &= f(t, x) + h(t), \ (t, x) \in I \times R^n \\ Lx &= r \ (r \in R^n) \end{aligned} \tag{2.1}$$

where $I \subset R$ is a compact interval, $L : C(I, R^n) \to R^n$ is a linear continuous map. Here, $C(I, R^n)$ is the space of continuous maps with the norm $\|x\| = \max\{|x(t)| \ |t \in I\}$.

Assume

(H1) $f : I \times R^n \to R^n$ is continuous in x for each fixed $t \in I$, and is summable with respect to t for each fixed $x \in R^n$.

(H2) A multivalued map $F : I \times R^n \to CK(R^n)$ is upper semicontinuous in x for each fixed $t \in I$, and for each compact set $K \subset R^n$ there is a function $m(t)$, summable on I, such that $\sup\{|u| \le m(t) \mid \text{for } u \in F(t,x), t \in I, x \in K\}$.

Let $\Gamma_0 = \{U_n, P_n\}$ be a projection scheme for $C(I, R^n)$. Then $\Gamma = \{U_n \times R^n, Q_n\}$ is a projection scheme for the Banach space $X = C(I, R^n) \times R^n$, where $Q_n(x, p) = (P_n x, p)$ and $\|(x, p)\| = \max\{\|x\|, \|p\|\}$.

Theorem 2.1 *Let* $(H1) - (H2)$ *hold and* f *satisfy a generalized Lipschitz condition*

$$f(t, q) - f(t, p) \in F(t, q - p) \text{ for } t \in I, p, q \in R^n. \tag{2.2}$$

Suppose that $x(t) \equiv 0$ *is the unique solution of* $x'(t) \in F(t, x)$ *with* $Lx = 0$. *Then BVP (2.1) is uniquely approximation solvable w.r.t.* Γ *for each* $h \in C$. *If* $F(t, p)$ *is also positively homogeneous in* $p \in R^n$, *then the approximate solutions satisfy (1.5)–(1.6).*

Proof. Define the map $N : X \to X$ by $N(x, p) = (y, p - Lx)$, where, for a fixed $t_0 \in I$,

$$y(t) = \int_{t_0}^{t} f(s, x(s)) ds + p.$$

Then BVP (2.1) is reduced to the study of the equation $u - Nu = v$ in X. Define the map $A : X \to CK(X)$ in a similar way using all measurable selections of the map F. Then N is continuous and compact and A is upper semicontinuous and compact (cf. [L-O]). Hence, the maps $I - N$ and $I - A$ are A-proper w.r.t. Γ (cf. [Mi-1]) and $u \equiv 0$ if $u \in Au$. Since $I - N$ is bijective by Theorem 1.3-(a), the unique approximation solvability follows by Theorem 1.3-(b). If $F(t, p)$ is positively homogeneous in p, then the error estimate assertions follow again from Theorem 1.3-(b). □

Condition (2.2) holds if f satisfies some type of a generalized Lipschitz condition in the second variable. For example, if

$$|f(t, p) - f(t, q)| \le w(t, |p - q|)$$

for a suitable function w, then (2.2) holds with $F(t, p) = \{q \mid |q| \le w(t, |p|)\}$. If $w(t, \cdot)$ is positively homogeneous for each $t \in I$, then so is $F(t, \cdot)$. The unique solvability in Theorem 2.1 is due to Lasota-Opial [L-O].

2.2 Strong solvability of nonresonant regular elliptic problems. Consider the following semilinear elliptic BVP without resonance

$$\begin{aligned} Au + G(x, u, Du, ..., D^{2m-1}u)u + F(x, u, Du, ..., D^{2m}u) &= f(x) \text{ in } Q \subset R^n \\ B_j(x, D)u &= 0 \text{ on } \partial Q, j = 0, 1, \ldots, m - 1, \end{aligned} \tag{2.3}$$

where the boundary ∂Q is smooth and $Au = \sum_{|\alpha| \le 2m} a_\alpha(x) D^\alpha u(x)$ is an elliptic operator acting on $V = \{u \in W_p^{2m}(Q) \mid B_j u = 0 \text{ on } \partial Q, 1 \le j \le m\}$, the space of functions satisfying "coercive" (i.e., Lopatinski-Schapiro) boundary conditions $B_j u = 0$ on ∂Q for some $p \in (1, \infty)$ with $a_\alpha \in C(\overline{Q})$ for $|\alpha| = |(\alpha_1, ..., \alpha_n)| = \alpha_1 + ... + \alpha_n = 2m$ and $a_\alpha \in L_\infty(Q)$ for $|\alpha| < 2m$. Let s_{2m} be the number of distinct derivatives of order $\le 2m$.

Regarding A, F and G, we assume

(A) There is a $\lambda \in R$ such that $A + \lambda I : L_p \to W_p^{2m}$ is continuously invertible.

(F1) $F : Q \times R^{s_{2m}} \to R$ satisfies the Caratheodory conditions and there are $M > 0$ and $h_1, h_2 \in L_p(Q)$ such that

$$|F(x,\xi)| \le h_1(x) + Mh_2(x) \sum_{|\alpha| \le 2m} |\xi_\alpha| \text{ for a.e. } x \in Q \text{ and each } \xi \in R^{s_{2m}}.$$

(F2) There is a k sufficiently small such that

$$\left|F(x,\eta,\zeta) - F(x,\eta,\zeta')\right| \le k \sum_{|\alpha|=2m} \left|\zeta_\alpha - \zeta'_\alpha\right|$$

for a.e. $x \in Q$ and all $\eta \in R^{s_{2m-1}}$, $\zeta, \zeta' \in R^{s'_{2m}}$, $s'_{2m} = s_{2m} - s_{2m-1}$.

(G) $G : Q \times R^{s_{2m-1}} \to R$ satisfies (F1) on $Q \times R^{s_{2m-1}}$ and, for each $u \in W_p^{2m-1}(Q)$, $B(u) = G(x,u,Du,...,D^{2m-1}u) : L_p \to L_p$ is a continuous linear map.

(H) There are $a < c = \|(A+\lambda I)^{-1}\|$, b and $R > 0$ such that for $\|u\|_{p,2m} \ge R$

$$\|G(x,u,\ldots,D^{2m-1}u)u + F(x,u,\ldots,D^{2m}u) - \lambda u\|_p \le a\|u\|_{p,2m} + b.$$

Define $M : V \to L_p$ by $Mu = F(x,u,...,D^{2m}u)$.

Theorem 2.2 ([Mi-5]) *Let $A_\lambda = A + \lambda I : V \to L_p$ have a continuous inverse and (A),(F1)-(F2), (G) and (H) hold. Then BVP (2.3) is approximation solvable in V for each $f \in L_p$.*

Proof. Define the scheme $\Gamma = \{X_n = A_\lambda^{-1}(Y_n), Y_n, Q_n\}$ for (V, L_p) and note that $\|Q_n A_\lambda u\| = \|A_\lambda u\| \ge \|A_\lambda^{-1}\|^{-1}\|u\|_{p,2m}$ on X_n. Let $T_1 u = Au + B(u)u$ on V and $T = T_1 + M$. Then $T_1, T_2 : V \to L_p$ are A-proper w.r.t Γ (cf. [Mi-5]) and $\|Tu - \lambda u\|_p \le a\|u\|_{p,2m} + b$ for $\|u\|_{p,2m} \ge R$. Hence, the conclusion follows by Theorem 1.6 (or Theorem 1.10 and Lemma 1.3). $\square$

Remark 2.1 Using the arguments similar to [G-N-K], we see that (G) and (H) hold if M is sufficiently small and

(i) Let α be the Sobolev embedding constant for $W_p^{2m} \hookrightarrow C(\overline{Q})$, where $p > n/2m$, μ and $\delta > 0$ be such that $\delta c\alpha < 1$ and, for all $u \in W_p^{2m}$,

$$c\alpha\left(\int_{\{|u(x)|>\mu\}} |G(x,u,\ldots,D^{2m-1}u) - \lambda|^p dx\right)^{1/p} \le 1 - \delta$$

(ii) There are a continuous nondecreasing function $c_1(t)$ on $[0,\infty)$, $h_3 \in L_p$, $\beta \in [0,1)$ and $c_2 \ge 0$ such that for each $x \in Q$, $\xi \in R^{s_{2m-1}}$

$$|G(x,\xi)\xi_0| \le c_1(|\xi_0|)(1 + |\xi|^\beta) \text{ and } |G(x,\xi)\xi_0| \le h_3(x) + c_2|\xi|.$$

By Remark 2.1, Theorem 2.2 includes a result of Gupta-Kwong-Necas [G-K-N] when F depends only on the derivatives of order $\le 2m - 1$.

2.3 Weak solvability of nonresonant quasilinear elliptic problems. Let us first look at (2.3) with $Au = \sum_{|\alpha|,|\beta| \le m}(-1)^{|\alpha|}D^\alpha(a_{\alpha\beta}(x)D^\beta u)$ uniformly strongly elliptic, where $a_{\alpha\beta} = a_{\beta\alpha} \in L_\infty(Q)$ for $|\alpha|,|\beta| \le m$, and $a_{\alpha\beta}$ are uniformly continuous on Q for $|\alpha| = |\beta| = m$.

Let $\lambda_1 \leq \lambda_2 \leq \ldots$ be the eigenvalues of : $Au = \lambda u$, $\partial^j u / \partial n^j = 0$ on ∂Q, $j = 0, 1, \ldots, m-1$.

We have the following extension of the result of Mawhin-Ward [M-W], where F was allowed to depend only on the derivatives of order $\leq 2m - 1$.

Theorem 2.3 *Let A be as above, (F1)-(F2) hold with M sufficiently small and G satisfy the Caratheodory conditions and*

(G') There are functions $\alpha, \beta \in L_\infty(Q)$ such that for some i,

$$\lambda_i \leq \alpha(x) \leq \liminf_{|\xi_0| \to \infty} G(x, \xi) \leq \limsup_{|\xi_0| \to \infty} G(x, \xi) \leq \beta(x) \leq \lambda_{i+1}$$

uniformly for $x \in Q$ and the non-ξ_0 components of $\xi \in R^{s_{2m-1}}$ and the end inequalities are strict on some sets of positive measure. Then, for $f \in L_2$, there is a weak solution $u \in W_2^m$ of Eq.(2.3).

Proof. Let $\mathcal{L}$ be the set of all linear maps $L : W_2^m \to L_2$ of the form $Lu = pu$ for $p \in L_\infty$ such that $\alpha(x) - \varepsilon \leq p(x) \leq \beta(x) + \varepsilon$ a.e. for some $\varepsilon > 0$ sufficiently small. Then, as in Mawhin-Ward [M-W], codition (1.16) is valid for the variational extension of A and each $L \in \mathcal{L}$ in place of $B(x)$. Reformulating the problem as in [M-W] and continuing as in Theorem 2.2, the conclusion follows from Theorem 1.10. $\square$

Next, consider the following formal differential equation

$$\sum_{|\alpha| \leq m} (-1)^{|\alpha|} D^\alpha A_\alpha(x, u, \ldots, D^m u) + \sum_{|\alpha| < m} (-1)^{|\alpha|} D^\alpha B_\alpha(x, u, \ldots, D^m u) = f \text{ in } Q \quad (2.4)$$

with $u \in V$, $f \in V^*$, where V is a closed subspace of W_p^m such that $\overset{\circ}{W}_p^m \subseteq V$. Suppose

(A1) For each $|\alpha| \leq m$, let $A_\alpha(x, \xi) = A_\alpha^{(0)}(x, \xi) + A_\alpha^{(1)}(x, \xi)$ be such that each $A_\alpha^{(i)} : Q \times R^{s_m} \to R$ satisfies the Caratheodory conditions and

$$|A_\alpha^{(i)}(x, \xi)| \leq c_i(1 + |\xi|)^{p-1}$$

for some constants c_0 and c_1.

(A2) For each $|\alpha| \leq m$, $A_\alpha^{(0)}(x, \xi)$ is odd and $p - 1$–homogeneous in ξ.

(A3) Let $c_1 : R^+ \to R$ be continuos and $r \to 0$ if $c_1(r) \to 0$, $c : R^+ \times R^+ \to R^+$ be such that $c(R, .)$ is weakly uppersemicontinuous at 0 and $c(R, 0) = 0$ for each $R > 0$ and for $w \in V$ and $u, v \in \overline{B}(0, R) \subset V$ we have for some $k \leq m - 1$, $i = 0, 1$

$$\sum_{|\alpha|=m} \int_Q \left[A_\alpha^{(i)}(x, D^\gamma w, D^m u) - A_\alpha^{(i)}(x, D^\gamma w, D^m v) \right] D^\alpha(u - v) dx \geq$$

$$\geq c_1 \left(\|u - v\|_{p,m} \right) - c \left(R, \|u - v\|_{p,k} \right).$$

(B1) For each $|\alpha| \leq m$, $B_\alpha : Q \times R^{s_m} \to R$ satisfies the Caratheodory conditions and for some $c_2 > 0$

$$|B_\alpha(x, \xi)| \leq c_2(1 + |\xi|)^{p_0}.$$

(B2) Let $c : R^+ \times R^+ \to R^+$ be continuous and $c(R, tr)/t \to 0$ as $t \to 0^+$ for each r and R and suppose that for each $R > 0$ and $u, v \in \overline{B}(0, R) \subset V$ the integral inequality in (A3) holds for the B_α's with $c_1 \equiv 0$.

The weak solvability of BVP (2.4) is equivalent to the solvability of the operator equation

$$A_0 u + A_1 u + Nu = f \tag{2.5}$$

where A_0, A_1 and N are the corresponding operators induced by the functions $A_\alpha^{(0)}, A_\alpha^{(1)}$ and B_α respectively in the weak formulation of BVP (2.4).

Using Theorem 1.5, we obtain the following result.

Theorem 2.4 *Let (A1)-(A3) and (B1)-(B2) hold with $c_1(t) = c(t) \equiv 0$ for $i = 1$ in (A3), c_1 and c_2 sufficiently small and $u = 0$ if $A_0 u = 0$. Then BVP (2.4) has a weak solution for each $f \in V^*$.*

2.4 Periodic solutions of semilinear hyperbolic equations. Let $Q \subset R^n$ be a bounded domain with smooth boundary, $\Omega = (0, T) \times Q$ and let $H = L_2(\Omega, R^m)$ with the inner product defined by

$$(u, v) = \int_0^T \int_Q (u(t, x), v(t, x)) dx dt$$

where $(u(t, x), v(t, x))$, $(t, x) \in \Omega$, is the inner product in R^m. Let L_1 be a second order linear selfadjoint elliptic operator in space variables $x \in R^n$ with coefficients independent of t such that the induced bilinear form $a(u, v)$ on the Sobolev space $W_2^1(Q, R^m)$ is continuous and symmetric. Let V be a closed subspace of $W_2^1(Q, R^m)$, containing the test functions, such that $a(u, v)$ is semi-coercive on V, i.e. there are constants $a_1 > 0$ and $a_2 \geq 0$ such that

$$a(u, v) \geq a_1 \|u\|_{2,1}^2 - a_2 \|u\|_{L_2}^2 \text{ for all } u \in V.$$

Define a linear map $L_0 : D(L_0) \subset L_2(Q, R^m) \to L_2(Q, R^m)$ by $(L_0 u, v) = a(u, v)$ for each $v \in V$, where $D(L_0) = \{u \in V \,|\, a(u, \cdot) \text{ is continuous on } V \text{ in the } L_2 - \text{norm}\}$. It is well known that L_0 is selfadjoint and has a compact resolvent, since W_2^1 is compactly embedded in L_2. Next, define a selfadjoint map with compact resolvent $L : D(L) \subset H_1 = L_1(Q, R^m) \to H_1$ by:

$$D(L) = [D(L_0)]^m \text{ and } L = diag(L_0, \ldots, L_0).$$

Since L has a compact resolvent, there is an orthonormal basis $\{\psi_j | j \in J\}$ in H_1 and a sequence of its eigenvalues $\{\mu_j | j \in J\}$ such that $|\mu_j| \to \infty$.

Let $F : R \times Q \times R^m \to R^m$ be a Caratheodory function and consider the system of semilinear wave equations:

$$\begin{cases} u_{tt} - L_1 u - F(t, x, y) = f(t, x) \\ u(t, .) = 0 \text{ for } (t, x) \in [0, T] \times \partial Q \end{cases} \tag{2.6}$$

where $f \in L_2(\Omega, R^m)$ is a T-periodic function in t variable and $\tau = 2\pi/T$ is rational.

By a T-periodic weak solution of the variational boundary value problem (2.6) for the system of semilinear wave equations we mean a solution of the nonlinear operator equation

$$Au - Nu = f, \, u \in D(A), \, f \in H \tag{2.7}$$

where $Au = \sum_{j,k} u_{j,k}(\mu_j - \tau^2 k^2)\psi_j(x)e^{i\tau kt}$, $u_{j,-k} = \bar{u}_{j,k}$, for

$$u \in D(A) = \left\{ u = \sum_{j,k} u_{j,k}\psi_j(x)e^{i\tau kt} \mid \sum_{j,k} |u_{jk}(\mu_j - \tau^2 k^2)|^2 < \infty \right\}$$

and $Nu = F(t,x,u)$ for $u \in L_2(\Omega, R^m)$.

Set $Gu = u_t$ for $u \in W_2^1$, $A_\epsilon = A + \epsilon G$ and suppose that the null-space $N(A_\epsilon) = \{0\}$. Then for a real number $\lambda \notin \sigma(L)$ and $f \in H$ we have

$$(A_\epsilon + \lambda I)^{-1} f = \sum_{j,k} \frac{f_{jk}}{\mu_j - \tau^2 k^2 - \lambda + i\epsilon \tau k} \psi_j(x) e^{i\tau kt},$$

$(A_\epsilon + \lambda I)^{-1} f \in W_2^1(\Omega, R^m)$ and ([B-N])

$$\|(A_\epsilon + \lambda I)^{-1} f\|_{2,1}^2 = \sum_{j,k} |f_{j,k}|^2 \frac{1 + |\mu_j| + \tau^2 k^2}{|\mu_j - \tau^2 k^2 - \lambda + i\epsilon\tau k|^2} \leq C(\lambda, \epsilon)^2 \|f\|^2 \quad (2.8)$$

Let $0 < \alpha < \beta$ be given such that $(\alpha, \beta) \cap \sigma(A) \neq \emptyset$ and consists of a finite number of eigenvalues of A of finite multiplicity. Let V be the finite dimensional subspace of H spanned by the eigenvectors of A corresponding to the eigenvalues in $(\alpha, \beta) \cap \sigma(A)$. Set $W = V^\perp$ in H, $A_1 = -A|D(A) \cap W$ and $\lambda = (\alpha + \beta)/2$. Then $A_1 : D(A) \cap W \to D(A) \cap W$ and $\sigma(A_1) \subset (-\infty, \alpha) \cup (\beta, \infty)$ so that $(A_1 - \lambda I)^{-1} : W \to D(A) \cap W \subset W$ exists and is bounded with $\|(A_1 - \lambda I)^{-1}\| \leq 1/(\beta - \lambda)$.

Regarding F, we assume

(F1) F is a Caratheodory function and there are $h \in L_2(\Omega, R)$, $\alpha_0 > 0$ and $\lambda \notin \sigma(L)$ such that

$$|F(t,x,y) - \lambda y| \leq \frac{1}{\|(A_1 - \lambda I)^{-1}\| + \alpha_0} |y| + h(t,x), \text{ for a.e.}(t,x) \in \Omega, y \in R^m.$$

(F2) $F(t,x,\cdot)$ is positively homogeneous and $(F(t,x,y) - F(t,x,z))(y-z) \geq c|y-z|^2$ for a.e. $(t,x) \in \Omega$, all $y, z \in R^m$ and some $c > 0$.

Theorem 2.5. Let $(F1)-(F2)$ hold with $c = 0$ if $n = 1$, and also $|F(t,x,y) - F(t,x,z)| \leq k|y-z|$ for a.e. $(t,x) \in \Omega$, $y,z \in R^m$ and k sufficiently small when $n > 1$. Let $\sigma(L) = \{\mu_j > 0 | j = 1,2,\ldots\}$. Then, if $u = 0$ is the only solution of $Au - Nu = 0$, Eq.(2.6) has a weak solution in L_2 for all f with $f_t \in H$ if $n > 1$ and for all $f \in H$ if $n = 1$.

Proof. Let $H = L_2(\Omega, R^m)$ and $X = W_2^1$. Suppose that $H_n = $ linspan $\{\psi_j(x)e^{i\tau kt} \mid |j|, |k| \leq n\}$, $P_n : H \to H_n$ is the orthogonal projection for each n and $\Gamma = \{H_n, P_n\}$. Then $A + N : D(A) \subset H \to H$ is pseudo A-proper w.r.t. Γ ([Mi-6]). Since $(A_\epsilon + \lambda I)^{-1} : X \to H$ is continuous by (2.8), the map $A - tN + \epsilon G + (1-t)\epsilon I : D(A) \cap X \subset X \to H$ is A-proper w.r.t. Γ by Proposition 1.2 for each $\epsilon > 0$ small and $t \in [0,1]$. By $(F1)$, we see that $(1.9)-(1.10)$ for $A_1 - \lambda I|W$ hold. Also, for $u \in D(A) \cap X$,

$$(Au, Gu) = \left(\sum_{j,k} u_{j,k}(\mu_j - \tau^2 k^2)\psi_j e^{i\tau kt}, \sum_{j,k} i k\tau u_{j,k}\psi_j e^{i\tau kt} \right)$$

$$= -\sum_{j,k} i|u_{j,k}|^2 \tau k(\mu_j - \tau^2 k^2) = 0$$

since $u_{j,-k} = \bar{u}_{j,k}$, and for each j fixed and $n \geq 1$

$$\sum_{|k|\leq n} k|u_{j,k}|^2 (\mu_j - \tau^2 k^2) = 0.$$

Integrating by parts and using the boundary conditions, we get that $(Nu, Gu) = 0$ for all $u \in D(A) \cap X$ and $(Gu, u) = 0$ for all $u \in D(A) \cap X$. For each $u \in X$ and $Gu = u_t$, we have

$$(Au, u) = \sum_{j,k} \lambda_j |u_{j,k}|^2 - \sum_{j,k} \tau^2 k^2 |u_{j,k}|^2 = (Lu, u) - \|Gu\|^2.$$

Since $(Lu, u) \geq \mu_1 \|u\|^2$ for each u, it follows that $(Au, u) \geq 0$ whenever $Gu = 0$. Hence, the theorem follows from Theorem 1.8. □

When $m = n = 1$ and A is the wave operator, Theorem 2.5 includes a result by Lazer-McKenna [L-McK] that was proved using very different arguments based heavily on a Fredholm Alternative for the system of ordinary differential equations.

The theory applies to more general systems of the form

$$\begin{cases} u_{tt} - L_1 u + F(t, x, u) = f(t, x), \ (t, x) \in (0, T) \times Q \\ u^{2\alpha_i}(t, \cdot) = 0 \ for \ (t, x) \in [0, T] \times \partial Q, \alpha_i = k \ for \ 0 \leq k \leq m-1, \ 1 \leq i \leq n \end{cases}$$

where $\alpha = (\alpha_1, ..., \alpha_n)$ and L_1 is an elliptic operator (see [Mi-8] for details).

2.5 Hamiltonian systems. Consider the existence of 2π-periodic solutions of asymptotically quadratic Hamiltonian systems

$$-J\dot{u} = H'(u, t) + f(t), \qquad (2.9)$$

where $J = \begin{pmatrix} 0 & -I \\ I & 0 \end{pmatrix}$, $I = I_n$ is the identity $n \times n$ matrix. Assume that the Hamiltonian $H \in C^1(R^{2n} \times R, R)$ and is 2π-periodic in t. Assume also that there is a constant symmetric matrix h such that

$$|H'(s, t) - hs| \leq a|s| + b \ for \ |s| \geq r, \qquad (2.10)$$

where H' denotes the gradient of H with respect to the first $2n$ variables. We shall work in the Hilbert space $H = W^{1/2}(S^1, R^{2n})$ defined as follows. Every function $u \in L_2(S^1, R^{2n})$ has a Fourier expansion

$$u(t) = a_0 + \sum_{m=1}^{\infty} a_m \cos mt + b_m \sin mt, \qquad (2.11)$$

where a_m, b_m are $2n$-vectors. Then, H is the set of such functions with

$$\|u\| = \left(|a_0|^2 + \sum_{m=1}^{\infty} m\left(|a_m|^2 + |b_m|^2\right)\right)^{1/2} < \infty.$$

H is a Hilbert space with this norm. For a function $u \in C^\infty(S^1, R^{2n})$, we define a linear functional $l : H \to R$ by

$$l(v) = \int_0^{2\pi} (-J\dot{u}, v) dt \ for \ v \in H,$$

where $(,)$ is the inner product in R^{2n}. Since $|l(v)| \leq C\|u\|\|v\|$ for each $v \in H$, by the Riesz representation theorem there is a linear bounded selfadjoint map $A : H \to H$ such that $(Au, v) = l(v)$ for each $v \in H$. Similarly, the symmetric $2n \times 2n$ matrix h defines a compact selfadjoint map $B_\infty = B_\infty(h)$ on H by

$$(N_\infty u, v) = -\int_0^{2\pi} (hu, v) dt \ \ for\ all\ \ u, v \in H.$$

The explicit expressions for these maps are

$$Au = \sum_{m=1}^{\infty} -Jb_m cosmt + Ja_m sinmt,$$

$$N_\infty u = -ha_0 - \sum_{m=1}^{\infty} m^{-1}(ha_m cosmt + hb_m sinmt),$$

where u is given by (2.11). Define a nonlinear map $N : H \to H$ by

$$(Nu, v) = \int_0^{2\pi} \bigl(H'(u,t), v\bigr) dt \ \ for\ all\ \ u, v \in H.$$

Hence, the existence of 2π-periodic solutions of (2.9) is reduced to the solvability of the operator equation in H

$$Au + Nu = f. \tag{2.12}$$

The set $\{u_{mj}(t) = a_{mj} cosmt + b_{mj} sinmt | j = 1, ..., 2n, m = 0; j = 1, ..., 4n, ; m = 1, 2, ...\}$ is a complete orthonormal basis in H. For each $m \geq 0$, define a finite dimensional subspace E_m of H by $E_m = \{a cosmt + b sinmt\ |a, b \in R^{2n}\}$. Set $H_n = E_0 + E_1 + ... + E_n$. Then the orthogonal projection P_n of H onto H_n is given by

$$P_n u = a_0 + \sum_{m=1}^{n} a_m cosmt + b_m sinmt.$$

Moreover, $P_n u \to u$ as $n \to \infty$ for each $u \in H$ and each P_n commutes with A and N_∞ since each E_m is invariant under A and N_∞ because

$$(A + N_\infty)(a cosmt + b sinmt) = (-ha/m - Jb) cosmt + (Ja - hb/m) sinmt.$$

Theorem 2.6 *Let $H : R^{2n} \times R \to R$ be a C^1-function, 2π-periodic in t and satisfy (2.10) with a sufficiently small. Then Eq.(2.12) has a solution $u \in H$ for each $f \in L_2$.*

Proof. Since N is compact and the null space of A is finite dimensional, it is easy to see that $A + N : H \to H$ is A-proper w.r.t. $\Gamma = \{H_n, P_n\}$. Moreover, $A + N_\infty$ is invertible and selfadjoint. By (2.10),

$$\|Nu - N_\infty u\| \leq C\|H'(u,t) - hu\|_{L_2} \leq C(a\|u\| + b).$$

Since a is sufficiently small, we have that $Ca\|(A - N_\infty)^{-1}\| < 1$. Hence, the conclusion follows from Theorem 1.6. □

Asymptotically linear Hamiltonian systems have been studied extensively (cf. [M-Wi]).

References

[A] H. Amann, On the unique solvability of semilinear operator equations in Hilbert spaces, J. Math. Pures Appl. 61 (1982), 149-175.

[B-N] H.Brezis, L. Nirenberg, Characterizations of the ranges of some nonlinear operators and applications to boundary value problems, Ann. Scuola Norm. Sup. Pisa, 5(1978), 225-326.

[D] E.N. Dancer, Order intervals of self-adjoint linear operators and nonlinear homeomorphisms, Pacific J. Math. 115 (1984), 57-72.

[F-M-P] P.M. Fitzpatrick, I.Massabo, J.Pejsachowicz, On the covering dimension of the set of solutions of some nonlinear equations, Trans. AMS, 296 (1986), 777-798.

[F-M] A. Fonda, J. Mawhin, Iterative and variational methods for the solvability of some semilinear equations in Hilbert spaces, J. Diff. Eq. (to appear); also this proceedings.

[G-K-N] C.P. Gupta, Y.C. Kwong, J. Necas, Nonresonance conditions for the strong solvability of a general elliptic partial differential operator, Nonlinear Analysis, TMA 17 (1991), 613–625

[H] P.Hess, On the Fredholm alternative for nonlinear functional equations in Banach spaces, Proc. Amer. Math. Soc. 33 (1972), 55-61.

[K] R.I.Kachurovskii, On the Fredholm theory for nonlinear operator equations, Dokl. Akad. Nauk SSSR, 192 (1970), 969-972.

[K-Z] M.A. Krasnoselskii, P.P. Zabreiko, Geometrical Methods in Nonlinear Analysis, Nauka, Moskva, 1975.

[L] A. Lasota, Une généralisation du premier théorème de Fredholm et ses applications à la théorie des èquations differentielles ordinaires, Ann. Polon. Math. 18 (1966), 65-77.

[L-O] A. Lasota, Z. Opial, On the existence and uniqueness of solutions of nonlinear functional equations, Bull. Acad. Pol. Sci. 15 (1967), 97-107.

[L-McK] A.C. Lazer, P.J. McKenna, Fredholm theory for periodic solutions of some semilinear P.D.E.s with homogeneous nonlinearities, Contemporary Math., vol. 107, 1990, 109-122.

[M-W] J. Mawhin, J.R. Ward, Jr, Nonresonance and existance for nonlinear elliptic boundary value problems, Nonlinear Anal., TMA, 6(1981), 677–684.

[M-Wi] J. Mawhin, M. Willem, Critical Point Theory and Hamiltonian systems, Springer, N.Y. 1989.

[Mi-1] P.S. Milojević, Some generalizations of the first Fredholm theorem to multivalued condensing and A-proper mappings, Boll. Unione Mat. Ital. 13-B (1976), 619-633.

[Mi-2] P.S. Milojević, Some generalizations of the first Fredholm theorem to multivalued A-proper mappings with applications to nonlinear elliptic equations, J. Math. Anal. Appl. 65 (1978), 468-502.

[Mi-3] P.S. Milojević, Fredholm alternatives and surjectivity results for multivalued A-proper and condensing mappings with applications to nonlinear integral and differential equations, Czechoslovak Math. J., 30(105) (1980), 387–417.

[Mi-4] P.S. Milojević, Continuation theory for A-proper and strongly A- closed mappings and their uniform limits and nonlinear perturbations of Fredholm mappings, Proc. Int. Seminar in

Funct. Anal., Holomorphy and Approx. Theory, Rio de Janeiro, August 1980, Math. Studies, v. 71 (A. Barroso, ed.) North-Holland, 1982, 299-372.

[Mi-5] P.S. Milojević, Fredholm theory and semilinear equations without resonance involving noncompact perturbations, I, II, Applications, Publications de l'Institut Math. 42 (1987), 71-82 and 83-95.

[Mi-6] P.S. Milojević, Solvability of semilinear operator equations and applications to semilinear hyperbolic equations, in Nonlinear Functional Analysis, P.S. Milojević ed., Marcel Dekker, NY, 1989, 95-178.

[Mi-7] P.S. Milojević, Implicit function theorems, approximate solvability of nonlinear equations and error estimates, to appear.

[Mi-8] P.S. Milojević, Fredholm theory for semilinear equations and applications, to appear.

[N-1] J. Nečas, Sur l'alternative de Fredholm pour les operateurs non lineaires avec applications aux problèmes aux limites, Ann. Scuola Norm. Sup. Pisa 23 (1969), 331-345.

[N-2] J. Nečas, Fredholm alternative for nonlinear operators and applications to partial differential equations and integral equations, Cas. Pest. Mat. 9 (1972), 15-21.

[P] A.I. Perov, On the principle of the fixed point with two-sided estimates, Dokl. Akad. Nauk SSSR 124 (1959), 756-759.

[Pe] W.V. Petryshyn, Fredholm alternative for nonlinear A-proper mappings with applications to nonlinear elliptic boundary value problems, J. Funct. Anal. 18 (1975), 288-317.

[S] M.W. Smiley, Eigenfunction methods and nonlinear hyperbolic boundary value problems at resonance, J. Math. Anal. Appl. 122 (1987), 129–151.

[T] S.A. Tersian, A minimax theorem and applications to nonresonance problems for semilinear equations, Nonl. Anal. TMA, 10(1986), 651-668.

Department of Mathematics & CAMS
New Jersey Institute of Technology,
Newark, NJ, USA

Activation waves and threshold phenomena in platelet aggregation

1. Introduction

Platelet aggregation is a major component of the blood's clotting response, and involves the clumping together of blood platelets along portions of a blood vessel wall in response to injury to the vessel. Overly aggressive platelet aggregation, particularly in the coronary arteries or associated with the use of blood-contacting prosthetic devices, causes severe medical problems. (For references, see [1].) To participate in the aggregation process, a platelet must be primed or 'activated' by exposure to certain chemicals which have been released into the blood plasma by other, already activated, platelets. The equations (1) which we study below concern the interplatelet chemical signaling that triggers activation. These equations are one-dimensional versions of the transport equations for the concentration φ_n of non-activated platelets and the concentration c of the activating chemical ADP. The larger, multidimensional system of equations from which these equations come is derived in [1].

2. Critical points and scaling

The critical points of the system:

$$\frac{\partial \varphi_n}{\partial t} = D_n \frac{\partial^2 \varphi_n}{\partial x^2} - R(c)\varphi_n$$

$$\frac{\partial c}{\partial t} = D_c \frac{\partial^2 c}{\partial x^2} + AR(c)\varphi_n \tag{1}$$

are solutions of the equation $H(c - c_T)\varphi_n = 0$. Below, we construct a travelling wave solution which connects $(\varphi_{-\infty}, c_{-\infty})$ and $(0, c_{+\infty})$, with $\varphi_{-\infty} > 0$ and $c_{-\infty} < c_T < c_{+\infty}$, under the condition $A\varphi_{-\infty} = c_{+\infty} - c_{-\infty}$. That this is a necessary condition for existence of a travelling wave follows upon integrating the ordinary differential equations satisfied by such waves.

We set $\Phi = \varphi_n/\varphi_{-\infty}$ and $C = (c - c_{-\infty})/(c_{+\infty} - c_{-\infty})$ so that the travelling wave (in Φ and C) connects the states $(1, 0)$ and $(0, 1)$. After an appropriate scaling of the independent variables, (1) becomes:

$$\frac{\partial \Phi}{\partial t} = \epsilon \frac{\partial^2 \Phi}{\partial x^2} - H(C - a)\Phi$$

$$\frac{\partial C}{\partial t} = \frac{\partial^2 C}{\partial x^2} + H(C - a)\Phi \tag{2}$$

where $\epsilon = D_n/D_c$ and $a = (c_T - c_{-\infty})/(c_{+\infty} - c_{-\infty})$.

3. Existence of travelling waves

We construct a travelling wave solution $(\Phi(\xi), C(\xi))$ to the system (2). We require that $(\Phi(\xi)$ and $C(\xi))$ have limits as $\xi \equiv x + vt \to \pm\infty$. It then follows that the wave *must* be a front connecting $(1,0)$ with $(0,1)$. The equations for the front are piecewise linear and can be solved explicitly. The condition $A\varphi_{-\infty} = c_{+\infty} - c_{-\infty}$ is seen to be sufficient for the existence of a travelling front. The front is uniquely determined by ϵ and a; it travels with speed $v = a'/(\epsilon + a')^{1/2}$ where $a' = a/(1-a)$, and it is given by:

$$\Phi_f(\xi) = \begin{cases} 1 - (1 - \frac{a'}{a'+\epsilon})e^{\frac{v\xi}{\epsilon}} & \xi \leq 0 \\ \frac{a'}{a'+\epsilon}e^{r_-\xi} & \xi \geq 0 \end{cases} \qquad C_f(\xi) = \begin{cases} ae^{v\xi} & \xi \leq 0 \\ 1 - (1-a)e^{r_-\xi} & \xi \geq 0 \end{cases}$$

where $r_- = -1/(a'+\epsilon)^{1/2}$. From the expression for the front one observes that each term in (2) is of the same order with respect to ϵ.

4. Linearized Stability

We linearize (2) about the travelling front (Φ_f, C_f). Differentiation of the nonlinear term yields δ-functions that we interpret as jump conditions for the second derivatives. Let (Φ_1, C_1) be an infinitesimal perturbation of the front. The solution of the following eigenvalue problem determines linear stability: Find $\lambda \in \mathbf{C}$ and $\Phi_1, C_1 \in L^\infty(\mathbf{R}) \cap C^1(\mathbf{R})$ such that:

$$(L) = \begin{cases} \epsilon\Phi_1'' - v\Phi_1' - \lambda\Phi_1 &= H(C_f(\xi) - a)\Phi_1 \\ C_1'' - vC_1' - \lambda C_1 &= -H(C_f(\xi) - a)\Phi_1 \\ \lim_{\xi \to \pm\infty}(\Phi_1(\xi), C_1(\xi)) &= (0,0) \\ \Phi_1'(0^+) - \Phi_1'(0^-) &= \frac{1}{\epsilon}\Phi_f(0)C_1(0)/C_f'(0) \\ C_1'(0^+) - C_1'(0^-) &= -\Phi_f(0)C_1(0)/C_f'(0) \end{cases}$$

has a nontrivial solution (Φ_1, C_1). This problem can be explicitly solved due to the piecewise linear form of the nonlinearity. We have derived an algebraic equation that gives the (possible) unstable modes. We find that, for $\epsilon < 1$, there is a critical value $a^*(\epsilon)$ such that there are two unstable modes for $a > a^*(\epsilon)$ and zero otherwise. For $\epsilon > 1$ there is always a real unstable mode. Finally, at $\epsilon = 1$ there are no unstable modes.

5. Existence of classical solutions

We consider the existence of solutions to the initial value problem (2) with initial conditions which resemble the front. In particular, we consider initial data C_0 that crosses the threshold level only once. We have shown that this property is maintained for a short time, so that there exists a *free boundary* $x = m(t)$ such that $C(x,t) < a \leftrightarrow x < m(t)$. The system (2) can therefore be converted into a system of integral equations:

$$\Phi(x,t) = \int_{-\infty}^{+\infty} Q(x-y,\epsilon t)\Phi_0(y)dy - \int_0^t \int_{m(s)}^{+\infty} Q(x-y,\epsilon(t-s))\Phi(y,s)dyds$$

$$C(x,t) = \int_{-\infty}^{+\infty} Q(x-y,t)C_0(y)dy + \int_0^t \int_{m(s)}^{+\infty} Q(x-y,t-s)\Phi(y,s)dyds. \quad (3)$$

These equations are solved, locally in time, by an iteration procedure.

6. Bounds on Φ and C.

The classical solutions just described are defined on $\mathbf{R} \times (0,T)$. From the integral equations (3) it is easy to show

$$0 \leq \Phi(x,t) \leq 1, \quad \text{and,} \quad 0 \leq C(x,t) \leq 1+T. \quad (4)$$

Therefore, the solutions do not blow up in finite time, and so classical solutions exist for all time. We have been unable to establish an upper bound for $C(x,t)$ that is independent of T. In [2], we provided substantial evidence for the following conjectured bound:

$$C(x,t) \leq \|C_0\|_\infty + \|\Phi_0\|_\infty. \quad (5)$$

This bound would follow from a bound on the exponential decay for Φ in the region $\{(x,t) : x > m(t)\}$. The latter is reasonable because Φ satisfies $\Phi_t = \epsilon \Phi_{xx} - \Phi$ there.

7. Numerical results

We computed approximate solutions to the initial value problem (2). For a calculation with a moderate threshold level $a = 0.6$ and with $\epsilon = .01$, the solution of the initial value problem evolved into the front solution. This is shown in Fig.1. We see, in this figure, numerical support for the conjectured bound on C. In Fig.2A, we show the location of the threshold $m(t)$ for this computation. Note that this curve is essentially linear. A least-squares fit to this data for the time interval [5,5C] gives a speed of -0.8101 compared with the speed of the front -0.8104 computed from the formula given in Section 3. In Fig.2B, we show the location of the threshold for a calculation with a high threshold level $a = 0.9$. In this case, the solution of the initial value problem did not approach the front. The value of Φ in the region in which $C > a$ became negligible, and the location of the threshold moved to the right because of linear diffusion.

8. A new threshold problem

We consider the initial value problem (2) for initial data such that C_0 crosses threshold twice and Φ_0 has limits at $\pm\infty$. In [3], we show global existence of a classical solution. We consider, in particular, initial data along a ray,

$$\Phi(x,0) = \alpha\Phi_0(x) \quad \text{and} \quad C(x,0) = C_0(x). \tag{6}$$

Here, we fix functions C_0, Φ_0 of the type just described and take $\alpha \in \mathbf{R}^+$ to be the scaling factor along the ray. The behavior of solutions for α small is easy to predict: the concentration $C(x,t)$ converges to 0. In this case we say that the initial data *collapses*. For α large, we have numerical evidence that, for each fixed x, $C(x,t)$ converges to a value above the threshold level a. This situation is called *expansion*. We have conjectured a weak comparison principle for the free boundaries which allows us to prove the existence of the numbers:

$$\alpha_*[\Phi_0, C_0] = \sup\{\alpha \in \mathbf{R}^+ : (\alpha\Phi_0, C_0) \text{ collapses}\} \tag{7}$$
$$\text{and} \quad \alpha^*[\Phi_0, C_0] = \inf\{\alpha \in \mathbf{R}^+ : (\alpha\Phi_0, C_0) \text{ expands}\}. \tag{8}$$

We call the remaining interval $[\alpha_*, \alpha^*]$ the gap. We have obtained strong numerical evidence to support the:

Conjecture: For each fixed pair of functions (Φ_0, C_0), we have

$$\alpha_*[\Phi_0, C_0] = \alpha^*[\Phi_0, C_0]. \tag{9}$$

Equivalently, there is no gap.

9. Conclusions

We study a system of reaction diffusion equations which describes the chemically - mediated spread of platelet activation during the platelet aggregation process. We have explicitly constructed travelling front solutions to this sytem and proven their linear stability. First we consider the class of initial data which cross threshold exactly once abd have the same limiting behavior as the front. We have presented numerical results which illustrate that convergence to the fronts occurs when the threshold level is low. We also have computational examples which show that convergence to the front does not occur for very high levels of the threshold.

We next consider a class of initial data in which the concentration of the chemical vanishes at infinity. We have extensive numerical evidence for the existence of a critical scaling that separates data that collapse below the threshold level a from those that expand above a.

The detailed description of these two problems will be the subject of future work.

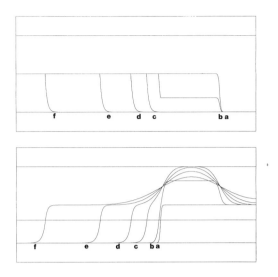

Figure 1: Evolution of Φ (top) and C (bottom).

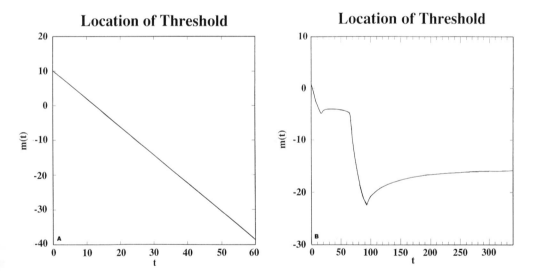

Figure 2: Location $m(t)$ at which C crosses the threshold level a: (A) $a = 0.6$. (B) $a = 0.9$.

References

[1] **A. Fogelson:** *Continuum models of platelet aggregation: formulation and mechanical properties*, to appear in SIAM Journal of Applied Math (1991).

[2] **V. Moll, A. Fogelson:** *Activation waves in a model of platelet aggregation: existence of solutions and stability of travelling fronts*, submitted to Journal of Mathematical Biology (1991).

[3] **V. Moll:** *Threshold Phenomena in a model of platelet aggregation*, in preparation (1991).

Victor H. Moll   Aaron L. Fogelson
Tulane University   University of Utah

J BERKOVITS AND V MUSTONEN
On the asymptotic bifurcation for a class of wave equations with sublinear nonlinearity

1. Introduction

In this note we shall deal with the existence of solutions (u, λ) of wave equations of the form

$$\begin{cases} u_{tt} - u_{xx} - \lambda u - g(\lambda, x, t, u) = 0 & \text{in } (0, \pi) \times \mathbb{R} \\ u(0, t) = u(\pi, t) = 0 & \text{for all } t \in \mathbb{R} \\ u(x, \cdot) \text{ is } 2\pi\text{-periodic} \end{cases} \quad (1)$$

where $\lambda \in \mathbb{R}_+$ and the nonlinear part g is sublinear with respect to u. In particular, we are interested in solutions where u is "large" which leads us to the study of asymptotic bifurcation points of (1). According to a general principle a necessary condition for λ_0 to be an asymptotic bifurcation point of (1) is that λ_0 belongs to the spectrum of the wave operator. This condition is not sufficient. The purpose of this note is to give some sufficient conditions. Using the abstract results of [Be2] we show first that bifurcation from infinity occurs at any eigenvalue λ_0 of odd multiplicity $m(\lambda_0)$. Secondly, under some further conditions for g we are able to show that bifurcation takes place also at eigenvalues of even multiplicity. Finally we indicate that two different branches of solutions may emanate from one eigenvalue.

2. Prerequisites

We start with giving the precise setting of problem (1) in $H = L_2(\Omega)$, $\Omega = (0, \pi) \times (0, 2\pi)$. We shall assume that the function $(\lambda, x, t, s) \to g(\lambda, x, t, s)$ from $\mathbb{R}_+ \times [0, \pi] \times \mathbb{R} \times \mathbb{R}$ to R is continuous in (λ, s) for allmost all (x, t) and measurable in (x, t) for all λ and s, 2π-periodic in t and that for each bounded interval $[a, b] \subset \mathbb{R}_+$ there exist constants $\beta > 0$ and $1 \leq \alpha < 2$ and a function $h_0 \in L_2(\Omega)$ such that

$$|g(\lambda, x, t, s)| \leq \beta |s|^{\frac{\alpha}{2}} + h_0(x, t) \quad \text{for a.a. } (x, t) \in \Omega \text{ and } \lambda \in [a, b]. \quad (2)$$

Then the Nemytskii operator $N(\lambda, u) = g(\lambda, \cdot, \cdot, u)$ from H to H is continuous and sublinear in the sense that

$$\frac{\|N(\lambda, u)\|}{\|u\|} \to 0 \quad \text{as } \|u\| \to \infty \text{ uniformly on } [a, b] \quad (3)$$

where $\|\cdot\|$ stands for the usual norm in H. Moreover we assume that for each bounded interval $[a, b] \subset \mathbb{R}_+$ there exists a constant $\mu > 0$ such that $\lambda s + g(\lambda, x, t, s)$ is strongly μ-monotone in s, i.e.,

$$\{g(\lambda, x, t, s_1) - g(\lambda, x, t, s_2)\}(s_1 - s_2) \geq (\mu - \lambda)|s_1 - s_2|^2 \quad \text{for all } s_1, s_2 \in \mathbb{R}, \lambda \in [a, b] \quad (4)$$
$$\text{and a.a. } (x, t) \in \Omega.$$

Denoting
$$\phi_{jk}(x,t) = \begin{cases} \frac{\sqrt{2}}{\pi}\sin(jx)\sin(kt), & j,k \in \mathbb{Z}_+ \\ \frac{1}{\pi}\sin(jx), & j \in \mathbb{Z}_+, k = 0 \\ \frac{\sqrt{2}}{\pi}\sin(jx)\cos(kt), & j, -k \in \mathbb{Z}_+ \end{cases}$$

the set $\{\phi_{jk} : j \in \mathbb{Z}_+, k \in \mathbb{Z}\}$ forms an orthonormal basis in H and the wave operator $\frac{\partial^2}{\partial t^2} - \frac{\partial^2}{\partial x^2}$ admits in H the abstract realization $L\colon D(L) \subset H \to H$ defined by

$$Lu = \sum_{j,k}(j^2 - k^2)(u, \phi_{jk})\phi_{jk} \qquad (5)$$

with the domain $D(L) = \{u \in H : Lu \in H\}$, $(\cdot,\cdot)$ standing for the usual inner product in H. A pair $(\lambda, u) \in \mathbb{R}_+ \times H$ is said to be a *weak solution* of the problem (1) if

$$Lu - \lambda u - N(\lambda, u) = 0, \qquad u \in D(L). \qquad (6)$$

It is well-known fact that L is densely defined closed self adjoint linear operator with $\operatorname{Im} L = (\operatorname{Ker} L)^{\perp}$. The inverse of $L_0 = L|_{\operatorname{Im} L \cap D(L)}$ is compact and L has a pure point spectrum of eigenvalues

$$\sigma(L) = \{\lambda_{jk} = j^2 - k^2 : j \in \mathbb{Z}_+, k \in \mathbb{Z}\} = \mathbb{Z} \setminus \{-1, -4, 4m+2, m \in \mathbb{Z}\}$$

with corresponding eigenvectors ϕ_{jk}. Each non-zero eigenvalue has finite multiplicity, but $\operatorname{Ker} L$ is infinite dimensional. Only the eigenvalues $\lambda = j^2$ have odd multiplicity $m(\lambda)$. For more details about the weak formulation of the wave equation we refer to [BM1] and the references mentioned therein. Finally we recall that λ_0 is called an asymptotic bifurcation point of (6) if there exists a sequence $\{(\lambda_n, u_n)\} \subset \mathbb{R}_+ \times D(L)$ of solutions such that $\lambda_n \to \lambda_0$ and $\|u_n\| \to \infty$.

3. Results on asymptotic bifurcation

In order to find asymptotic bifurcation points for the problem (6) we introduce first the abstract result of [Be2] which we employ in suitable subspaces of H. Indeed, let V be a closed subspace of H and let P_V and $P_{V^{\perp}}$ stand for the projections to V and $V^{\perp}$, respectively. We assume that L is *completely reduced* by V, i.e., $P_V(D(L)) \subset D(L)$ and $P_V L = LP_V$ on $D(L)$. Then the restrictions L_V and $L_{V^{\perp}}$ inherit the properties of L (cf. [TL] p. 288) and for the spectra we have $\sigma(L) = \sigma(L_V) \cup \sigma(L_{V^{\perp}})$. Of course we have also to ensure that $N(V) \subset V$ which makes it possible to define N_V as the restriction of N to V. Then we consider the reduced problem

$$L_V u - \lambda u - N_V(\lambda, u) = 0, \qquad u \in V \cap D(L). \qquad (7)$$

Obviously any solution of (7) solves also the original problem (6). From [Be2] we adopt

Theorem 1. *Assume that the function g satisfies the conditions (2) and (4) and that V is a nontrivial closed subspace of H such that L is completely reduced by V and $N(V) \subset V$. Then if λ_0 is a positive eigenvalue of L_V with odd multiplicity, the problem (7) admits a sequence of solutions $\{(\lambda_n, u_n)\} \subset \mathbb{R}_+ \times D(L_V)$ such that $\lambda_n \to \lambda_0$ and $\|u_n\| \to \infty$.*

The proof of Theorem 1 is based on the classical 'change of degree argument' for the topological degree constructed in [BM1] and [Be1].

Since only the eigenvalues $\lambda_0 = j^2$ ($j \in \mathbb{Z}_+$) of L have odd multiplicity we obtain by choosing $V = H$ above the following

Corollary 1. *Assume that g satisfies (2) and (4). If $\lambda_0 = j^2$ for some $j \in \mathbb{Z}_+$, λ_0 is an asymptotic bifurcation point of (6).*

In order to deal with eigenvalues $\lambda_0 \in \sigma(L)$ of even multiplicity we use reductions which enables us to manipulate the multiplicity as an eigenvalue of the reduced operator. Main difficulty encountered here is to find subspaces which are invariant under N. To overcome this we impose on g some further symmetry conditions. We start with

Corollary 2. *Assume that g satisfies (2) and (4) and that $g = g(\lambda, x, t, s)$ is even in t. If λ_0 is a positive eigenvalue of L with multiplicity $m(\lambda_0) = 4n - 2$ for some $n \in \mathbb{Z}_+$, then λ_0 is an asymptotic bifurcation point of (6).*

Proof. Define $V = \{u \in H : u(x, 2\pi - t) = u(x, t) \text{ a.e. in } \Omega\}$. It is easy to see that $V = \overline{\mathrm{sp}}\{\phi_{jk} : j \in \mathbb{Z}_+, -m \in \mathbb{N}\}$. Clearly L is completely reduced by V and $N(V) \subset V$. Now $\lambda_0 \in \sigma(L_V)$ and its reduced multiplicity $m_V(\lambda_0) = 2n - 1$. Hence we can apply Theorem 1 to obtain the desired result.

Remark. Note that for each $n \geq 2$ there are infinitely many positive eigenvalues with multiplicity n.

Corollary 3. *Assume that g satisfies (2) and (4) and that $g = g(\lambda, x, s)$. If λ_0 is a positive eigenvalue of L with multiplicity $m(\lambda_0) = 2n$ for some $n \in \mathbb{Z}_+$, then λ_0 is an asymptotic bifurcation point of (6).*

Proof. We can write $\lambda_0 = j_m^2 - (\pm k_m)^2$, $j_m, k_m \in \mathbb{Z}_+$, $m = 1, 2, \ldots, n$, with $j_1 < j_2 < \cdots < j_n$ and $k_1 < k_2 < \cdots < k_n$. Define

$$V = \left\{ u \in H : u(x, 2\pi - t) = u(x, t) \text{ a.e. in } \Omega \text{ and } \right.$$
$$\left. u\left(x, t + \frac{2\pi}{k_n}\right) = u(x, t) \text{ for a.a. } x \in (0, \pi), t \in \left(0, 2\pi - \frac{2\pi}{k_n}\right) \right\}.$$

Then $V = \overline{\mathrm{sp}} \left\{ \phi_{jk} : j \in \mathbb{Z}_+, -\frac{k}{k_n} \in \mathbb{N} \right\}$, and the conditions of Theorem 1 are met.

We close this paper by an example where two different branches bifurcate from infinity at an eigenvalue.

Example. Assume that $g = g(\lambda, x, s)$ satisfies (2) and (4) and that g is odd in s. We take $\lambda_0 = 32$. Clearly $32 = 9^2 - (\pm 7)^2 = 6^2 - (\pm 2)^2$ and hence $m(32) = 4$. Define for $q = 2$ and $q = 7$ the subspaces

$$W_q = \Big\{ u \in H : u(x, 2\pi - t) = u(x, t) \text{ a.e in } \Omega,$$
$$u(x, t + \frac{\pi}{q}) = -u(x, t) \text{ for a.a. } x \in (0, \pi),\, t \in (0, 2\pi - \frac{\pi}{q}) \Big\}$$
$$= \overline{\text{sp}} \Big\{ \phi_{jk} : j \in \mathbb{Z}_+,\, -\frac{k}{q} \in \mathbb{Z}_+,\, \frac{k}{q} \text{ odd} \Big\}.$$

Applying Theorem 1 for $V = W_2$ and $V = W_7$ respectively, we conclude that $\lambda_0 = 32$ is an asymptotic bifurcation point for (7). Since W_2 and W_7 are orthogonal subspaces, two different branches of solutions of (6) emanate from infinity at $\lambda_0 = 32$. Clearly these braches are not caused by a shift $t \to t + \tau$ for any $\tau \in (0, 2\pi)$.

Remark. The method of reductions described above has been used for a long time in the connection of periodic solutions for ordinary differential equations. For the use of reductions in the case of wave equations we refer to [Ve], [Co] and [BM2].

References

[Be1] Berkovits, J., *Topological degree and multiplication theorem for a class of nonlinear mappings*, Bull. London Math. Soc. (to appear).

[Be2] Berkovits, J., *Some bifurcation results for a class of semilinear equations via topological degree method*, Preprint August 1991, Mathematics, Univ. of Oulu, 10 p.

[BM1] Berkovits, J., Mustonen, V., *An extension of Leray-Schauder degree and applications to nonlinear wave equations*, Differential and Integral Equations **3** (1990), 945–963.

[BM2] Berkovits, J., Mustonen, V., *On the existence of multiple solutions for semilinear equations with monotone nonlinearities crossing a finite number of eigenvalues*, Nonlinear Analysis **17** (1991), 399–412.

[Co] Coron, J.M., *Periodic solutions of a nonlinear wave equation without assumption of monotonicity*, Math. Annln **262** (1983), 273–285.

[TL] Taylor, A.E., Lay, D.C., *Introduction to Functional Analysis*, John Wiley & Sons, New York, 1980.

[Ve] Vejvoda, O., *Partial Differential Equations*, Noordhoff, Sijhoff, 1981.

University of Oulu
Department of Mathematics
SF-90570 Oulu
Finland

O A OLEINIK
Some mathematical problems of elasticity and Korn's inequalities

This lecture is dedicated to mathematical problems of linear elasticity. We consider boundary-value problems for the linear stationary system of elasticity in bounded and unbounded domains. It is well known that Korn's inequalities are important to study boundary-value problems for the system of elasticity. Using these inequalities, one can prove existence theorems, uniqueness and stability of solutions of boundary-value problems for the system of elasticity in bounded and unbounded domains, estimate approximate solutions and so on.

In this lecture

1) Korn's inequality with asymptotically sharp constants is proved;

2) Korn's inequality with a sharp constant is proved for thin domains (shells).

3) Hardy's and Korn's type inequalities are formulated for unbounded domains and stability theorems are given for solution of boundary-value problems in unbounded domains for the linear stationary system of elasticity.

4) The Boussinesq boundary-value problem is considered.

The linear stationary elasticity system has the form

$$\sum_{i,h,j=1}^{n} \frac{\partial}{\partial x_i}\left(a_{ij}^{kh}(x) \frac{\partial u_h}{\partial x_j}\right) = f_k, \quad k = 1, \ldots, n, \tag{1}$$

$$u = (u_1, \ldots, u_n), \quad x = (x_1, \ldots, x_n), \quad f = (f_1, \ldots, f_n),$$

$$a_{ij}^{kh}(x) \equiv a_{ji}^{hk}(x) \equiv a_{kj}^{ih}(x), \quad \forall x \in \Omega, \tag{2}$$

$$\lambda_1 |\eta|^2 \leq \sum_{i,k,j,h=1}^{n} a_{ij}^{kh}(x) \eta_i^k \eta_j^h \leq \lambda_2 |\eta|^2 \tag{3}$$

for any matrix $\{\eta_i^k\}$,

$$\eta_i^k = \eta_k^i, \quad \lambda_1, \lambda_2 = \text{const} > 0, \quad |\eta|^2 \equiv \sum_{i,k=1}^{n} (\eta_k^i)^2.$$

$$\lambda_3 \sum_{i,k=1}^{n} \left(\xi_i^k + \xi_k^i\right)^2 \leq \sum_{i,k,j,h=1}^{n} a_{ij}^{kh}(x)\xi_i^k \xi_j^h \leq \lambda_4 \sum_{i,k=1}^{n} \left(\xi_i^k + \xi_k^i\right)^2, \tag{4}$$

$\lambda_3, \lambda_4 = \text{const} > 0,$ for any matrix $\xi = \{\xi_i^k\}$.

The property (4) of coefficients of system (1) is a base for applications of Korn's inequalities. Consider the main boundary-value problems for system (1) in a domain Ω with a boundary $\partial\Omega$.

The boundary condition

$$u = 0 \quad \text{on} \quad \partial\Omega \tag{5}$$

corresponds to the Dirichlet problem.

The boundary condition

$$\sigma_i(u) = 0, \quad i = 1, \ldots, n, \quad \text{on} \quad \partial\Omega, \tag{6}$$

where

$$\sigma_i(u) \equiv \sum_{j,k,h=1}^{n} a_{ij}^{kh}(x) \frac{\partial u_h}{\partial x_j} \nu_k,$$

ν is an exterior normal vector to $\partial\Omega$, corresponds to the Neumann problem.

The boundary condition

$$u = 0 \quad \text{on} \quad \Gamma_1, \quad \sigma_i(u) = 0, \, i = 1, \ldots, n, \quad \text{on} \quad \Gamma_2, \tag{7}$$

where $\Gamma_1 \cup \Gamma_2 = \partial\Omega$, $\Gamma_1 \cap \Gamma_2 = \emptyset$, corresponds to the mixed problem.

The first existence theorem for the Dirichlet problem was proved by I. Fredholm in 1906, using his theory of integral equations [3].

Almost in the same time (1906–1909) boundary value problems for the elasticity system (the Dirichlet problem and the Neumann problem) were considered by A. Korn in several papers. In one of these papers at the first time Korn's inequalities appeared [10]. He proved the Korn inequality for vector-functions $u(x) = (u_1(x), \ldots, u_n(x))$ with the boundary condition $u(x) = 0$ on $\partial\Omega$ (the first case) and for vector-functions $u(x)$ in a domain Ω with the condition

$$\int_{\Omega} \left(\frac{\partial u_i}{\partial x_j} - \frac{\partial u_j}{\partial x_i}\right) dx = 0, \quad i,j = 1, \ldots, n, \tag{8}$$

(the second case). Very often they are called the first and the second Korn inequality correspondingly. (Korn's proof of the second Korn inequality takes 200 pages).

K.O. Friedrichs in his famous paper [4], published in 1947 and dedicated to mathematical problems of elasticity, writes: "The author of the present paper has been unable to verify Korn's proof for the second case". He notes that the Korn proof of Korn's inequality for the first case is very simple, that for the second case is very complicated.

Similar remarks one can find also in other papers [1]. In 1947 K.O. Friedrichs [4] gave a clear proof of Korn's inequality for the second case and solved the Dirichlet problem and the Neumann problem for the linear system of elasticity, using variational methods and Korn's inequalities. (The Friedrichs proof of Korn's inequality for the second case takes 18 pages).

G. Fichera [2] solved the mixed problem in 1950, using functional analysis methods. After 1947 many various proofs of Korn's inequality for the second case appeared. (J.L. Lions, G. Fichera, J. Nitsche, J. Nečas and others). Now, after more than 80 years from the first proof of Korn's inequality for the second case, there exists a proof of this inequality which takes less than 2 pages and uses only a simple calculus. This proof does not use any modern technique and could be found in the 19 century when the mathematical theory of elasticity just started to develop and the problem arose.

Since this proof is short and simple, one can see, how constants in Korn's inequality depend on the geometry of a domain Ω. Such kind of Korn's inequalities with explicit constants have many applications in the theory of homogenization, in numerical analysis, in mathematical theory of elasticity itself.

There are many different equivalent forms of the Korn inequality for the second case. Let us introduce some notations. We set

$$\mathcal{D}(u,\Omega) = \int_\Omega \sum_{i,j=1}^n \left|\frac{\partial u_i}{\partial x_j}\right|^2 dx, \quad E(u,\Omega) = \int_\Omega \sum_{i,j=1}^n |e_{ij}(u)|^2 dx,$$

where $u = (u_1, \ldots, u_n)$, $x = (x_1, \ldots, x_n)$,

$$e_{ij}(u) \equiv \frac{\partial u_i}{\partial x_j} + \frac{\partial u_j}{\partial x_i}.$$

The first Korn inequality is

$$\mathcal{D}(u,\Omega) \leq \frac{1}{2} E(u,\Omega) \tag{9}$$

for any bounded domain Ω and any vector-function u which belongs to $H_0^1(\Omega)$,

$$\left(u_j, \frac{\partial u_i}{\partial x_j} \in L_2(\Omega), \quad u = 0 \quad \text{on} \quad \partial\Omega, \ i,j = 1,\ldots,n\right).$$

It is of interest to note that (9) is also valid for any bounded domain and for a more wide class of vector-functions. For example, it is valid for a vector-function u, such that $u_j \in H^1(\Omega)$, $j = 1,\ldots,n$, $u_j = 0$ on $\partial\Omega$, $j = 1,\ldots,n-1$, and u_n is any function of $H^1(\Omega)$.

A more general proposition can be proved.

Theorem 1. Assume that $\partial\Omega = S_1 \cup \ldots \cup S_n$,

$$S_i \cap S_j = \emptyset \quad \text{for} \quad i \neq j, \quad u_j = 0 \quad \text{on} \quad \partial\Omega \setminus S_j, \quad j = 1,\ldots,n.$$

Then

$$\mathcal{D}(u,\Omega) \leq \frac{1}{2} E(u,\Omega).$$

The proof of this theorem is similar to the proof of (9) in the classical case.

The second Korn inequality is

$$\mathcal{D}(u,\Omega) \leq C(\Omega) E(u,\Omega),$$

where the constant $C(\Omega)$ depends on Ω and does not depend on u. It is supposed that $\partial\Omega$ is sufficiently smooth and $u(x)$ satisfies the condition (8), $u \in H^1(\Omega)$.

Some other forms of the second Korn inequality:

1) $\mathcal{D}(u,\Omega) \leq C_1(\Omega,\Omega') \left[E(u,\Omega) + \|u\|_{L_2(\Omega')}^2\right],$

   $u \in H^1(\Omega), \quad \Omega' \subset \Omega, \quad \Omega$ is smooth;

2) $\mathcal{D}(u,\Omega) \leq C_2(\Omega,\Omega') \left[E(u,\Omega) + \mathcal{D}(u,\Omega')\right],$

   $u \in H^1(\Omega), \quad \Omega' \subset \Omega, \quad \Omega$ is smooth;

3) For any $u \in H^1(\Omega)$ there exists a skew-symmetric matrix A with constant elements such that

$$\mathcal{D}(u - Ax, \Omega) \leq C_3(\Omega) E(u,\Omega),$$

   Ω is a smooth domain.

Due to the property (4) of the elasticity system one can get a priori estimates for solutions of systems (1), using Korn's inequalities.

A domain Ω is called a starshaped domain with respect to a ball Q, if $Q \subset \Omega$ and the segment between any point of Q and any point of Ω belongs to Ω.

Theorem 2. Let the domain Ω be starshaped with respect to the ball $Q = \{x : |x| < R_1\}$. Assume that Ω has the diameter R, $u_j \in H^1(\Omega)$, $j = 1, \ldots, n$. Then

$$\mathcal{D}(u, \Omega) \leq C_1 \left(\frac{R}{R_1}\right)^n [E(u, \Omega) + \mathcal{D}(u, Q)], \quad n \geq 3, \tag{10}$$

$$\mathcal{D}(u, \Omega) \leq C_2 \left[\left(\frac{R}{R_1}\right)^2 \ln\left(\frac{3R}{R_1}\right) E(u, \Omega) + \left(\frac{R}{R_1}\right)^2 \mathcal{D}(u, Q)\right], \quad n = 2, \tag{11}$$

where the constants C_1, C_2 depend on n only.

The estimate (10) is sharp in the sense that $C_1 \left(\frac{R}{R_1}\right)^n$ cannot be substituted by $C_1 \left(\frac{R}{R_1}\right)^n \alpha\left(\frac{R}{R_1}\right)$, where $\alpha(t) \to 0$ as $t \to \infty$.

Theorem 3. Assume that the conditions of Theorem 2 are satisfied. Then there exists a skew-symmetric matrix A with constant elements such that

$$\mathcal{D}(u - Ax, \Omega) \leq C_3 \left(\frac{R}{R_1}\right)^n \left(1 + \frac{R_1^2}{\gamma^2}\right) E(u, \Omega), \quad n \geq 3$$

$$\mathcal{D}(u - Ax, \Omega) \leq C_4 \left(\frac{R}{R_1}\right)^2 \left(\ln\left(\frac{3R}{R_1}\right) + \frac{R_1^2}{\gamma^2}\right) E(u, \Omega), \quad n = 2,$$

where the constants C_3, C_4 depend on n only, γ is the distance between Q and $\partial \Omega$, $u_j \in H^1(\Omega)$, $j = 1, \ldots, n$.

Theorem 4. Assume that the conditions of Theorem 2 are valid and

$$\int_\Omega \left(\frac{\partial u_i}{\partial x_j} - \frac{\partial u_j}{\partial x_i}\right) dx = 0, \quad i, j = 1, \ldots, n.$$

Then

$$\mathcal{D}(u, \Omega) \leq C_5 \left(\frac{R}{R_1}\right)^n \left(1 + \frac{R_1^2}{\gamma^2}\right) E(u, \Omega), \quad n \geq 3,$$

$$\mathcal{D}(u, \Omega) \leq C_6 \left(\frac{R}{R_1}\right)^2 \left(\ln\frac{3R}{R_1} + \frac{R_1^2}{\gamma^2}\right) E(u, \Omega), \quad n = 2,$$

where the constants C_5, C_6 depend on n only, $u_j \in H^1(\Omega)$, $j = 1, \ldots, n$.
This is the classical Korn inequality for the second case with explicit constants.

Theorems 2–4 are proved in [12], [7].

Here we give a proof of the Korn inequality in the form

$$\mathcal{D}(u,\Omega) \leq C_1 \left(\frac{R}{R_1}\right)^{n+1} E(u,\Omega) + C_2 \left(\frac{R}{R_1}\right)^n \mathcal{D}(u,Q), \quad n \geq 2, \tag{12}$$

for any domain Ω which is starshaped with respect to the ball Q, where the constants C_1, C_2 depend on n only.

This proof is based on the relation

$$\frac{\partial^2 V_i}{\partial x_p \partial x_q} \equiv \frac{1}{2}\left[\frac{\partial}{\partial x_p} e_{iq}(V) + \frac{\partial}{\partial x_q} e_{ip}(V) - \frac{\partial}{\partial x_i} e_{pq}(V)\right], \tag{13}$$

where

$$e_{ij}(V) \equiv \frac{\partial V_i}{\partial x_j} + \frac{\partial V_j}{\partial x_i}.$$

Suppose firstly that $R_1 = 1$. Consider $\theta = (\theta_1, \ldots, \theta_n)$, where

$$\Delta \theta_i = \Delta u_i \quad \text{in} \quad \Omega, \tag{14}$$

$$\theta_i = 0 \quad \text{on} \quad \partial \Omega. \tag{15}$$

Due to the identity (13), for solutions θ_i of the problem (14), (15) we have an energy estimate

$$\mathcal{D}(\theta, \Omega) \leq C_3 E(u, \Omega),$$

where C_3 depends on n only. Set $w = u - \theta$. Then $\Delta w_i = 0$,

$$\Delta e_{ij}(w) = 0, \quad i,j = 1,\ldots,n, \quad \text{in} \quad \Omega.$$

Since $\Delta e_{ij}(w) = 0$ in Ω and $\rho(x)$, which is the distance from x to $\partial \Omega$, satisfies the Lipschitz condition with the constant equal to 1, we obtain the estimate

$$\int_\Omega \rho^2(x) \sum_{i,j=1}^n |\nabla e_{ij}(w)|^2 dx \leq C_4 E(u,\Omega), |\nabla \phi|^2 \equiv \sum_{i,j=1}^n \left|\frac{\partial \phi_i}{\partial x_j}\right|^2,$$

and due to (13)

$$\int_\Omega \rho^2(x) \sum_{i,j,k=1}^n \left|\frac{\partial^2 w_i}{\partial x_j \partial x_k}\right|^2 dx \leq C_5 E(u,\Omega),$$

where the constants C_4, C_5 depend on n only.

Consider the elementary inequality

$$\int_0^a f^2(t)\,dt \le 4\left[\int_0^a t^2(f')^2\,dt + af^2(a)\right], \quad a = \text{const} > 0.$$

We apply this inequality to functions $f = \dfrac{\partial w_i}{\partial x_j}$ and to the segment AP of the ray OP, where P is a point on the boundary of Ω, O is the origin, $Q = \{x : |x| < 1\}$, $A \in Q$, $|A| = \lambda$, $1/2 \le \lambda \le 1$. The constant λ is chosen in such a way that

$$\int_{|A|=\lambda} |\nabla w(A)|^2\,d\omega \le C_6 \int_Q |\nabla w(x)|^2\,dx,$$

where $d\omega$ is an element of the area of the unit sphere, the constant C_6 does not depend on w. So we get

$$\int_{|A|}^{|P|} |\nabla w|^2 d|x| \le C_7\left[\int_{|A|}^{|P|} |P-x|^2 \sum_{i,k,l=1}^n \left|\frac{\partial^2 w_i}{\partial x_k \partial x_l}\right|^2 d|x| + |P-A||\nabla w(A)|^2\right],$$

C_7 depends on n only. From this inequality it follows that

$$\int_{|A|}^{|P|} \frac{|x|^{n-1}}{R^{n-1}} |\nabla w|^2 d|x| \le C_7\, 2^{n-1}\left[\int_{|A|}^{|P|} |P-x|^2 |x|^{n-1} \sum_{i,k,l}^n \left|\frac{\partial^2 w_i}{\partial x_k \partial x_l}\right|^2 d|x|\right.$$

$$+ |P-A||\nabla w(A)|^2, \text{ since } \frac{|x|^{n-1}}{R^{n-1}} \le 1,\ |x|^{n-1} 2^{n-1} \ge 1 \text{ for } |x| \ge |A|.$$

Integrating this inequality over the unit sphere and taking into account that for domains Ω which are starshaped with respect to the ball Q of radius 1, the inequality

$$|P - x| \le \rho(x) R$$

is valid, we obtain the inequality (12) for $R_1 = 1$.

For any $R_1 > 0$ the inequality (12) can be deduced from (12) for $R_1 = 1$ by changing of variables $y = \dfrac{x}{R_1}$. Thus the inequality (12) is proved.

Since any domain Ω which satisfies the Lipschitz condition or the cone condition can be represented as a union of a finite number of starshaped domains, from theorems 2–4 the Korn inequality for the second case follows for such domains. (The cone condition means that any point of $\bar{\Omega}$ can be considered as a vertex of a cone with the fixed size, which belongs to $\bar{\Omega}$.

Let us consider now thin domains (shells).

Theorem 5. Let $\Omega = \{x : F_1(x') < x_n < F_1(x') + \varepsilon F_2(x')\}$, $x' \in \Omega'$, $x' = (x_1, \ldots, x_{n-1})$, $x = (x_1, \ldots, x_n)$, $u_j \in H^1(\Omega)$, $j = 1, \ldots, n$, Ω' is a bounded domain with the Lipschitz boundary, $F_2(x') \geq 1$, $\varepsilon = $ const, $0 \leq \varepsilon \leq 1$, functions $F_1(x')$, $F_2(x')$ satisfy the Lipschitz condition. Then there exists a constant skew-symmetric matrix A such that

$$\mathcal{D}(u - Ax, \Omega) \leq C_1 \varepsilon^{-2} E(u, \Omega), \tag{16}$$

where C_1 does not depend on ε and u.

Proof. Let us consider $\theta = (\theta_1, \ldots, \theta_n)$, where θ_i is a solution of the problem

$$\Delta \theta_i = \Delta u_i, \quad i = 1, \ldots, n,$$

in Ω and

$$\theta_i = 0 \quad \text{on} \quad \partial \Omega, \; \theta_i \in H_0^1(\Omega).$$

It is easy to see that

$$\mathcal{D}(\theta_i, \Omega) \leq C_2 E(u, \Omega), \tag{17}$$

where the constant C_2 does not depend on u, θ_i, Ω. We set

$$w = u - \theta.$$

It is obvious that $\Delta w_i = 0$, $\Delta(e_{ij}(w)) = 0$, $i, j = 1, \ldots, n$. Multiplying the equation $\Delta(e_{ij}(w)) = 0$ by $e_{ij}(w)\rho^2(x)$, where $\rho(x)$ is the distance from x to $\partial\Omega$, and integrating the obtained equality over Ω, after the integration by parts we get

$$\int_\Omega \rho^2(x) |\nabla e_{ij}(w)|^2 \, dx \leq C_3 E(u, \Omega), \; C_3 = \text{const}, \tag{18}$$

C_3 does not depend on u and Ω.

Let us denote by G_ε the domain $\{x : F_1(x') + \frac{\varepsilon}{4} < x_n < F_1(x') + \frac{3}{4}\varepsilon F_2(x')$, $x' \in \Omega'$, $\rho(x', \partial\Omega') > \kappa\varepsilon\}$, $\kappa = $ const $ > 0$.

It follows from (18) that

$$\int_{G_\varepsilon} |\nabla e_{ij}(w)|^2 dx \leq \varepsilon^{-2} C_4 E(u,\Omega), \quad C_4 = \text{const} > 0, \tag{19}$$

where C_4 does not depend on ε and u.

Since according to (13)

$$\frac{\partial^2 w_i}{\partial x_k \partial x_l} = \frac{1}{2}\left[\frac{\partial}{\partial x_k} e_{il}(w) + \frac{\partial}{\partial x_l} e_{ik}(w) - \frac{\partial}{\partial x_i} e_{lk}(w)\right],$$

we deduce from (19) that

$$\int_{G_\varepsilon} \sum_{i,j,k=1}^n \left|\frac{\partial^2 w_i}{\partial x_j \partial x_k}\right|^2 dx \leq \varepsilon^{-2} C_5 E(u,\Omega), \quad C_5 = \text{const}, \tag{20}$$

C_5 does not depend on u and ε. From the Poincaré inequality it follows that

$$\int_{G_\varepsilon} \left(\frac{\partial w_i}{\partial x_j} - \tilde{a}_{ij}\right)^2 dx \leq C_6 \int_{G_\varepsilon} \left|\nabla \frac{\partial w_i}{\partial x_j}\right|^2 dx, \tag{21}$$

where constant C_6 does not depend on ε. Indeed, let us change variables $y_n = (x_n - F_1(x'))\varepsilon^{-1}$, $y' = x'$. Then the domain G_ε turns into the domain $G = \left\{y : \frac{1}{4} < y_n < \frac{3}{4} F_2(y'), y' \in \Omega'\right\}$ which does not depend on ε. From the Poincaré inequality for the domain G, passing to coordinates x, we obtain the Poincaré inequality for the domain G_ε and apply it to the function $\frac{\partial w_i}{\partial x_j}$. From (21), (17), (20) it follows that

$$\mathcal{D}(u-\tilde{A}x, G_\varepsilon) = \mathcal{D}(w+\theta-\tilde{A}x, G_\varepsilon) \leq 2\mathcal{D}(w-\tilde{A}x, G_\varepsilon) + 2\mathcal{D}(\theta, G_\varepsilon) \leq \varepsilon^{-2} C_7 E(u,\Omega), \tag{22}$$

where C_7 does not depend on ε, the matrix $\tilde{A} = \{\tilde{a}_{ij}\}$ is defined by (21).

Let us prove that the matrix $\tilde{A}$ can be taken as a skew-symmetric matrix. Since due to (22)

$$E(\tilde{A}x, G_\varepsilon) \leq 2E(u-\tilde{A}x, G_\varepsilon) + 2E(u, G_\varepsilon) \leq C_8\, \varepsilon^{-2} E(u,\Omega), \tag{23}$$

we have from (23) that

$$|\tilde{a}_{ij} + \tilde{a}_{ji}|^2 \text{ mes } G_\varepsilon \leq C_8\, \varepsilon^{-2} E(u,\Omega).$$

Therefore
$$\mathcal{D}\left(u - \frac{1}{2}(\tilde{A} - \tilde{A}^*)x - \frac{1}{2}(\tilde{A} + A^*)x, G_\varepsilon\right) \le \varepsilon^{-2} C_7 E(u, \Omega),$$

$$\mathcal{D}\left(u - \frac{1}{2}(\tilde{A} - \tilde{A}^*)x, G_\varepsilon\right) \le C_9[\varepsilon^{-2} E(u, \Omega)$$
$$+ \mathcal{D}((\tilde{A} + \tilde{A}^*)x, G_\varepsilon)] \le C_{10}\,\varepsilon^{-2} E(u, \Omega),$$

where the constants C_{10}, C_9 do not depend on ε.

The domain Ω can be covered by sets $\bar{d}_i$, $i = 1, \ldots, N$, where d_i is a starshaped domain with respect to a ball Q_i of radius ρ_i, Q_i belongs to G_ε, diameter of Q_i is equal to R_i and $R_i/\rho_i \le \kappa_1$, the constant κ_1 does not depend on ε and i. Every d_i can have an intersection with a finite number of d_j, which is less than κ_2 and κ_2 does not depend on ε and i. It follows from the fact that functions $F_1(x')$, $F_2(x')$ and the boundary of Ω' satisfy the Lipschitz condition. To domains d_i and vector function $u - \frac{1}{2}(\tilde{A} - \tilde{A}^*)x$ we apply inequality (12). We get

$$\mathcal{D}\left(u - \frac{1}{2}(\tilde{A} - \tilde{A}^*)x, d_i\right) \le C_{11}\, E(u, d_i) + \mathcal{D}\left(u - \frac{1}{2}(\tilde{A} - \tilde{A}^*)x, Q_i\right), \qquad (24)$$

where the constant C_{11} does not depend on i, ε, u.

Summing inequalities (24) over all i, we obtain (16) with a matrix $A = \frac{1}{2}(\tilde{A} - \tilde{A}^*)$. Theorem is proved.

Theorem 6. Assume that conditions of Theorem 5 are satisfied. Then for elements a_{ij} of the matrix A in (16) we have

$$\sum_{i,j=1}^{n} a_{ij}^2\, \operatorname{mes} \omega \le C_1\, \varepsilon^{-2} \left(E(u, \Omega) + \int_{\Omega \setminus \omega} |u|^2 dx \right), \qquad (25)$$

where the constant C_1 does not depend on ε, u. The domain ω is a subdomain of Ω.

Proof. It is evident that

$$\mathcal{D}(Ax, \omega) \le 2\mathcal{D}(u - Ax, \omega) + 2\mathcal{D}(u, \omega). \qquad (26)$$

Let us take $\omega \subset \Omega$ and a function ψ in Ω, such that $\psi = 1$ in ω, $\psi = 0$ on $\partial \Omega$, $\psi \in C^\infty(\mathbf{R}^n)$, $0 \le \psi \le 1$. According to the classical Korn inequality

$$\mathcal{D}(\psi u, \Omega) \le \frac{1}{2} E(\psi u, \Omega) \le \frac{1}{2} E(u, \Omega) + C_2\, \varepsilon^{-2} \int_{\Omega \setminus \omega} |u|^2 dx. \qquad (27)$$

It follows from (27) that

$$D(u,\omega) \leq \frac{1}{2} E(u,\Omega) + C_2\, \varepsilon^{-2} \int_{\Omega\setminus\omega} |u|^2 dx. \qquad (28)$$

From (26), (28) and Theorem 5 we get

$$\sum_{i,j=1}^{n} |a_{ij}|^2\, \operatorname{mes} \omega \leq C_3\, \varepsilon^{-2} \left[E(u,\Omega) + \int_{\Omega\setminus\omega} |u|^2 dx \right].$$

The Theorem is proved.

Theorem 7. Assume that conditions of Theorem 5 are satisfied and $u = 0$ on $S' = \{x : x' \in \partial\Omega'\}$. Then for $u_j \in H^1(\Omega)$, $j = 1, \ldots, n$,

$$D(u,\Omega) \leq C_1\, \varepsilon^{-2} E(u,\Omega), \qquad (29)$$

where the constant C_1 does not depend on ε and u.

Proof. We can suppose that functions $F_1(x')$ and $F_2(x')$ are given in a domain Ω'' such that $\Omega'' \supset \Omega'$ and the distance between $\partial\Omega''$ and $\partial\Omega'$ is positive. Let us denote by Ω_1 the domain $\{x : x' \subset \Omega'', F_1(x') < x_n < F_1(x') + \varepsilon F_2(x')\}$. We set $u = 0$ in $\Omega_1\setminus\Omega$. It is evident that the vector-function $u(x) \in H^1(\Omega_1)$. According to Theorem 5 we have

$$D(u - Ax, \Omega_1) \leq C_2\, \varepsilon^{-2} E(u,\Omega_1) = C_2\, \varepsilon^{-2} E(u,\Omega),\ C_2 = \text{const}, \qquad (30)$$

where $A = \{a_{ij}\}$ is a constant skew-symmetric matrix, C_2 does not depend on ε, u. It is easy to see that

$$D(Ax, \Omega_1\setminus\Omega) \leq 2D(Ax - u, \Omega_1\setminus\Omega) + 2D(u, \Omega_1\setminus\Omega)$$
$$\leq 2D(u - Ax, \Omega_1). \qquad (31)$$

From (31) and (30) it follows that

$$\operatorname{mes}(\Omega_1\setminus\Omega) \sum_{i,j=1}^{n} |a_{ij}|^2 \leq 2C_2\, \varepsilon^{-2} E(u,\Omega). \qquad (32)$$

Using the estimates (30), (32), we obtain

$$D(u,\Omega) \leq 2D(u - Ax, \Omega) + 2D(Ax, \Omega)$$
$$\leq 2C_3\, \varepsilon^{-2} \left[E(u,\Omega) + \operatorname{mes}\Omega (\operatorname{mes}(\Omega_1\setminus\Omega))^{-1} E(u,\Omega) \right] \leq C_4\, \varepsilon^{-2} E(u,\Omega),$$

since mes $\Omega \leq C_5$ mes $(\Omega_1 \setminus \Omega)$, where the constants C_4, C_3 do not depend on ε. Theorem is proved.

Remarks.

1) In a similar way the inequality (29) can be proved, when $u = 0$ on some set

$$\{x : x' \in \alpha,\ F_1(x') < x_n < F_1(x') + \varepsilon F_2(x'),\ \alpha \subset \partial\Omega'\},$$

where α is an open set of $\partial\Omega'$.

2) The theorem 5 can be proved also for a case when the domain Ω is bounded by two surfaces S_1 and S_2, which satisfy the Lipschitz condition and $\kappa_1\varepsilon \leq \rho(x^1, S_2) \leq \kappa_2\varepsilon$ for any point x^1 of S_1; $\rho(x^1, S_2)$ is the distance between x^1 and S_2, the constants κ_1, κ_2 do not depend on ε.

Let us consider Korn's and Hardy's type inequalities in unbounded domains. K.O. Friedrichs in his paper [4] wrote that it is of interest to consider Korn's inequalities in unbounded domains and boundary value problems for the elasticity system in unbounded domains but he restricts himself by consideration of bounded domains, since in unbounded domains inequalities which he uses, are not valid.

Here we formulate the Korn type inequalities and the Hardy type inequalities for unbounded domains and use them to study boundary-value problems in these domains. Let κ be a cone in $\mathbf{R}^n$ with a vertex at the origin. (The domain κ is called a cone, if λx belongs to κ for all positive λ and all $x \in \kappa$). We assume that the boundary of κ satisfies the Lipschitz condition.

We say that a vector-function $\theta \in H^1_{\text{loc}}(\Omega)$, if $\theta_j \in H^1(\omega)$, $j = 1, \ldots, n$, for any bounded domain $\omega \subset \Omega$.

Theorem 8. Let Ω be an unbounded domain in $\mathbf{R}^n$ and mes $(\mathbf{R}^n \setminus \Omega) = \infty$, $u \in H^1_{\text{loc}}(\Omega)$, $u = 0$ on $\partial\Omega$. Then

$$\mathcal{D}(u, \Omega) \leq \frac{1}{2} E(u, \Omega).$$

Theorem 9. Let the domain $\Omega = \kappa \setminus G$, where G is a bounded set with the Lipschitz boundary, κ is a cone, $u = (u_1, \ldots, u_n)$, $u_j \in H^1_{\text{loc}}(\Omega)$. Then there exists a skew-symmetric matrix A with constant elements such that

$$\mathcal{D}(u - Ax, \Omega) \leq C(\Omega) E(u, \Omega),$$

where the constant $C(\Omega)$ does not depend on u.

Theorem 10. (Hardy's type inequality). Let $n \geq 3$, $\kappa_R = \kappa \cap \{x : |x| > R\}$, $R \geq 0$; κ is a cone, $\theta \in H^1_{\text{loc}}(\kappa_R)$. Then there exists a constant vector M such that

$$\int_{\kappa_R} |\theta - M|^2 |x|^{-2} dx \leq C\mathcal{D}(\theta, \kappa_R),$$

where the constant C does not depend on θ.

Theorem 11. Let $n = 2$, κ is an angle, $\theta \in H^1_{\text{loc}}(\kappa)$, $\kappa_R = \kappa \cap \{x : |x| > R\}$, $\theta = 0$ for $\kappa \cap \{x : |x| \leq R\}$, $R > 0$. Then

$$\int_{\kappa_R} |\theta(x)|^2 \left(|x| \ln \frac{2|x|}{R}\right)^{-2} dx \leq 4\mathcal{D}(\theta, \kappa_R).$$

Theorem 12. Let $n = 2$, a domain $\Omega = \kappa \backslash G$, κ is an angle, Ω has the Lipschitz boundary. G is a bounded set, $G \subset \{x : |x| < R\}$, $\theta \in H^1_{\text{loc}}(\Omega)$. Then there exists the constant M such that

$$\int_\Omega |\theta - M|^2 \left(1 + \left(|x| \ln \frac{2|x|}{R}\right)^2\right)^{-1} dx \leq C_1 \mathcal{D}(\theta, \Omega), \quad C_1 = \text{const},$$

C_1 does not depend on θ.

Theorems 8–12 and some more general theorems of this type are proved in [5], [9], [12].

We note that Korn's inequality in the classical sense is not valid for some unbounded domains. For example, Korn's inequality is not valid for a layer $\Omega = \{x : 0 < x_n < l\}$, $l = $ const, (see [5], [8]). Boundary value problems in such unbounded domains are studied in [8], [9].

Stability theorems for the Dirichlet boundary-value problem, for the Neumann and the mixed boundary-value problems in unbounded domains are proved in [12], [11].

Let us consider here the Dirichlet problem in an exterior domain $\Omega \subset \mathbf{R}^n$, $n \geq 3$, using Korn's and Hardy's inequalities. Let a domain $\Omega = \mathbf{R}^n \backslash G$, where G is a bounded set. The vector-function $u(x)$ is called a weak solution of the Dirichlet problem for the system (1) with the boundary condition $u = 0$ on $\partial \Omega$, if $u \in H^1_{\text{loc}}(\Omega)$, $u = 0$ on

$\partial\Omega$, and for any vector-function $\theta(x)$ such that $\theta = 0$ on $\partial\Omega$, θ is equal to zero in the neighbourhood of infinity, $\theta \in H^1_{\text{loc}}(\Omega)$, $u(x)$ satisfies the integral identity

$$-\sum_{i,k,h,j=1}^{n} \int_\Omega a_{ij}^{kh}(x) \frac{\partial u_h}{\partial x_j} \frac{\partial \theta_k}{\partial x_i}\, dx = \sum_{k=1}^{n} \int_\Omega f_\kappa\, \theta_\kappa\, dx. \tag{33}$$

Let w^i be a weak solution of the system (1) in Ω with $f_\kappa = 0$ in Ω, $\kappa = 1,\ldots,n$, and with the boundary condition $w^i = A_i x$ on $\partial\Omega$, where

$$A_1,\ldots, A_{n(n-1)/2}$$

are linearly independent skew-symmetric matrices with constant elements, and with condition $\mathcal{D}(w^i,\Omega) < \infty$. Let W^i be a weak solution of the elasticity system (1) with $f_\kappa = 0$ in Ω, $\kappa = 1,\ldots,n$, and with the boundary condition $W^i = e^i$ on $\partial\Omega$, where $e^i = (e_1^i,\ldots,e_n^i)$,

$$e_j^i = 0 \text{ for } i \neq j, \ e_i^i = 1, \text{ and } \mathcal{D}(W^i,\Omega) < \infty, \ i = 1,\ldots,n.$$

Solutions W^i and w^i can be constructed by the variational method.

Theorem 13. Let $u(x)$ be a weak solution of the exterior Dirichlet problem in $\Omega = R^n \backslash G$ for the system (1) with the boundary condition $u = 0$ on $\partial\Omega$, G is a bounded set. Assume that $E(u,\Omega) < \infty$ and $n \geq 3$. Then

$$\mathcal{D}(u - V, \Omega) \leq C_1 \int_\Omega |f|^2 |x|^2\, dx, \tag{34}$$

$$\int_\Omega |u - V|^2 |x|^{-2} \leq C_2 \int_\Omega |f|^2 |x|^2\, dx, \tag{35}$$

where

$$V = Ax + B - \sum_{i=1}^{n(n-1)/2} M_i w^i - \sum_{i=1}^{n} \kappa_i W^i,$$

A is a skew-symmetric matrix with constant elements, B is a constant vector, constants C_1, C_2 do not depend on f and u, the constants M_i and κ_i are chosen in such a way that $V = 0$ on $\partial\Omega$.

Proof. According to Theorem 9 the Korn inequality

$$\mathcal{D}(u - Ax, \Omega) \leq C_3\, E(u,\Omega) \tag{36}$$

is valid, where A is a constant skew-symmetric matrix, C_3 is a constant. Suppose that $G \subset \{x : |x| < R\}$. Then due to Theorem 10 there exists a constant vector B such that

$$\int_{R^n \setminus \{x : |x| < R\}} |u - Ax - B|^2 |x|^{-2} \, dx \leq C_4 \, \mathcal{D}(u - Ax, \Omega). \tag{37}$$

We define

$$V = Ax + B - \sum_{i=1}^{n(n-1)/2} M_i \, w^i - \sum_{i=1}^{n} \kappa_i \, W^i,$$

$$A = \sum_{i=1}^{n(n-1)/2} M_i \, A_i, \quad B = \sum_{i=1}^{n} \kappa_i \, e^i.$$

Therefore $V = 0$ on $\partial\Omega$, V satisfies system (1) with $f_k = 0$, $\kappa = 1, \ldots, n$, $U = u - V$ is a solution of system (1) with the boundary condition $U = 0$ on $\partial\Omega$. Setting $u = U$, $\theta = U$ in integral identity (33) and using (36), (37) we get (34), (35).

The Dirichlet problem for the system (1) for any unbounded domain is studied in [5]. The dimension of the kernel for the Dirichlet problem in the class of solutions u with $E(u, \Omega) < \infty$ is defined there in terms of g-capacity, introduced in [5].

Using Korn's and Hardy's inequalities one can study the classical I. Boussinesq and V. Cerruti problems. I. Sneddon [15] applies the Fourier transform to study these problems in the case of constant coefficients of the system (1). We consider here one of these problems, which was posed and considered by I. Boussinesq in 1888 in the case of homogeneous isotropic media.

The problem is to find a solution of system (1) satisfying the boundary conditions:

$$u_1 = \phi_1, \quad u_2 = \phi_2, \quad \sigma_3(u) = 0 \quad \text{on} \quad \partial\Omega, \tag{38}$$

where $\Omega \subset \mathbf{R}^3$, $\Omega = \{x : x_3 < 0\}$.

The vector-function $u(x)$ is called a weak solution of problem (1), (38), if $u \in H^1_{\text{loc}}(\Omega)$, $u_1 = \phi_1$, $u_2 = \phi_2$ on $\partial\Omega$ and $u(x)$ satisfies the integral identity (33) for any vector-function $\theta(x)$ such that $\theta_1 = 0$, $\theta_2 = 0$ on $\partial\Omega$, $\theta \in H^1_{\text{loc}}(\Omega)$, $\theta = 0$ in a neighbourhood of infinity.

Theorem 14. There exists a weak solution $u(x)$ of system (1) with $f \equiv 0$ in Ω and with the boundary condition (38) such that

$$\mathcal{D}(u, \Omega) < \infty, \quad \int_\Omega |u|^2 |x|^{-2} \, dx < \infty,$$

where $\phi_1 \in H^1_{\text{loc}}(\Omega)$, $\phi_2 \in H^1_{\text{loc}}(\Omega)$, $\phi_1 = 0$, $\phi_2 = 0$ in a neighbourhood of infinity.

Theorem 15. Let $u(x)$ be a weak solution of the problem (1), (38) with $\phi_1 = 0$, $\phi_2 = 0$, satisfying the condition $E(u, \Omega) < \infty$. Then there exists a skew-symmetric matrix $A = \{a_{ij}\}$ with $a_{ij} = 0$ for $i, j = 1, 2$ and constant vector $B = (b_1, b_2, b_3)$ with $b_i = 0$ for $i = 1, 2$ such that

$$\mathcal{D}(u - Ax - B, \Omega) \leq C_1 \int_\Omega |f|^2 |x|^2 dx,$$

$$\int_\Omega |u - Ax - B|^2 |x|^{-2} dx \leq C_2 \int_\Omega |f|^2 |x|^2 dx,$$

where the constants C_1, C_2 do not depend on u, f. If $\mathcal{D}(u, \Omega) < \infty$, $f \equiv 0$ in Ω, then

$$u(x) = B.$$

The proof of this theorem and a detailed study of the Boussinesq and the Cerruti problems one can find in [13].

References

[1] Bernstein B, Toupin R.A. (1960), *Korn inequalities for the sphere and circle*. Arch. Rat. Mech. Anal. 6.

[2] Fichera G. (1984), *Existence theorems in elasticity*. Mechanics of solid, II, Springer Verlag, p.347–389.

[3] Fredholm I. (1906), *Solution d'un problème fondamentale de la théorie de l'élasticité*. Ark. Mat. Astron. Fys., 2. 28.

[4] Friedrichs K.O. (1947), *On the boundary value problems of the theory of elasticity and Korn's inequality*, Ann. Math. v. 48, p.441–471,

[5] Kondratiev V.A., Oleinik O.A. (1988), *Boundary-value problems for the system of elasticity theory in unbounded domains. Korn's inequalities*. Russian Math. Survey, v. 43, No 5, p.65–119.

[6] Kondratiev V.A., Oleinik O.A. (1989), *On Korn's inequalities*, C.R. Acad. Sci. Paris, v.308, Ser. 1, p.483–487.

[7] Kondratiev V.A., Oleinik O.A. (1989), *On the dependence of constants in Korn's inequalities on parameters, characterizing the geometry of a domain*. Russian Math. Survey, v.44, No 6, p.157–158.

[8] Kondratiev V.A., Oleinik O.A. (1990), *Korn's type inequalities for a class of unbounded domains and applications to boundary-value problems in elasticity.* In: Elasticity. Mathematical methods and applications. Ed. by G. Eason and R.W. Ogden, Ellis Horwood Limited, West Sussex, England, p.211–233.

[9] Kondratiev V.A., Oleinik O.A. (1990), *Hardy's and Korn's type inequalities and their applications.* Rendiconti di Matematica, Seri VII, v. 10, Fac. 3, p.641–666.

[10] Korn A. (1908), *Solution générale du problème d'équilibre dans la théorie de l'élasticité dans le cas où les efforts sont donnés à la surface.* Annales, Toulouse Université, (1990).

[11] Michailova T.V. (1990), *On the mixed problem for the system of elasticity in an unbounded domain.* Russian Math. Survey. v. 45, No 4, p.115–116.

[12] Oleinik O.A. (1991), *Korn's type inequalities and applications to elasticity.* Convegno Internationale in memoria di Vitto Volterra (Roma 6–11, Oct. 1990), Atti dei convegni Lincei, 92.

[13] Oleinik O.A. (1991), *On Boussinesq and Cerruti problems in elasticity.* Rend. Sem. Mat. Univers. Politech. Torino. vol. 48, 1.

[14] Payne L.E., Weinberger H.F. (1961), *On Korn's inequality.* Arch. Rat. Mech. Anal. 8.

[15] Sneddon I.N. (1989), *The problem of Boussinesq and Cerruti: a new approach to an old problem.* Continuum Mechanics and its applications. Hemisphere compl. Corp., New York, p.27–50.

C D PAGANI, M VERRI AND A ZARETTI
On a problem of seakeeping

1. Position of the problem

This paper deals with the unsteady motion of a rigid body floating in the water under the action of the gravity, of the liquid pressure and of other given external forces (wind, constraints on the body motions etc .....).

The problem is treated within the framework of the linearised theory near equilibrium. Let us summarise the model to be adopted (see e.g. [1], [3]). A cartesian rectangular coordinate system (x, y, z) is considered with the y-axis positive upwards. When the system is at rest, the region occupied by water is a bounded domain Ω (see figure below) whose boundary is composed by: Γ_F on $y = 0$ (the undisturbed free surface), Γ_I (the immersed surface of the body in its rest position), Γ' (the fixed fluid bottom and lateral walls).

The fluid is assumed to be inviscid and incompressible (with constant density ρ) and its motion to be irrotational, hence the fluid velocity field can be represented by the gradient of a velocity potential $\phi(x, y, z, t)$. In response to this interaction with the fluid, the rigid body moves with six degrees of freedom and its position is completely specified by $\underline{\xi}(t) = (\underline{\xi}_T(t), \underline{\xi}_R(t))$ where: $\underline{\xi}_T(t) = (\underline{\xi}_i(t)) (i = 1, 2, 3)$ are the coordinates of the center of mass and $\underline{\xi}_R(t) = (\underline{\xi}_i(t)) (i = 4, 5, 6)$ are the angles of rotation about the axes through the center of mass.

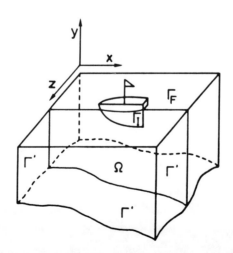

In the absence of any external force apart from gravity g, water and body are both in a rest position of equilibrium. At the initial time t = 0 some exterior forces acting on the body are considered to be switch on and the system evolves accordingly. It is assumed that their resultant is a first order quantity $\varepsilon \underline{f}(t)$ (ε suitable small parameter) and that a perturbation expansion in powers of ε is permitted for the quantities ϕ, $\underline{\xi}_T$, $\underline{\xi}_R$. Then, to first order, the governing evolution equations of the system are the following:

$$\Delta \phi = 0 \quad \text{in } \Omega \quad t > 0 \qquad (1.a)$$

$$M_1 \underline{\ddot{\xi}}(t) = \underline{F}(t) - M_2 J\underline{\xi}(t) - \int_{\Gamma_I} \phi_t \underline{q} \qquad (1.b)$$

where we assumed: $\rho = g = 1$ and:

$$\underline{F} = (\underline{f}, \underline{g}) \,, \quad \underline{q} = (\underline{\nu}, \underline{p})$$
$$J\underline{\xi} = (0, \xi_2, 0, \xi_4, 0, \xi_6)$$

ν is the unit normal pointing into the body, $\underline{g}$ and $\underline{p}$ respectively are the moments of $\underline{f}$ and $\underline{\nu}$ about the center of mass of the body, M_1 is the symmetric positive-definite matrix of body-inertia coefficients, M_2 is the matrix of the hydrostatic restoring coefficients of the body.

The appropriate boundary conditions must be imposed on the surfaces of equilibrium; they are: the classical Fourier condition on the calm-water surface:

$$\frac{\partial^2 \phi}{\partial t^2} + \frac{\partial \phi}{\partial y} = 0 \text{ on } \Gamma_F \qquad t > 0 \qquad (1.c)$$

$$\frac{\partial \phi}{\partial \nu} = 0 \quad \text{on} \quad \Gamma' \qquad t > 0 \qquad \text{(no-slip condition)} \qquad (1.d)$$

and the kinematic condition on the wetted surface of the body:

$$\frac{\partial \phi}{\partial \nu} = \underline{q} \cdot \underline{\dot{\xi}} \text{ on } \Gamma_I \qquad t > 0 \qquad (1.e)$$

The initial data are:

$$\phi(x,y,z,0) = 0 \quad \text{in } \Omega$$

$$\phi = 0, \frac{\partial \phi}{\partial t} = 0 \quad \text{on } \Gamma_F \quad \text{for } t = 0 \qquad (1.f)$$

$$\underline{\xi}(0) = \underline{\dot{\xi}}(0) = 0$$

Uniqueness of the solution of the problem (1.a)-(1.f) follows easily from a standard energy-argument. Let us define the total energy E(t) of the system as the sum of the kinetic energies of the body and the liquid plus the potential energies of the free surface and the hydrostatic restoring forces acting on the body:

$$E(t) = \frac{1}{2} \int_\Omega |\nabla \phi|^2 + \frac{1}{2} \int_{\Gamma_F} \left(\frac{\partial \phi}{\partial t}\right)^2 + \frac{1}{2} M_1 \dot{\underline{\xi}} \cdot \dot{\underline{\xi}} + M_2 J \underline{\xi} \cdot \underline{\xi}$$

Then equations (1.a) - (1.f) yield:

$$\dot{E}(t) = \underline{F} \cdot \dot{\underline{\xi}} \quad , \quad E(0) = 0$$

If no energy is supplied from outside i.e., $\underline{F} = 0$, $E(t)$ is conserved, hence it vanishes identically and a simple argument yields the conclusion that the system stays at rest for all time.

The existence of the solution will be proved by using the Laplace transform. This approach enables us to represent the velocity potential as the convolution of the body-velocity with the solution of a differential problem depending only on the geometry of the fluid domain. Furthermore the body velocity turns out to be simply computable by solving a linear system.

For existence and other results for a model concerning a free floating body in a unbounded domain see [2].

2. Method of solution

Let s denote a complex variable and:

$$\tilde{\phi}(x,y,z,s) = \mathcal{L}\, \phi(x,y,z,t)$$
$$\underline{\theta}(s) = \mathcal{L}\, \dot{\underline{\xi}}(t)$$
$$\underline{\tilde{F}}(s) = \mathcal{L}\, \underline{F}(t)$$

We obtain from Problem 1 the following transformed problem:

$$\Delta \tilde{\phi} = 0 \qquad (2.1.a)$$

$$(M_1 + \frac{1}{s^2} M_3) \underline{\theta} = \frac{1}{s} \underline{\tilde{F}} - \int_{\Gamma_I} \tilde{\phi}\, \underline{q} \qquad (M_3 = M_2 J) \qquad (2.1.b)$$

$$\frac{\partial \tilde{\phi}}{\partial \nu} = -s^2\, \tilde{\phi} \quad \text{on } \Gamma_F \qquad (2.1.c)$$

$$\frac{\partial \tilde{\phi}}{\partial \nu} = 0 \quad \text{on } \Gamma' \qquad (2.1.d)$$

$$\frac{\partial \tilde{\phi}}{\partial \nu} = \underline{q} \cdot \underline{\theta} \quad \text{on } \Gamma \qquad (2.1.e)$$

Let us look for a solution $\tilde{\phi}$ of the form:

$$\tilde{\phi}(x,y,z,s) = \underline{u}(x,y,z,s) \cdot \underline{\theta}(s) \qquad (2.2)$$

Then, in order $\tilde{\phi}$ (given by (2.2)) and θ to satisfy Pb. (2.1) it suffices that $\underline{u}$ solves the following differential boundary value problem:

$$\Delta \underline{u} = 0 \quad \text{in } \Omega \tag{2.3.a}$$

$$\frac{\partial \underline{u}}{\partial \nu} = -s^2 \underline{u} \quad \text{on } \Gamma_F \tag{2.3.c}$$

$$\frac{\partial \underline{u}}{\partial \nu} = 0 \quad \text{on } \Gamma' \tag{2.3.d}$$

$$\frac{\partial \underline{u}}{\partial \nu} = \underline{q} \quad \text{on } \Gamma_I \tag{2.3.e}$$

and θ solves the linear system:

$$(M_1 + \frac{1}{s^2} M_2)\, \underline{\theta} = \frac{1}{s}\, \tilde{\underline{F}} - \int_{\Gamma_I} (\underline{u} \cdot \underline{\theta})\, \underline{q} \quad . \tag{2.3.b}$$

Let q be any component of $\underline{q}$ and u the corresponding component of $\underline{u}$. We introduce on $H^1(\Omega) \times H^1(\Omega)$ the continuous sesquilinear form:

$$a_s(u,v) = s^2 \int_{\Gamma_F} u\bar{v} + \int_\Omega \nabla u \cdot \nabla \bar{v} \quad .$$

The problem (2.3) admits the following variational formulation:

Given $s \in \mathbb{C}$, Re $s > 0$, *and* $q \in L^2(\Gamma_I)$, *find* $u \in H^1(\Omega)$ *such that*:

$$a_s(u,v) = \int_{\Gamma_I} q \cdot \bar{v} \quad \forall\ v \in H^1(\Omega) \tag{2.4}$$

Notice that $a_s(u,v)$ is continuous but non coercive for all $s \in \mathbb{C}$ varying in a right half plane.

We prove the following theorem:

Proposition 1. *Problem* (2.4) *admits a unique solution* $u \in H^1(\Omega)$ *and the estimate*:

$$\|u\|_{H^1(\Omega)} \leq \frac{c}{|s|^3}\, \|q\|_{L^2(\Gamma_I)} \tag{2.5}$$

holds: the constant c *depends only on the geometry of* Ω.

3. Conclusions

Let us now write the equation (2.3.b) in the following form:

$$(M_1 + Q + \frac{1}{s^2} M_3)\, \underline{\theta} = \frac{1}{s}\, \tilde{\underline{F}}$$

where $Q_{ij}(s) = s^3 \int_{\Gamma_I} q_i v_j$

By using estimate (2.5) we prove that the matrix $M_1 + Q + s^{-2}M_3$ is invertible for s fixed (Re s > 0) and we obtain the following estimate:

$$|\underline{\theta}| \leq c(s) \frac{|\underline{\tilde{E}}|}{|s|} \qquad (3.1)$$

We observe that the eingenvalues of Q are bounded (with respect to s): they may tend to the real negative semiaxis when $s \to \infty$. Then, keeping in mind that M_1 is a positive definite matrix and using estimate (2.5) we can draw the following conclusions:

i) If Re s is large enough and M_1 is large compared to Q, then c(s) in (3.1) is bounded for $s \to \infty$ and we get:

$$|\underline{\theta}| \leq \frac{c}{|s|} |\underline{\tilde{F}}|$$

$$\|u\|_{H^1(\Omega)} \leq \frac{c}{|s|} |\underline{\tilde{F}}| \quad \|q\|_{L^2(\Gamma_I)}$$

ii) Otherwise, compatibility conditions on $\underline{F}$ must be satisfied, in order a solution to exist. E.g., assume $c(s) \approx s^{k-1}$, k integer ≥ 2; a solution still exists if $\underline{\tilde{F}}(s) \approx s^{-k}$, i.e., $\underline{F}$ and its first k-2 derivatives must vanish for t = 0.

References

[1] P. Bassanini, U. Bulgarelli, F. Pitolli, (1989). A numerical method for studying the interaction between a partially submerged body and the surrounding fluid, Meccanica, 24.

[2] J.T. Beale, (1977). Eigenfunction expansions for objects floating in an open sea. Communications on pure and applied mathematics, 30.

[3] D. Euvrard, G. Fernandez, A. Jami, (1984). Methodes numeriques en hydrodynamique navale, Report ENSTA, 31, Paris.

E AVGERINOS AND N S PAPAGEORGIOU
Optimal control and relaxation of nonlinear evolution equations

1. Introduction.

In a recent paper, Flytzanis-Papageorgiou [4], proved the existence of optimal controls for a class of Lagrange optimal control problems, governed by nonlinear evolution equations of the subdifferential type. Their work extended earlier ones by Ahmed [1], Ganesh-Joshi [5], Hou [6], Lions [7] and Papageorgiou [8], which either examined semilinear systems (see [5] and [8]) or considered nonlinear systems, with the control and state variables separated, with open loop control constraints and with more restrictive overall hypotheses on the data (see [5], [6], and [7]). Flytzanis-Papageorgiou [4] removed some of those limitations. However, their system is linear in the control variable, the subdifferential operator is time independent and the control constraint set is state independent (open loop). In this paper, we drop all those restrictions and using the Cesari reduction technique, we establish the existence of an optimal "state-control" pair. We also introduce the relaxed system and prove that under some regularity hypotheses on the data, it has the same value as the original problem. An example is also included.

2. Existence theorem.

Our notation and terminology follows the book of Brezis [3]. Let $T \in [0,b]$, h a separable Hilbert space (state space) and Y a separable Banach space (control space). The problem under consideration is the following:

$$\left| \begin{array}{l} J(x,u) = \int_0^b L(t,x(t),\ u(t,x(t),u(t)))dt \to \inf = m \\ \text{s.t.} \ -\dot{x}(t) \in \partial\varphi(t,x(t)) + f(t,x(t),u(t)) \ \text{a.e.} \\ x(0) = x_0, \ u(t) \in U(t,x(t)) \ \text{a.e.}, \ u(\cdot) \ \text{measurable.} \end{array} \right| \quad (*)$$

$H(\varphi)$: $\varphi: T \times H \to \overline{\mathbb{R}} = \mathbb{R} \cup \{+\infty\}$ is a measurable function s.t. for all $t \in T$, $\varphi(t,\cdot)$ is proper, convex l.s.c. and of compact type and for any $r \in \mathbb{N}_+$, there exists $K_r > 0$, an absolutely continuous function $g_r: T \to \mathbb{R}$ with $\dot{g}_r \in L^\beta(T)$ and a function of bounded variation $h_r: T \to \mathbb{R}$ s.t. if $t \in T$, $x \in \text{dom}\varphi(t,\cdot)$, $\|x\| \le r$ and $s \in [t,b]$, then there exists $\hat{x} \in \text{dom}\varphi(s,\cdot)$ satisfying $\|\hat{x} - x\| \le |g_r(s) - g_r(t)| (\varphi(t,x) + K_r)^\alpha$ and $\varphi(s,\hat{x},) \le \varphi(t,x) + |h_r(s) - h_r(t)| (\varphi(t,x) + K_r)$ where $\alpha \in [0,1]$ and

185

$\beta = 2$ if $\alpha \in [0, 1/2]$ or $\beta = \frac{1}{1-\alpha}$ if $\alpha \in [1/2, 1]$, (see Yotsutani [11]).

<u>$H(f)$</u>: $f: T \times H \times Y \to H$ is a function s.t. $t \to f(t, x, u)$ is measurable, $(x, u) \to f(t, x, u)$ is sequentially continuous from $H \times Y_w$ into H_w, $|f(t, x, u)| \leq a(t) + b(t)|x|$ a.e. for all $u \in W$, with $a, b \in L^2_+$.

<u>$H(U)$</u>: $U: T \times H \to P_{wk}(Y) = \{A \subseteq Y:$ nonempty and w-compact$\}$ is graph measurable, $GrU(t, \cdot) = \{(x, u) \in H \times Y: u \in U(t, x)\}$ is sequentially closed in $H \times Y_w$ and $U(t, x) \subseteq W \in P_{wkc}(Y)$ (i.e. $W \in P_{wk}(Y)$, and is convex).

<u>$H(L)$</u>: $L: T \times H \times Y \to \bar{\mathbb{R}} = \mathbb{R} \cup \{+\infty\}$ is measurable, $(x, u) \to L(t, x, u)$ is sequentially l.s.c. on $H \times Y_w$ and $\psi(t) - M(\|x\| + \|u\|) \leq L(t, x, u)$ a.e. with $\psi(\cdot) \in L^1$, $M > 0$.

<u>H_τ</u>: $Q(t, x) = \{(\eta, v) \in \mathbb{R} \times H: L(t, x, u) \leq \eta, u \in U(t, x), v \in \partial \varphi(t, x) + f(t, x, u)\}$ is convex for every $(t, x) \in T \times H$. (This hypothesis is satisfied if for example u enters linearly in the dynamics, $U(\cdot, \cdot)$ has convex values and $L(t, x, \cdot)$ is convex).

<u>H_0</u>: there exists admissible "state-control" pair (y, v) s.t. $J(y, v) < \infty$ (thus m is finite).

Theorem 1: *If hypotheses $H(\varphi)$, $H(f)$, $H(U)$, $H(L)$, H_τ, H_0 hold and $x_0 \in H$, then* $(*)$ *admits a solution.*

Proof: Let $M: T \times H \times H \to 2^Y$ be defined by $M(t, x, v) = \{u \in U(t, x): v \in \partial \varphi(t, x) + f(t, x, u)\}$. Set $h(t, x, v) = \inf_{u \in Y}[L(t, x, u) + \delta_{M(t, x, v)}(u)]$ (here $\delta_{M(t, x, v)}(u) = 0$ if $u \in M(t, x, v)$ and $+\infty$ otherwise; also $\delta_\emptyset(\cdot) = +\infty$).

Claim #1: $(t, x, v) \to h(t, x, v)$ is measurable: Note that $GrM = \{(t, x, v, u) \in T \times H \times H \times W: (t, x, u) \in GrU, (v - f(t, x, u), x) = \varphi(t, x) + \varphi^*(t, v - f(t, x, u))\}$. Let $\theta(t, x, v, u) = \varphi(t, x) + \varphi^*(t, v - f(t, x, u)) - (v - f(t, x, u), x)$ and $\hat{U}(t, x) = H \times U(t, x)$. Then θ is measurable and $\hat{U}$ is graph measurable. Therefore $GrM = \{(t, x, v, u) \in T \times H \times H \times W: \theta(t, x, v, u) = 0, (t, x, v, u) \in Gr\hat{U}\} \in B(T) \times B(H) \times B(H) \times B(W)$. Also using the demiclosedness of the subdifferential operator, we can show that M has weakly compact values. To establish claim #1, we need to show that for every $\lambda \in \mathbb{R}$, $H(\lambda) = \{(t, x, v) \in T \times H \times H: h(t, x, v) \leq \lambda\}$ belongs to the Borel σ-field $B(T \times H \times H) = B(T) \times B(H) \times B(H)$. But since M has w-compact values $H(\lambda) = proj_{T \times H \times H}\hat{H}(\lambda)$, $\hat{H}(\lambda) = \{(t, x, v, u) \in T \times H \times H \times W: q(t, x, v, u) \leq \lambda\}$ and $q(t, x, v, u) = L(t, x, u) + \delta_{M(t, x, v)}(y)$. Note that because Y is separable, $B(Y) = B(Y_w) \Rightarrow B(W) = B(W_w)$ and W_w is compact, metrizable. Also $\hat{H}(\lambda) \in B(T) \times B(H) \times B(H) \times B(W_w)$ (hypothesis $H(L)$ and the graph measurability of M). Thus via Novikov's theorem, we get $H(\lambda) = proj_{T \times H \times H}\hat{H}(\lambda) \in B(T) \times B(H) \times B(H)$.

Claim #2: $(x, v) \to h(t, x, v)$ is l.s.c: We can easily check this claim by using the demiclosedness of the subdifferential operator and hypotheses $H(f)$, $H(U)$ and $H(L)$.

Claim #3: $v \to h(t,x,v)$ is convex: Note that $epih(t,x,\cdot) = \{[v,\lambda] \in H \times \mathbf{R}: h(t,x,v) \leq \lambda\} = Q(t,x)$ and the latter is convex for every $(t,x) \in T \times H$ (see hypothesis H_τ).

Let $\{(x_n, u_n)\}_{n \geq 1}$ be a minimizing sequence of admissible "state-control" pairs; i.e. $J(x_n, u_n) \downarrow m$. Let $v(\cdot) \in C(T,H)$ be the unique solution of $-\dot{v}(t) \in \partial\varphi(t,v(t))$ a.e., $v(0) = v_0$ (see [11]). Then exploiting the monotonicity of the subdifferential operator, we get

$$\tfrac{1}{2} \| v(t) - x_n(t) \|^2 \leq \int_0^t \| f(s, x_n(s), u_n(s)) \| \cdot \| v(s) - x_n(s) \| ds$$

$$\Rightarrow \| v(t) - x_n(t) \| \leq \int_0^t (a(s) + b(s) \| x_n(s) \|) ds \text{ (see Brezis [3], lemma A.5, p. 157)}$$

$$\Rightarrow \| x_n(t) \| \leq \| v \|_\infty + \| a \|_1 + \int_0^t b(s) \| x_n(s) \| ds$$

$$\Rightarrow \| x_n(t) \| \leq M_1 \text{ for all } t \in T \text{ and all } n \geq 1 \text{ (Gronwall's inequality).}$$

Set $\gamma(t) = a(t) + b(t)M_1 \in L_+^2$ and $B(\gamma) = \{h \in L^2(H): \|h(t)\| \leq \gamma(t) \text{ a.e.}\}$ and let $V = p(B(\gamma))$ with $p: L^2(H) \to C(T,H)$ be the solution map of $-\dot{x}(t) \in \partial\varphi(t,x(t)) + g(t)$ a.e., $x(0) = x_0$. We claim that $V \subseteq C(T,H)$ is compact. Indeed by lemma 6.1 of Yotsutani [11], $\|x(t') - x(t)\| \leq M_2(t'-t)^{1/2} M_2 > 0$ for all $x(\cdot) \in V \Rightarrow V$ is equicontinuous. Furthermore, using inequality 7.9 of [11], we have that $\|x(t)\|^2 + \varphi(t,x(t)) \leq M_3$, $M_3 > 0$. Thus $V(t) = \{x(t): x \in V\} \subseteq \{x \in H: \|x\|^2 + \varphi(t,x) \leq M_3\}$ and the latter is compact, since by hypothesis $H(\varphi)$, $\varphi(t,\cdot)$ is of compact type. So invoking the Arzela-Ascoli theorem, we have that $V \subseteq C(T,H)$ is compact. Also $\|\dot{x}_n\|_{L^2(H)} \leq M_2$ (see [11]). Thus, we may assume that $x_n \xrightarrow{s} x$ in V and $\dot{x}_n \xrightarrow{w} \dot{x}$ in $L^2(H)$. From claims #1, #2, and #3 and from Theorem 2.1 of Balder [2], we get $-\infty < \int_0^b h(t, x(t), -\dot{x}(t)) dt \leq \underline{\lim} \int_0^b h(t, x_n(t), -\dot{x}_n(t)) dt = m < \infty \Rightarrow h(t, x(t), -\dot{x}(t))$ is a.e. finite. Via a straightforward application of Aumann's selection theorem, we can get $u: T \to Y$ measurable, $u(t) \in M(t, x(t), -\dot{x}(t))$ a.e. and $h(t, x(t), -\dot{x}(t)) = L(t, x(t), u(t))$ a.e. $\Rightarrow J(x,u) = m$, and (x,u) is admissible.

<u>Q.E.D.</u>

3. Relaxed problem.

In the previous theorem, the convexity hypothesis H_τ is crucial in establishing the existence of an optimal pair. If hypothesis H_τ is not present, then the Cesari reduction technique employed in the previous proof, collapses. Nevertheless, it is interesting to study the behavior of the minimizing sequences, by characterizing their limit points as minimizers, of a new augmented system, known as the "relaxed system". There is no unique approach to relaxation. Here we propose a relaxed version of (∗), based on the "Gamkrelidze-Warga" approach that utilizes transition probabilities (parametrized measures). Let $W \in P_{wkc}(Y)$ as in $H(U)$ (Section 2). Equipped with the relative weak topology, W becomes a compact metric space. Let $M^1_+(W)$ be the space of probability measures defined on W. Endowed with the weak (narrow) topology, $M^1_+(W)$ becomes a compact metric space. Let $\Sigma(t) = \{\mu \in M^1_+(W) : \mu(U(t)) = 1\}$ (note here we will assume U to be state independent). Our version of the relaxed problem is the following:

$$J_r(x,\lambda) = \int_0^b \int_W L(t,x(t),u)\lambda(t)(du)dt \to \inf = m_r$$

s.t. $-\dot{x}(t) \in \partial\varphi(t,x(t)) + \int_W f(t,x(t),u)\lambda(t)(du)$ a.e. $\quad\quad (*)_r$

$x(0) = x_0, \lambda \in S_\Sigma = \{$measurable selectors of $\Sigma(\cdot)\}$.

$\underline{H(L)_1}$: $L:T \times H \times W \to \mathbb{R}$ is measurable in t, continuous in (x,u) and $|L(t,x,u)| \leq \theta_n(t)$ a.e. for all $[x,u] \in H \times W$, $\|x\| \leq n$ and with $\theta_n(\cdot) \in L^1_+$.

$\underline{H(U)_1}$: $U:T \to P_{wkc}(Y)$ is a measurable multifunction s.t. $U(t) \subseteq W$ μ-a.e., $W \in P_{wkc}(Y)$.

$\underline{H(f)_1}$: $f:T \times H \times Y \to H$ is a map satisfying hypothesis $H(f)$ in Section 2 and in addition $|f(t,x',u) - f(t,x,u)| \leq k(t)|x' - x|$ a.e. with $k(\cdot) \in L^1_+$.

Theorem 2: _If hypotheses $H(\varphi)$, $H(f)_1$, $H(U)_1$, $H(L)_1$, H_0 hold and $x_0 \in H$, then $(*)_r$ has a solution and $m = m_r$._

Proof: Since the control enters linearly in the dynamics and in the cost functional of $(*)_r$, the existence part follows easily by using the direct method (note that S_Σ is $w(L^\infty(M_b(W)), L^1(C(W))$-compact). So let $[x,\lambda] \in C(T,H) \times S_\Sigma$ s.t. $J_r(x,\lambda) = m_r$. Invoking corollary 4, p. 377 of Saint Beuve [10], we can find $\{u_n\}_{n \geq 1} \subseteq S^1_U$ s.t. $\delta_{u_n} \xrightarrow{w^*} \lambda$ (δ_{u_n} is the Dirac transition probability concentrated on u_n). Also, we can easily check that the corresponding states $x_n(\cdot)$ converge to $x(\cdot)$ in $C(T,H)$. Then we have:

$$J_r(x_n, \delta_{u_n}) = J(x_n, u_n) = \int_0^b L(t,x_n(t),u_n(t))dt = \int_0^b \int_W L(t,x_n(t),u)\delta_{u_n(t)}(du)dt$$
$$= ((\hat{L}_n, \delta_{u_n}))_0$$

where $\hat{L}_n(t,\cdot) = L(t,x_n(t),\cdot) \in C(W)$ and $((\cdot,\cdot))_0$ denotes the duality brackets of the pair $(L^1(C(W)), L^\infty(M_b(W)))$. Then if $\hat{L}(t,\cdot) = L(t,x(t),\cdot) \in C(W)$, we have for some $u_n \in W$, $u_n \xrightarrow{w} u$:

$$\|\hat{L}_n(f,\cdot) - \hat{L}(t,\cdot)\|_\infty = |L(t,x_n(t),u_n) - L(t,x(t),u_n)| \to 0 \text{ as } n\to\infty.$$

$$\Rightarrow ((\hat{L}_n, \delta_{u_n}))_0 \to ((\hat{L}, \lambda))_0$$

$$\Rightarrow m \le \int_0^b \int_W L(t,x(t),u)\lambda(t)(du)dt = J_r(x,\lambda) = m_r; \text{ i.e. } m = m_r.$$

Q.E.D.

4. An example.

Let $Z \subseteq \mathbf{R}^N$ be a bounded domain with smooth boundary $\Gamma = \partial Z$, and $T = [0,r]$. We consider the following nonlinear parabolic optimal control problem:

$$\left| \begin{array}{l} \int_0^r \int_Z L(t,z,x(t,z),u(t,z))dzdt \to \inf = \hat{m} \\[2mm] \text{s.t.} \dfrac{\partial x(t,z)}{\partial t} - \sum\limits_{i,j=1}^N \dfrac{\partial}{\partial z_j}\left(a_{ij}(t,z)\dfrac{\partial x(t,z)}{\partial z_i}\right) + \beta(x(t,z)) \ni f(t,z,x(t,z), \int_Z K(z,z')u(z')dz') \\[2mm] x(0,z) = x_0(z),\ x\big|_{T\times\Gamma} = 0,\ u(t,z) \in U(t,z) \text{a.e.},\ u(\cdot,\cdot)=\text{measurable}. \end{array} \right| \quad (**)$$

$H(\alpha):$ $\quad a_{ij} \in L^\infty(T\times Z),\ a_{ij} = a_{ji},\ \sum\limits_{i,j=1}^N a_{ij}(t,z)\theta_i\theta_j \ge c\|\theta\|^2\ (t,z| \in T\times Z, \theta \in \mathbf{R}^N$ and

$\qquad |a_{ij}(t,z) - a_{ij}(t',z)| \le k|t'-t|$ on Z, with $k > 0$.

$H(\beta):$ $\quad \beta = \partial j$, where $j:\mathbf{R}\to\mathbf{R}_+$ is proper, convex, l.s.c.

$H(f)_2:$ $\quad f:T\times Z\times\mathbf{R}\times\mathbf{R}\to\mathbf{R}$ is a function s.t. $(t,z)\to f(t,z,x,u)$ is measurable, $(x,u)\to f(t,z,x,u)$ is continuous, $|f(t,z,x,u)| \le a(t,z) + b(t,z)(|x|+|u|)$ a.e. with $a,b \in L^2(T\times Z)$.

$H(U)_2:$ $\quad U(t,z) = \{\theta \in \mathbf{R}: m_1(t,z) \le \theta \le m_2(t,z)\},\ m_1,\ m_2 \in L^\infty(T\times Z),\ m_1 < m_2$ and $K \in L^2(T\times Z)$.

$H'_r:$ $\quad \{f(t,z,x,\int_Z K(z,z')udz'): u \in U(t,z')\}$ is convex.

$H(L)_2:$ $\quad L:T\times Z\times\mathbf{R}\times\mathbf{R}\to\overline{\mathbf{R}}$ is measurable, l.s.c. in (x,u), convex in u and $\psi(t,z) - M(z)(|x|+|u|) \le L(t,z,x,u)$ a.e., $\psi \in L^1(T\times Z),\ M(\cdot) \in L^1$.

$H'_0:$ $\quad \hat{m} < \infty$ and there exists $g \in L^2(Z)$ s.t. $g(z) \in \partial j(x_0(z))$ a.e.

Let $\varphi(t,x) = \frac{1}{2}\sum\limits_{i,j=1}^n \int_Z a_{ij}(t,z)D_ix(z)D_jx(z)dz + \int_Z j(x(z))dz$ if $x \in H_0^1(Z),\ j(x(\cdot)) \in L^1(Z)$ and $+\infty$ otherwise. As in [9], we can check that $\varphi(\cdot,\cdot)$ satisfies hypothesis $H(\varphi)$ and $\partial\varphi(t,x) = \{-\sum\limits_{i,j=1}^N D_j(a_{ij}(t,\cdot)D_ix(\cdot)) + g(\cdot):\ g \in L^2(Z),\ g(z) \in \partial j(x(z)) = \beta(x(z))$ a.e.$\}$. Also set $\hat{f}(t,x,u)(z) = f(t,z,x(z),\int_Z K(z,z')u(z')dz'),\ \hat{U}(t) = \{u \in L^2(Z): m_1(t,z) \le u(z) \le m_2(t,z)$ a.e.$\} \subseteq W = \{u \in L^2(Z):\ \|u\|_2 \le max[\|m_1\|_\infty, \|m_2\|_\infty]\} \in P_{wkc}(L^2(Z))$ and $\hat{L}(t,x,z) = \int_Z L(t,z,x(z),u(z))dz$. It is easy to check that the above items satisfy hypotheses $H(f)$,

$H(U)$ and $H(L)$. Furthermore, hypothesis H'_r, the convexity of $\hat{L}(t,x,\cdot)$ and the convexity of the values of $U(\cdot)$, imply that hypothesis H_r is satisfied.

Rewrite problem (**) in the following equivalent abstract form:

$$\left| \begin{array}{l} \hat{J}(x,u) = \displaystyle\int_0^b \hat{L}(t,x(t),u(t))dt \to \inf = \hat{m} \\[2mm] s.t. -\dot{x}(t) \in \partial\varphi(t,x(t)) + \hat{f}(t,x(t),u(t)) \text{ a.e.} \\[2mm] x(0) = \hat{x}_0 = x_0(\cdot) \in L^2(Z) = H, \ u(t) \in U(t) \text{ a.e.} \end{array} \right. \quad (**)'$$

Using Theorem 1, we get:

Theorem 3: _If_ hypotheses $H(a)$, $H(\beta)$, $H(f)_2, H'_r, H(L)_2$, H'_0 hold and $x_0(\cdot) \in H_0^1(Z) \cap H^2(Z)$

then problem (**) admits an optimal pair $(x,u) \in C(T, L^2(Z)) \times L^2(T \times Z)$ s.t.

$$\frac{\partial x}{\partial t} \in L^2(T \times Z), \sqrt{t}\frac{\partial x}{\partial t} \in L^2(T \times Z).$$

Our formulation is general enough to incorporate also problems of the form $-\dot{x}(t) \in N_{K(t)}(x(t)) + f(t,x(t),u(t))$ a.e., $x(0) = x_0$, with $K(\cdot)$ h-Lipschitz (here $N_{K(t)}(\cdot)$ is the normal cone to the set $K(t)$). Such inclusions are known in the literature as "differential variational inequalities" and appear in mathematical economics and in theoretical mechanics.

References:

[1] N. Ahmed, "Optimal control of a class of strongly nonlinear parabolic systems", *J. Math. Anal. Appl.* **61**, (1977), pp. 188-207.

[2] E. Balder, "Necessary and sufficient conditions for a L_1-strong-weak lower semicontinuity of integral functionals", *Nonl. Analysis-TMA* **11**, (1987), pp. 1399-1404.

[3] H. Brezis, "*Operateurs Maximaux Monotones*", North Holland, Amsterdam (1973).

[4] E. Flytzanis, N.S. Papageorgiou, "On the existence of optimal controls for a class of nonlinear infinite dimensional systems", *Math. Nachrichten* **150**, (1991), pp. 203-217.

[5] M. Ganesh, M. Joshi, "Optimality of nonlinear control systems", *Nonl. Anal.-TMA* **16**, (1991), pp. 553-566.

[6] S.-H. Hou, "Existence theorems of optimal control problems in Banach spaces", *Nonl. Anal.-TMA* **7**, (1983), pp. 239-257.

[7] J.-L. Lions, "Optimisation pour certaines classes d'equations d'evolution nonlineaires", *Ann. Mat. Pura ed Appl.* **72**, (1966), pp. 275-294.

[8] N.S. Papageorgiou, "Existence of optimal controls for nonlinear systems in Banach space", *J. Optim. Th. Appl.* **53**, (1987), pp. 451-459.

[9] N.S. Papageorgiou, "On evolution inclusions associated with time dependent convex subdifferentials", *Comm. Math. Univ. Carol.* **31**, (1990), pp. 517-527.

[10] M.-F. Saint-Beuve, "Some topological properties of vector measures of bounded variation and its applications", *Ann. Mat. Pura ed Appl.* **t. CXVI**, (1978), pp. 317-379.

[11] S. Yotsutani, "Evolution equations associated with subdifferentials", *J. Math. Soc. Japan* **31**, (1978), pp. 623-646.

University of the Agean	and	Florida Institute of Technology
Department of Mathematics		Department of Applied Mathematics
Karlovassi, Samos 83000		150 W. University Blvd.
GREECE		Melbourne, Florida 32901-6988, USA

F ROTHE

A priori estimates for nonlinear Poisson equations and phase plane analysis of the Emden–Fowler equation

Abstract. The bootstrap and feedback method to derive uniform bounds from weaker a priori bounds was developed by the author for parabolic equations. In the present note, it is applied to the boundary value problem for the nonlinear Poisson equation in a bounded domain. The Emden-Fowler equation with different exponents gives illustrative counterexamples which allow us to confirm that rather optimal results are obtained. The phase plane analysis for the Emden-Fowler equation yields bifurcations at two values β_1 and β_2. For $\beta < \beta_1$, by the bootstrap and feedback method, finite L_1-norm implies finite L_∞-norm, hence a uniform bound. On the other hand, for $\beta \geq \beta_1$, the phase plane analysis of the Emden-Fowler equation yields unbounded weak solutions of finite mass. Analogously, for $\beta < \beta_2$, the bootstrap and feedback method shows that finite Dirichlet integral implies a uniform bound. On the other hand, for $\beta > \beta_2$, the phase plane analysis of the Emden-Fowler equation yields a sequence of classical solutions with bounded Dirichlet integrals but maximum norms growing to infinity.

Key words. Poisson equation, strong and distributional solutions, a priori bounds, phase plane, Hopf- and saddle-node bifurcation, negative Bendixon criterion

AMS(MOS) subject classifications: 34C25, 34A25, 34A55

1. Introduction.

It is well known that the nonlinear Poisson equation

$$\Delta\phi + f(x,\phi) = 0 \quad \text{for all } x \in \Omega \tag{1.1}$$

occurs in many different contexts. In this introduction, we consider our motivating example and derive (1.1) in the context of an equilibrium configuration for a self-gravitating gas cloud. In this context, the a priori estimates of section 2 correspond to estimates of mass and gravitational energy.

In hydrostatic equilibrium, the potential U of the gravitational accelaration and the density ϱ and pressure p of a self-gravitating gas satisfy the Poisson equation

$$\Delta U = 4\pi G \varrho \tag{1.2}$$

and the hydrostatic equation

$$\nabla p + \varrho \nabla U = 0 \tag{1.3}$$

Note that $-\varrho \nabla U$ is the gravitional force per volume. In order to solve (1.2)(1.3), we need an equation of state $p = p(\varrho)$ relating pressure and density. We introduce the pressure potential

$$\phi(\varrho) = \int_0^\varrho \frac{1}{\varrho'} \frac{dp}{d\varrho'} d\varrho' \tag{1.4}$$

(1.3) and (1.4) imply $U + \phi = $ const. Because of thermodynamic stability, it is reasonable to assume that the functions $p = p(\varrho)$ and hence $\phi = \phi(\varrho)$ are increasing. The inverse function $\varrho = \varrho(\phi)$ exists and (1.2) yields the nonlinear Poisson equation

$$\Delta\phi + 4\pi G \varrho(\phi) = 0 \qquad (1.5)$$

The astrophysical problem corresponds to a free boundary value problem with given total mass

$$M = \int_{B(R)} \varrho\, dx$$

inside the ball $B(R) = \{\, x \in \mathbb{R}^3\,;\, |x| \leq R\,\}$ and the boundary condition $p = 0$ at the unknown radius R. The nonlinear Poisson equation (1.5) occurs in the construction of special solutions of the Vlasov-Poisson system, too (see Batt and Pfaffelmoser [1]).

Shapiro and Teukolsky [10] give detailed derivations of equations of states over the whole range of pressures occuring in astrophysical problems. In some important limiting cases, this leads to explicit equations of state of the *polytropic* form

$$p = K\varrho^\Gamma \qquad (1.6)$$

For the polytropic equations of state, the nonlinear Poisson equation is

$$\Delta\phi + a^{-2}\phi^\beta = 0 \qquad (1.7)$$

with $\beta = (\Gamma - 1)^{-1}$. The phase plane analysis of the radially symmetric solutions of (1.7) leads to the bifurcation values $\beta_1 = 3$ and $\beta_2 = 5$ mentioned in the abstract. For all $\beta \in [0, 5)$, the Dirichlet problem for (1.7) in a ball has a unique radially symmetric bounded solution. Besides this unique classical solution, the radially symmetric problem has a whole continuum of solutions with a singularity at the center. For $\beta \in [0, 3)$, these solutions are not weak solutions of the Poisson equation in the sense of distributions. Instead, they correspond to a physical system with a point mass in the center, the attracting force of which causes the singularity.

The situation changes for $\beta \in [3, 5)$: the solutions with a singularity are weak distributional solutions of (1.7). Physically, this means that gravity is strong enough to produce a singularity in the center. We prove in a forthcoming paper that for $\beta \in [3, 5)$, this is an unstable equilibrium for the time evolution of the Poisson-Euler system (see [10], too).

2. Some a priori Estimates for the Nonlinear Boundary Value Problem.

In this section, we consider the nonlinear Poisson equation (1.1) in a bounded domain $\Omega \subset \mathbb{R}^N$ of arbitrary dimension N with the Dirichlet boundary condition

$$\phi(x) = 0 \quad \text{for all } x \in \partial\Omega \qquad (2.1)$$

and construct a priori estimates. We derive uniform bounds for the potential ϕ starting from some less restrictive a priori bounds. In the context of the introduction, we think of bounds for the total mass or energy. The estimates holds for classical as well as mild solutions, which are defined to be solutions of the integral equation corresponding to (1.1)(2.1). The complete proofs will be published elsewhere. They use similar "bootstrap and feedback" methods as applied to parabolic problems in Rothe [9].

For any measurable function $f = f(x)$ on the domain $x \in \Omega$ and any $q \in [1, \infty]$, we introduce the L_q-norm $\|f\|_q$ and uniform norm $\|f\|_\infty = \sup\{|\phi(x)| : x \in \Omega\}$. As usual, we denote by $L_q(\Omega)$ the set of all measurable function such that $\|f\|_q < \infty$.

For the nonlinear term $f = f(x, \phi)$, we need the following assumptions with $q \in [1, \infty]$, $\beta \in (0, \infty)$ and $U \geq 0$ to be specified lateron.

(f_0) The function $f : (x, \phi) \in \overline{\Omega} \times \mathbb{R} \mapsto f(x, \phi) \in \mathbb{R}$ is C^1 in both variables.
($f_{\beta,q,U}$) There exists a nonnegative function $c \in L_q(\Omega)$ such that

$$|f(x, \phi)| \leq c(x) \max(U^\beta, |\phi|^\beta) \quad \text{for all } x \in \overline{\Omega} \text{ and } \phi \in \mathbb{R}$$

To derive uniform estimates for the solutions of the nonlinear Dirichlet problem (1.1)(2.1), we need two kinds of assumptions: firstly a growth restriction for the nonlinear term f in (1.1), secondly a primary (rather weak a priori estimate). In Proposition 2.1, we assume a primary estimate for the righthand side $f = f(x, \phi(x))$ and in Proposition 2.2, this is an estimate for the solution ϕ.

PROPOSITION 2.1. *We assume that the right-hand side $f = f(x, \phi(x))$ of the Dirichlet boundary value problem (1.1) (2.1) satisfies the primary estimate*

$$\|f\|_1 < \infty \tag{2.2}$$

and that the source term satisfies (f_0) and ($f_{\beta,q,U}$) for some $U \geq 0$. For space dimension $N \geq 2$, we have to assume that the exponents $q \in [1, \infty], \beta \in (0, \infty)$ satisfy

$$\frac{1}{q} < \frac{2}{N} \quad \text{and} \quad \beta < \frac{1 - 1/q}{1 - 2/N} \tag{2.3}$$

Then there exist exponents σ and ϱ and a constant $K(\Omega, \sigma, \varrho)$ such that the solution ϕ can be estimated uniformly by

$$\|\phi\|_\infty \leq \max\left(U, K(\Omega, \sigma, \varrho) \|c\|_q^\sigma \|f\|_1^\varrho\right) \tag{2.4}$$

PROPOSITION 2.2. *We assume that a (classical or mild) solution ϕ of the nonlinear Dirichlet problem (1.1) (2.1) satisfies the primary estimate*

$$\|\phi\|_r < \infty \tag{2.5}$$

for some $r \in (0, \infty]$ and that the source term satisfies (f_0) and ($f_{\beta,q,U}$) for some $U \geq 0$. If the exponents $q \in [1, \infty], r \in (0, \infty], \beta \in (0, \infty)$ satisfy the assumption

$$\begin{aligned}\frac{1}{q} &< \min\left(\frac{2}{N}, 1\right) \\ \beta &< 1 + r\left[\min\left(\frac{2}{N}, 1\right) - \frac{1}{q}\right]\end{aligned} \tag{2.6}$$

then there exist σ, ϱ and a constant $K(\Omega, \sigma, \varrho)$ such that the solution ϕ can be estimated uniformly by

$$\|\phi\|_\infty \leq \max\left(U, K(\Omega, \sigma, \varrho) \|c\|_q^\sigma \|\phi\|_r^\varrho\right) \tag{2.7}$$

It is rather instructive to apply Proposition 2.1 and 2.2 to the special astrophysical problems from the introduction. Indeed, this allows us to show that the assumption (2.3) and (2.6) made for the exponents are optimal. Following Batt and Pfaffelmoser [1], we consider the Dirichlet boundary value problem

$$\begin{aligned}\Delta\phi + r^\alpha \phi |\phi|^{\beta-1} &= 0 \quad \text{for all } x \in B(R) \\ \phi(x) &= d \quad \text{for all } x \in \partial B(R)\end{aligned} \tag{2.8}$$

in the ball $\Omega = B(R) = \{x \in \mathbb{R}^N; |x| \le R\}$ of arbitrary dimension $N \ge 3$ and assume that $\alpha > -2$ and $\beta > 1$. Usually, we take zero boundary condition $d = 0$. We use the functionals

$$D_1(\phi, r) \equiv \int_{B(r)} |x|^\alpha |\phi|^\beta \, dx \quad \text{and} \quad D_2(\phi, r) \equiv \int_{B(r)} (\nabla \phi)^2 \, dx \qquad (2.9)$$

We take as a primary estimate either a bound for the total mass $D_1(\phi, R)$ or the Dirichlet integral $D_2(\phi, R)$. We get the following two critical exponents. Let $\overline{\beta}_1 = \overline{\beta}_2 = \infty$ for dimension $N = 1$ or $N = 2$. For dimension $N \ge 3$, we define

$$\overline{\beta}_1 = \frac{N + \overline{\alpha}}{N - 2} \quad \text{and} \quad \overline{\beta}_2 = \frac{N + 2 + 2\overline{\alpha}}{N - 2} \quad \text{with} \quad \overline{\alpha} = \begin{cases} \alpha & \text{for } \alpha \in (-2, 0] \\ 0 & \text{for } \alpha \in [0, \infty) \end{cases} \qquad (2.10)$$

PROPOSITION 2.3. *We consider a classical or mild solution of the Dirichlet value problem (2.8). We assume that $\alpha > -2$ and consider case (1) or (2). Let $i = 1$ or $i = 2$ depending on the case considered.*

(1) $\beta < \overline{\beta}_1$, $\phi \ge 0$ and the total mass $D_1(\phi, R) < \infty$ is finite.
(2) $\beta < \overline{\beta}_2$ and the Dirichlet integral $D_2(\phi, R) < \infty$ is finite.

Then ϕ is a classical solution. There exist ϱ_i and $K_i(R, \alpha, \beta)$ such that we get the uniform estimate

$$\|\phi\|_\infty \le K_i(R, \alpha, \beta) D_i(\phi, R)^{\varrho_i} \qquad (2.11)$$

PROOF: We consider the case $i = 2$, from which the proof for $i = 1$ becomes obvious. By the Sobolev inequality, the estimate for the Dirichlet integral (2.9) implies the estimate (2.5) with $r = \infty$ for space dimension $N = 1$, all $r < \infty$ for space dimension $N = 2$ and $r \le 2^*$ for $N \ge 3$, where the critical exponent is

$$2^* = \frac{2N}{N-2} \qquad (2.12)$$

In assumption $(f_{\beta,q,U})$ take $c(x) = |x|^\alpha$. An elementary calculation yields $c \in L_q(B(R))$ for $1/q > -\alpha/N$ if $\alpha < 0$ or $1/q = 0$ if $\alpha \ge 0$. Inserting r and q into (2.6) yields the assertion. $\square$

3. Phase Plane Analysis of the Emden-Fowler Equation.

The radially symmetric solutions of the nonlinear Poisson equation (2.8) satisfy the ordinary differential equation

$$\frac{d^2 \phi}{dr^2} + \frac{N-1}{r} \frac{d\phi}{dr} + r^\alpha \phi |\phi|^{\beta-1} = 0. \qquad (3.1)$$

Equation (3.1) is reduced to a two-dimensional autonomous system. We introduce the new independent variable t by

$$r = e^t, \quad \frac{d}{dt} = r \frac{d}{dr} \qquad (3.2)$$

and new dependent variables

$$x = r^\mu \phi \quad \text{and} \quad y = -r^{\mu+1} \frac{d\phi}{dr} \qquad (3.3)$$

with the parameter μ given by

$$\mu = \frac{2+\alpha}{\beta - 1}. \qquad (3.5)$$

We get the system
$$\frac{dx}{dt} = \mu x - y$$
$$\frac{dy}{dt} = x|x|^{\beta-1} + (2 - N + \mu)y \qquad (3.6)$$

Several similar reductions occurs in the literature. See van den Broek and Verhulst [4] for a transformation involving additional denominators as well as Joseph and Lundgren [6]. For system (3.6) with parameters $N \in (2, \infty)$, $\alpha \in (-2, \infty)$, $\beta \in (1, \infty)$ and μ given by formula (3.5), we do a phase plane analysis. Especially, we have to find the bifurcation values for the parameter β, at which the topology of the flow changes.

At first, one determines the equilibria and their topological types and bifurcations. The trivial equilibrium $(x_0, y_0) = (0, 0)$ exists for all parameters. The flow in a neighbourhood of the trivial equilibrium is determined by the Jakobi matrix at $(0,0)$, which has the eigenvalue $2 - N + \mu > 0$ with eigenvector $(1, N - 2)^T$ and the eigenvalue $\mu > 0$ with eigenvector $(1, 0)^T$.

There exists a unique solution $(x_E(t), y_E(t))$ which is tangent to the side eigenvector $(1, 0)^T$ and satisfies
$$\lim_{t \to -\infty} e^{-\mu t} x_E(t) = 1$$
$$\lim_{t \to -\infty} e^{-\mu t} y_E(t) = 0 \qquad (3.7)$$

The equilibrium $(0, 0)$ changes from an unstable two tangent node into a saddle as the parameter β increases beyond $\beta = \beta_1$ given by $2 - N + \mu = 0$ and hence
$$\beta_1 = 1 + \frac{2 + \alpha}{N - 2} \qquad (3.8)$$

Note that the solution (x_E, y_E) depends continuounsly on β in a neighbourhood of $\beta = \beta_1$. For $\beta > \beta_1$, it is the unstable seperatrix of the saddle $(0, 0)$.

At $\beta = \beta_1$, a symmetric pair $(x_e, \mu x_e)$ and $(-x_e, -\mu x_e)$ of nontrivial equilibria bifurcate. The nontrivial equilibria exist if and only if the equation
$$|x_e|^{\beta-1} + \mu(2 - N + \mu) = 0 \qquad (3.9)$$

has a positive solution $x_e > 0$, hence for $\mu(2 - N + \mu) < 0$ which implies $\beta > \beta_1$.

The Jakobi matrix at the nontrivial equilibrium $(x_e, \mu x_e)$ has the determinant and trace
$$\text{Det } J_e = -(2 + \alpha)(2 - N + \mu) > 0 \quad \text{and} \quad \text{Tr } J_e = 2 - N + 2\mu \qquad (3.10)$$

A further change of stability occurs for Tr $J_e = 0$, hence for $\beta = \beta_2$ given by
$$\beta_2 = 1 + 2\frac{2 + \alpha}{N - 2} \qquad (3.11)$$

at which parameter we get a Hamiltonian system with a pair of homoclinic loops.

We turn to the global properties of system (3.6). System (3.6) is indeed a Hamiltonian system with friction and a Lyapunov functional can be constructed explicitly. One can show that all trajectories can be extended globally for all $t \in (-\infty, \infty)$. No limit cycles or homoclinic loops exist and in the limits $t \to \pm\infty$, all trajectories converge either to one of the at most three equilibrium points or escape to infinity. Positive solutions of (3.6) correspond to trajectories in the positive half plane
$$H = \{(x, y) \in \mathbb{R}^2 \, ; \, x > 0\} \qquad (3.12)$$

4. The Different Types of Solutions of the Emden-Fowler Equation.

We use the formula
$$\phi(r) = r^{-\mu} x(\ln r) \tag{4.1}$$
to translate the solutions $(x(t), y(t))$ of system (3.6) back to solutions of (2.8). From section 3, we get results about the the different types of nonnegative solutions of (2.8). Following general usage, we define

DEFINITION 4.1. *A nonnegative continuous solution* $r \in I \mapsto \phi(r) \in [0, \infty)$ *of the Emden-Fowler equation (2.8) on a finite or infinite interval* $I \subset [0, \infty)$ *is called*
(E)- *solution iff* $I = [0, r_2]$ *or* $I = [0, \infty)$ *and* $\phi(0)$ *is finite.*
(M)- *solution iff* $I = (0, r_2]$ *or* $I = (0, \infty)$ *and* $\lim_{r \to 0} \phi(r)$ *is* ∞ *or does not exist.*
(F)- *solution iff* $I = [r_1, r_2]$ *or* $I = [r_1, \infty)$ *where* $r_1 > 0$ *and* $\phi(r_1) = 0$.

The solution of the Emden-Fowler equation (2.8) corresponding to the unstable seperatrix (x_E, y_E) of the trivial equilibrium $(0, 0)$ in the (x, y)-phase plane is an (E)-solution. The potential ϕ_E is continuous at $r = 0$ and $\phi_E(0) = \lim_{r \to 0} \phi_E(r) = 1$. Let the index E denote quantities corresponding to ϕ_E. All other (E)-solutions can be produced from ϕ_E by application of the homology transformation. They are
$$\phi(r; \varrho) = \varrho^{-\mu} \phi_E(r/\varrho) \tag{4.2}$$
with arbitrary homology parameter $\varrho > 0$.

For $\beta \in (1, \beta_2)$, there exists a least "time" T_0 such that
$$x_E(T_0) = 0 \quad \text{and} \quad y_E(T_0) > 0 \tag{4.3}$$
To satisfy the Dirichlet boundary condition with $d = 0$, the corresponding trajectory (x, y) has to leave the positive half plane H in forward time. For $\beta < \beta_2$, the unstable seperatrix (x_E, y_E) leaves the half plane H for $t > T_0$. Hence, it is possible to satisfy the zero boundary condition $\phi(R; \varrho) = 0$ by appropriate choice of the parameter ϱ in (4.2).

The nontrivail equilibrium $(x_e, \mu x_e)$ become stable for $\beta > \beta_2$ via the degenerate Hopf- and homoclinic bifurcation. Hence the trajectory (x_E, y_E) does no longer leave the half-plane H and there exist no solutions of (2.8) with boundary value $d = 0$.

With boundary value $d = 0$, similar nonexistence results follow from the identity of Pohozaev [8]. The situation is different for $d > 0$ occuring in Proposition 5.3 below. More general existence- and nonexistence results for bounded and unbounded domains related to the crital Sobolev exponent $\beta = \beta_2$ are given by Peletier and Serrin [7]. Starting in [3], Brézis and Nirenberg have considered the borderline case with $\beta = \beta_2$.

For the (M)-solutions, we do more detailed study of their singularity for $r \to 0$. Going back to the (x, y)-system, we have to study the trajectories in the bounded subdomain $D \subset H$ enclosed by $t \in (-\infty, T] \mapsto (x_E(t), y_E(t))$ and the straight line segment $l = \{(0, y); 0 \leq y \leq y_E(T_0)\}$. For all $\beta < \beta_2$, these trajectories leave the half plane H in forward time by crossing l. Hence a zero Dirichlet boundary condition $\phi(R) = 0$ can be satisfied.

Concerning the asymptotic behavior for $t \to -\infty$, the situation is different depending whether $\beta < \beta_1$ or $\beta > \beta_1$. Note that the change of behavior at $\beta = \beta_1$ is related to the pitchfork bifurcation creating the nontrivial equilibra. For $\beta \in (1, \beta_1]$, the behavior of (M)-solutions is determined by the (x, y)-system (3.6) near the trivial equilibrium $(0, 0)$; whereas for $\beta > \beta_1$, the nontrivial equilibrium $(x_e, \mu x_e)$ bifurcates.

For $\beta < \beta_1$, their asymptotic behavior is given by
$$\lim_{t \to -\infty} e^{-(2-N+\mu)t} x(t) = k \quad \text{and} \quad \lim_{t \to -\infty} e^{-(2-N+\mu)t} y(t) = (N-2)k \tag{4.4}$$

where $k > 0$ depends on the individuel trajectory.

For $\beta \in (\beta_1, \beta_2)$, the nontrivial equilibrium $(x_e, \mu x_e)$ exists inside the domain D. Since it is totally unstable,
$$\lim_{t \to -\infty} x(t) = x_e \quad \text{and} \quad \lim_{t \to -\infty} y(t) = \mu x_e \tag{4.5}$$
holds for all trajectories in the open domain D. For the Emden-Fowler equation, the equilibrium $(x_e, \mu x_e)$ gives rise to the power law solution
$$\phi_M(r) = r^{-\mu} x_e \sim r^{-\frac{2+\alpha}{\beta-1}} \tag{4.6}$$
This is the basic (M)-solution for $\beta > \beta_1$ denoted by the index M in the following.

For $\beta > \beta_2$, (4.6) corresponds to the isolated unique (M)-solution ϕ_M. Indeed, in the (x,y)-phase plane, the nontrivial equilibrium has turned stable by a degenerate Hopf bifurcation. Thus the trajectories in a neighbourhood $N \ni (x_e, \mu x_e)$ escape N for $t \to -\infty$. Hence their asymptotic behavior for $t \to -\infty$ respectively $r \to 0$ prevents them to produce (M)-solutions.

PROPOSITION 4.2. *Whole continua of (M)-solutions exist for $\beta \in (1, \beta_2]$. For $\beta > \beta_2$, there exists only the single exceptional (M)-solution (4.6). For $r \to 0$, the singularities of the potential $\phi(r)$ are:*
$$\phi(r) \sim \begin{cases} r^{-(N-2)} & \text{for } \beta \in (1, \beta_1) \\ (\ln r)^{-1} r^{-(N-2)} & \text{for } \beta = \beta_1 \\ r^{-\frac{2+\alpha}{\beta-1}} & \text{for } \beta \in (\beta_1, \infty) \end{cases} \tag{4.7}$$

5. Counterexamples to A Priori Estimates from the Emden-Fowler Equation.

To obtain counterexamples proving the a priori bounds for strong and mild solutions of Dirichlet boundary value problems obtained in Proposition 2.3 are optimal, we consider the Dirichlet boundary value problem (2.8). Assume $d \geq 0$ and $N > 2$ throughout. Besides the potential $\phi(r)$, we consider the functionals (2.9). Because of (3.3), we get in the radially symmetric case
$$\begin{aligned} D_1(\phi, r) &= \Sigma_{N-1} r^{N-2-\mu} y(\ln r) \\ D_2(\phi, r) &= \Sigma_{N-1} \int_{-\infty}^{\ln(r)} e^{(N-2-2\mu)t} y^2(t)\, dt \end{aligned} \tag{5.1}$$

The meaning of the singularity of (M)-solutions for the original nonlinear Poisson problem become obvious in the following Lemma based on a well-known calculation from the theory of distributions.

LEMMA 5.1. *Assume that $\phi = \phi(r)$ is a (M)-solution such that $D_1(R) < \infty$ for some $R > 0$. Define the point mass at zero by*
$$D_1(\phi, 0) = \lim_{R \to 0} D_1(\phi, R) \tag{5.2}$$
In the sense of distributions, the function ϕ satisfies the following partial differential equation:
$$\Delta \phi + r^\alpha \phi^\beta = D_1(\phi, 0)\, \delta \tag{5.3}$$
where δ denotes the N-dimensional Dirac distribution supported at $x = 0 \in \mathbb{R}^N$.

As can be seen from (4.7), a singular point mass
$$D_1(\phi, 0) = \Sigma_{N-1} \lim_{t \to -\infty} e^{(N-2-\mu)t} y(t) > 0 \tag{5.4}$$

at the origin occurs for (M)-solutions ϕ if $\beta \in (1, \beta_1)$. In these cases, $\phi(r)$ is not a weak solution of the physical meaningfull nonlinear Poisson equation (2.8), but a weak solution of (5.3). Thus it describes a gas sphere with an additional attracting point mass at its center.

For $\beta > \beta_1$, the basic (M)-solution ϕ_M appears, for which the functional

$$D_1(\phi_M, r) = \Sigma_{N-1} \mu x_e r^{-\mu} \sim r^{\frac{(N-2)(\beta-\beta_1)}{\beta-1}} \quad (5.5)$$

is easy to calculate. Hence $D_1(\phi_M, 0) = 0$.

Let $N > 2$, $\beta \in (\beta_1, \beta_2)$. Since $\phi(r) \simeq \phi_M(r)$ as $r \to 0$ holds for all (M)-solutions, we get $D_1(\phi, r) \simeq D_1(\phi_M, r)$ and hence $D_1(\phi, 0) = 0$. The case $\beta = \beta_1$ needs as more detailed analysis, which we skip in this short note. For $\beta \geq \beta_2$, we have ϕ_M as unique isolated (M)-solution.

Hence for $\beta \in [\beta_1, \infty)$, all the (M)-solutions $\phi(r)$ satisfy the nonlinear Poisson equation (2.8) in the sense of distributions. They have bounded mass in any ball $B(R)$, but they are not uniformly bounded because of their singularity as $r \to 0$. Thus, we have a counterexample showing that Proposition 2.1 does not hold without the assumption (2.3) saying that $\beta < \beta_1$.

Remark: Such a counterexample does necessarily have to use weak or mild solutions. Indeed, Gidas and Spruck [5] give a priori bounds for classical solutions in case $\beta \in [\beta_1, \beta_2)$.

PROPOSITION 5.2. *Let $d = 0$ and $N > 2$, $\beta \in (1, \beta_2)$. Then the Dirichlet problem (2.8) has a unique classical solution.*

For $\beta \in [\beta_1, \beta_2)$, besides the classical, there exists a continuum of weak solutions of (2.8). They have finite total mass $D_1(\phi, R) < \infty$ but infinite Dirichlet integral $D_2(\phi, R) = +\infty$.

There exist neither classical nor weak solutions of (2.8) for $\beta \in (\beta_2, \infty)$.

The next Proposition considers a case with infinitely many solutions of (2.8). This arises for $\beta > \beta_2$ and boundary value

$$d = R^{-\mu} x_e \quad (5.6)$$

The argument exploits the situation when the nontrivial equilibrium $(x_e, \mu x_e)$ is a stable vortex. This idea goes back to Joseph and Lungren [6], see also Bebernes [2].

PROPOSITION 5.3. *We assume $N > 2$, $\beta > \beta_2$ and $\beta - \beta_2$ small.*

Then there exists a sequence ϕ_k with $k \in \mathbb{N}$ of classical solutions of the boundary value problem (2.8) with boundary value d given by (5.6).

In the limit $k \to \infty$, these solutions converge weakly to the (M)-solution ϕ_M defined by (4.6). The central values $\phi_k(0)$ and hence the maximum norms goes to infinity, whereas mass and Dirichlet integral approach those of ϕ_M and hence are bounded.

PROOF: For $\beta > \beta_2$, the unstable seperatrix (x_E, y_E) of the equlibrium $(0,0)$ converges to the stable vortex $(x_e, \mu x_e)$ as $t \to +\infty$. For $\beta - \beta_2$ small, we have complex eigenvalues and hence trajectories are spiraling. Hence there exists a sequence $t_k \to \infty$ such that

$$x_E(t_k) = x_e \quad \text{for all } k \in \mathbb{N} \quad (5.7)$$

The solution corresponding to $x_E(t + t_k - T)$ via (4.1) is

$$\phi_k(r) = r^{-\mu} x_E (\ln r + t_k - T)$$

With $T = \ln R$, the boundary value is $\phi_k(R) = d$ from (5.6). The central values are $\phi_k(0) = R^{-\mu} e^{\mu t_k}$ and hence the uniform norms diverge:

$$\lim_{k \to \infty} \|\phi_k\|_\infty = \infty$$

Total mass and Dirichlet integral satisfy

$$D_1(\phi_k, R) = \Sigma_{N-1} R^{N-2-\mu} y_E(t_k)$$

$$D_2(\phi_k, R) = \Sigma_{N-1} \int_{-\infty}^{0} e^{(N-2-2\mu)s} y_E^2(s + t_k)\, ds$$

Since $\lim_{k\to\infty} y_E(s) = \mu x_e$ and $(N-2-2\mu) > 0$, we can take the limit $k \to \infty$ and get

$$\lim_{k\to\infty} D_i(\phi_k, R) = D_i(\phi_M, R)$$

with $i = 1, 2$. Mass and Dirichlet integral converge to the values for the (M)-solution ϕ_M and hence are bounded. □

6. References.

[1] Batt, J. and Pfaffelmoser, K., *On the radius continuity of the models of polytropic gas spheres which correspond to the positive solutions of the generalized Emden-Fowler equation*, Math. Meth. in the Appl. Sci., **10**(1988), pp. 499-516.

[2] Bebernes, J. *Mathematical Problems from Combustion Theory*, Applied Mathematical Sciences, Vol. **83**, Springer-Verlag 1989.

[3] Brézis, H. and Nirenberg, L., *Positive solutions of nonlinear elliptic equations involving critical Sobolev exponents*, Communications on Pure and Applied Mathematics, **36**(1983), pp. 437-477.

[4] van den Broek, W.J. and Verhulst, F., *A generalized Lane-Emden-Fowler equation*, Math. Meth. in the Appl. Sci., **4**(1982), pp. 259-271.

[5] Gidas, B. and Spruck, J., *A priori bounds for positive solutions of nonlinear elliptic problems*, Comm. in Partial Differential Equations, **6**(1981), pp. 883-901.

[6] Joseph, D.D. and Lundgren, T.S., *Quasilinear Dirichlet problems driven by point sources*, Arch. Rat. Mech. Anal., **49**(1973), pp. 241-269.

[7] Peletier, L.A. and Serrin, J., *Uniqueness of positive solutions of semilinear equations in* $\mathbb{R}^n$, Arch. Rat. Mech. Anal., **81**(1983), pp. 181-197.

[8] Pohozaev, S.I., *Eigenfunctions of the equation* $\Delta u + \lambda f(u) = 0$, Soviet Math. Doklady, **6**(1963), pp. 1408-1411.

[9] Rothe, F. *Global Existence for Reaction-Diffusion Systems*, Lecture Notes in Mathematics 1072, (1984).

[10] Shapiro, S.L. and Teukolsky, S.A., *Black Holes, White Dwarfs, and Neutron Stars*, The Physics of Compact Objects, John Wiley & Sons 1983.

Mathematics Department,

University of North Carolina at Charlotte

Charlotte NC 28223 USA

… RUF

Butterflies in Banach space and forced secondary bifurcation

Introduction

In this note we illustrate on an example how singularity theory in Banach space may be applied to obtain information on the solution structure of a nonlinear elliptic boundary value problem. The example we will consider is

$$-\Delta u - \lambda u + u^3 = f(x), \quad x \in \Omega,$$
$$\frac{\partial u}{\partial \nu} = 0, \quad x \in \partial\Omega. \tag{1}$$

Here Ω denotes a bounded and smooth domain in R^n, λ is a real parameter, and $f(x)$ is a given forcing term.

To study (1), we consider the nonlinear mapping

$$\Phi = -\Delta - \lambda + (\cdot)^3 \; : \; E \to F, \tag{2}$$

where E and F denote suitable Banach spaces, e.g. $E = \{u \in C^{2,\alpha}(\Omega), \frac{\partial u}{\partial \nu}|_{\partial\Omega} = 0\}$, $F = C^{0,\alpha}(\Omega)$, with $\alpha \in (0,1)$ fixed. The idea is to study the mapping Φ locally in *every* point in E, and then try to piece together this information to obtain a global description of the solution structure of (1). By a local study of Φ we mean the foling: For a given $u \in E$ we consider the Fréchet derivative of Φ, $\Phi\prime(u) \in L(E,F)$. Now, if $\Phi\prime(u)$ has a continuous inverse, then the mapping Φ is locally invertible, i.e. we have the desired local characterization. More interesting from the point of view of singularity theroy are the points where Φ is not invertible, i.e. the *singular points*. Recently, the classification of singularities in finite dimensions due to R. Thom [11] has been extended (in the simplest cases) to Banach spaces. In section 1 we will describe the classification in Banach space of the first four of the so-called Morin singularities, the fold, the cusp, the swallow tail and the butterfly.

Applications of singularity theory to differential equations have been given by numerous authors. The pioneering work is a paper by Ambrosetti - Prodi [1], in which a class of nonlinear elliptic operators which have only fold singularities is described. In a paper by Cafagna - Donati [4] a more complicated singularity structure was proved for a first order Riccati equation. In short words, one may say that this structure is given by a *global cusp*. This means in particular that only fold and cusp singularities occur.

In a recent paper [8] the author has shown that for λ in a certain parameter range the operator given by (2) has a similar *global cusp structure*. This result is described in section 2 below. For other results in this direction, cf. [6], [7]. In section 3 the result of a forthcoming paper is presented, in which it is proved that for λ above the parameter range of section 2 the operator given by (2) develops other singularities than folds and cusps, namely swallow tails and butterflies. This can be interpreted as the occurence of a secondary bifurcation.

1. Singularities in Banach space

Here we assume that $\Phi : E \to F$ is a smooth, proper Fredholm operator of index zero, where E and F are Banach spaces. We assume furthermore that $E \subset F \subset H$, where H denotes a Hilbert space, and that we have $(\Phi\prime(u)v, w) = (v, \Phi\prime(u)w)$ for all $v, w \in E$, where $(.,.)$ denotes the inner product in H. For example, the operator Φ in the introduction satisfies this assumption with $H = L^2(\Omega)$.

Definition: A point $u \in E$ is a *singular point* of Φ if there exists a $v \in E \setminus \{0\}$ such that $\Phi\prime(u)[v] = 0$. The *singular set* S is the set of singular points.

We will make here the general assumption that the dimension of the kernel of $\Phi\prime(u)$ is at most one, for all $u \in E$. This means that we restrict ourselves to Morin-singularities. Note that by the implicit function theorem S is locally a codimension 1 manifold in E if there exists a $w \in E$ which is transverse to S in u. In this case we say that Φ is *1-transverse* in u. Let S_1 denote the set of 1-transverse singular points.

Definition: A point $u \in S_1$ is a *fold point* for Φ if $v(u) \notin T_u S_1$, where $v(u)$ is the unique, normalized solution of $\Phi\prime(u)[v(u)] = 0$ and $T_u S_1$ denotes the tangent space to the manifold S_1 in u.

One has the following proposition.

Proposition, cf. [2,5]: If u is a fold point for Φ then Φ is locally diffeomorphic to the map $\phi_1 : R \times X \to R \times X$ given by $\phi_1(t, x) = (t^2, x)$. Here X denotes a suitable Banach space.

Consider now the set of fold points as a subset of S_1. If in a fold point u there exists a $w \in T_u S_1$ which is transverse to the set of fold points, then this set forms locally a codimension 1 manifold in S_1. We call this set $S_{1,1}$, the set of *1-1-transverse singularities*.

Definition: A point $u \in S_{1,1}$ is a *cusp singularity*, if $v(u) \notin T_u S_{1,1}$.

Proposition, cf. [3,5]: If u is a cusp, then Φ is locally diffeomorphic to the mapping $\phi_2 : R^2 \times X \to R^2 \times X$, given by $\phi_2(t_1, t_2, x) = (t_1 t_2 + t_1{}^3, t_2, x)$. X denotes a suitable Banach space.

One now continues in the same pattern. Considering the cusp points as subset of $S_{1,1}$, we say that a cusp point u is *1-1-1-transverse*, if there exists a vector w in $T_u S_{1,1}$ which is transverse to the set of cusps. We denote this set by $S_{1,1,1}$.

Definition: A point $u \in S_{1,1,1}$ is a *swallow tail singularity*, if $v(u) \notin T_u S_{1,1,1}$.

Proposition, cf. [10]: If $u \in S_{1,1,1}$ is a swallow tail, then Φ is locally diffeomorphic to the map $\phi_3 : R^3 \times X \to R^3 \times X$ given by $\phi_3(t_1, t_2, t_3, x) = (t_1 t_2 + t_1{}^2 t_3 + t_1{}^4, t_2, t_3, x)$. X is a suitable Banach space.

Finally, considering the swallow tails as a subset of $S_{1,1,1}$, we say that a swallow tail is *1-1-1-1-transverse*, if there exists a vector w in $T_u S_{1,1,1}$ which is transverse to the set of swallow tails. We denote this set by $S_{1,1,1,1}$.

Definition: A point $u \in S_{1,1,1,1}$ is a *butterfly singularity*, if $v(u) \notin T_u S_{1,1,1,1}$.

Proposition, cf. [10]: If u is a butterfly, then the mapping Φ is locally diffeomorphic to the mapping $\phi_4 : R^4 \times X \to R^4 \times X$, given by $\phi_4(t_1, t_2, t_3, t_4, x) = (t_1 t_2 + t_1^2 t_3 + t_1^3 t_4 + t_1^5, t_2, t_3, t_4, x)$. X denotes a suitable Banach space.

2. A Global Characterization

We now apply these concepts to the study of equation (1). That is, the operator Φ is now given by (2). In [8] the following proposition was proved.

Proposition: Suppose that $0 = \lambda_1 < \lambda < \lambda_2$. Then the singluar set S is given by $\{u \in E; \mu_1(u) = 0\}$, and it is a starshaped, smooth manifold of codimension 1 in E. In particular, all singular points are 1-transverse, i.e. $S = S_1$.

Remark: In one dimension, i.e. if $\Omega \subset R$, one has in addition that S is diffeomorphic to the unit sphere in E provided that $0 < \lambda < \frac{\lambda_2}{4}$.

Also in [8] the following theorem was proved:

Theorem: Suppose that $0 < \lambda < \frac{\lambda_2}{7}$. Then the singluar set S contains a smooth submanifold C of codimension 1 (with respect to S) which consists entirely of cusp points. All the other points in S, i.e. $S \setminus C$ are fold points.

Based on this result, one is able to prove, cf. [8]:

Theorem: Suppose that $0 < \lambda < \frac{\lambda_2}{12}$. Then the set C is a connected submanifold in S such that $S \setminus C = S^1 \cup S^2$ consists of exactly two components. Furthermore, $\Phi|S$ is a diffeomorphism, and $F \setminus \Phi(S) = F_1 \cup F_3$ consists of exactly two connected components, with
- if $f \in F_1 \cup \Phi(C)$, then (1) has exactly 1 solution
- if $f \in F_3$, then (1) has exactly 3 solutions
- if $f \in \Phi(S \setminus C)$, then (1) has exactly 2 solutions

Furthermore, the forcing term $f \equiv 0$ lies in F_3.

Remark: This result should be interpreted as a *global bifurcation* which takes place when the parameter λ crosses the value 0. In fact, in that moment a singular set starts to "blow up" in the space E, which in turn produces the set $\Phi(S)$ in the space F, with the described 3-1 solution alternative.

3. A Secondary Bifurcation

The above results lead to the question whether the given restrictions on the parameter λ are necessary. In fact they are, even though we do not know whether they are sharp. We first mention a result in one dimension, cf. [10]:

Theorem: Suppose that Ω is an interval in R. Then there exists a $\epsilon > 0$ such that for

$$\frac{2}{\pi^2}\lambda_2 < \lambda < \frac{2}{\pi^2}\lambda_2 + \epsilon$$

the singular set S contains butterfly singularities.

Remarks: 1. Note that this result implies by the structure of a butterfly singularity that S contains also swallow tail singularities.

2. Again by the structure of the butterfly, we have that equation (1) has for certain forcing terms *at least five* solutions. This result can be interpreted as the occurence of a *secondary bifurcation*.

3. The existence of butterfly singularities is only proved near the value $\frac{2}{\pi^2}\lambda_2$. However, in [9] it is shown that for parameter values λ with

$$\frac{2}{\pi^2} < \lambda < \lambda_2$$

equation (1) has at least five solution. This implies the existence of other singularities than folds and cusps in this whole range.

Similar results can be proved for certain domains in higher dimensions. We mention here the following

Theorem: 1. Assume that Ω is the unit disk in R^2. Then there exists a $\epsilon > 0$ such that the singular set contains butterfly singularities provided that λ satifies

$$\frac{1}{4.84} < \frac{\lambda}{\lambda_2} < \frac{1}{4.84} + \epsilon .$$

2. Assume that Ω is a rectangle in R^2. Then there exists a $\epsilon > 0$ such that the singular set contains butterfly singularities provided that λ satisfies

$$\frac{1}{\pi^2} < \frac{\lambda}{\lambda_2} < \frac{1}{\pi^2} + \epsilon .$$

Remark 3 of the previous theorem applies here also.

We close with the following open problem:
Do there exist other Morin–singularities, maybe of any order, for parameter values $\lambda < \lambda_2$?

References

[1] Ambrosetti, A., Prodi, G., *On the inversion of some differentiable mappings between Banach spaces*, Ann. Math. Pura Appl. 93 (1973), 231-247.

[2] Berger, M.S., Church, P.T., *Complete integrability and perturbation of a nonlinear Dirichlet problem I*, Indiana Univ. Math. J. 28 (1979), 935-952; *II*, ibida 29 (1980), 715-735.

[3] Berger, M.S., Church, P.T., Timourian, J.G., *Folds and cusps in Banach spaces with applications to nonlinear partial differential equations*, Indiana Univ. Math. J. 34 (1985), 1-19.

[4] Cafagna, V., Donati, F., *Un résultat global de multiplicité pour un problème différentiel non linéaire du premier ordre*, C.R. Acad. Sci. Paris, Ser. I. Math. t. 300 (1985), 523-526.

[5] Cafagna, V., Donati, F., *Singularity theory of Fredholm maps and the number of solutions to some nonlinear first order differential equations*, preprint.

[6] Church, P.T., Timourian, J.G., *The singular set of a nonlinear elliptic operator*, Michigan Math. J. 35, (1988), 197-213.

[7] Church, P.T., Dancer, T.N., Timourian, J.G., *The structure o a nonlinear operator*, preprint.

[8] Ruf, B., *Singularity theory and the geometry of a nonlinear elliptic operator*, Ann. Sc. Norm. Sup. Pisa, C.Sc. (4) 17 (1990), 1-33.

[9] Ruf, B., *Forced secondary bifurcation in a nonlinear elliptic equation*, Diff. Int. Equ., to appear.

[10] Ruf, B., *Singularity theory and forced secondary bifurcation*, preprint.

[11] Thom, R., *Les singularités des applications diffé- rentiables*, Ann. Inst. Fourier, t.6 (1055-56), 43-87.

B. Ruf
Dip. di Matematica
Università degli Studi
Via Saldini 50
20133 Milano, Italia

N G MEDHIN AND M SAMBANDHAM
Numerical treatment of random integro-differential equations

Abstract. In this article a numerical method for a random integro-differential equation is presented. We illustrate the method by an example.

1. Introduction

It is well known that the numerical treatment of random equations is essential for all subjects in science and engineering. Recently we developed numerical methods for system of random Volterra integral equations by an application of successive approximation method and Newton's method [3, 4]. In this article we present a numerical method for a random integro-differential equation. For some related results we refer to [1-5].

We organize this article as follows. In Section 2, we list the numerical method and in Section 3, we present an example to illustrate the numerical method.

2. Numerical Method

In this section we state our assumptions and present the numerical method. Consider the random integro-differential equation

$$x'(t) = g(t,x(t)) + \int_a^t K(t,s,x(s))ds + L(t,w),$$
$$x(a) = \alpha + \xi(w). \tag{2.1}$$

Equation (2.1) can be written as a system as follows.

$$x(t) = \alpha + \xi(w) + \int_a^t [L(s,w) + g(s,x(s)) + z(s)]ds$$
$$z(t) = \int_a^t K(t,s,x(s))ds \tag{2.2}$$

We make the following assumptions:

A1. The function g is continuous and continuously differentiable in the second argument

A2. The function K is continuous and continuously differentiable in the third argument

A3. The random variable ξ is essentially bounded and

$$\sup_t \|L(t,\cdot)\|_{L_\infty(\Omega)} < \infty$$

and $L(\cdot,w)$ is piecewise continuous and α is any real variable

A4. (2.2) have a unique solution with probability one.

We solve (2.2) numerically to find the solution at equally spaced points $t_r = rh$, $r = 0,1,2,\ldots$ Approximating (2.2) by a suitable quadrature rule, we write

$$x_n = x_0 + h \sum_{i=0}^{r} w_{ri}(L(t_i,w) + g(t_i,x_i) + z_i), \quad r = k, k+1, \ldots \quad (2.3)$$

$$z_i = h \sum_{j=0}^{i} w_{ij} K(t_i, t_j, x_j), \quad i = 1, 2, \ldots, r, \quad (2.4)$$

where $x_0 = x(a)$, $x_r = x(t_r)$, $z_r = z(t_r)$.

If x_0, x_1, x_{r-1} are known, then we can use (2.3) and (2.4) to determine (x_r, z_r). We rewrite (2.3) and (2.4) as follows

$$F_1(x_r, z_r) = x_r - hw_{rr}\{L(t_r,w) + g(t_r,x_r) + z_r\} - G_{1,r}, \quad (2.5)$$

$$F_2(x_r, z_r) = z_r - hw_{rr}K(t_r,t_r,x_r) - G_{2,r}, \quad (2.6)$$

where

$$G_{1,r} = x_0 + h \sum_{i=0}^{r-1} w_{ri}(L(t_i,w) + g(t_i,x_i) + z_i) \quad (2.7)$$

$$G_{2,r} = h \sum_{j=0}^{r-1} w_{rj} K(t_r, t_j, x_j). \quad (2.8)$$

Let

$$F(x,z) = \begin{bmatrix} F_1(x,z) \\ F_2(x,z) \end{bmatrix},$$

$$DF(x_r,z_r) = \begin{bmatrix} 1 - hw_{rr}g_x(t_r,x_r) & -hw_{rr}z_r \\ -hw_{rr}K_x(t_r,t_r,x_r) & 1 \end{bmatrix}$$

$$= (1 + O(h))I,$$

$$A(t_r,x_r,h) = \sum_{n=1}^{\infty} [g_x(t_r,x_r) + K_x(t_r,t_r,x_r)z_r]^n h^n = O(h),$$

and

$$B(t_r,x_r,h) = 1 - hw_{rr}g_x(t_r,x_r) - hw_{rr}g_r(t_r,x_r)A(t_r,x_r,h).$$

Then

$$DF(x_r,z_r)^{-1} = \begin{bmatrix} 1 + A(t_r,x_r,h) & -hw_{rr}z_r A(t_r,x_r,h) \\ hw_{rr}K_x(t_r,t_r,x_r)A(t_r,x_r,h) & B(t_r,x_r,h) \end{bmatrix}$$

$$= (1 + O(h))I.$$

If $(x,z)^T, (\bar{x},\bar{z})^T$ are two points in a neighborhood of $(x_r,z_r)^T$, choose h small enough so that

$$\max_{0 \leq s \leq 1} \left\| \begin{bmatrix} -hw_{rr}(g_x(t_r,x_s) - g_x(t_r,x_r)) & -hw_{rr}(z_s - z_r) \\ -hw_{rr}[K_x(t_r,t_r,x_s) - K_x(t_r,t_r,x_r)] & 0 \end{bmatrix} \right\|$$

$$\leq r \left\| \begin{bmatrix} 1 + A(t_r,x_r,h) & hw_{rr}A(t_r,x_r,h) \\ hw_{rr}K_x(t_r,t_r,x_r)A(t_r,x_r,h) & B(t_r,x_r,h) \end{bmatrix} \right\| \quad (2.9)$$

where $r < 1$, $(x_s,z_s) = (\bar{x},\bar{z}) + s(x-\bar{x}, z-\bar{z})$, then

$$\left\| T\begin{pmatrix} x \\ z \end{pmatrix} - T\begin{pmatrix} \bar{x} \\ \bar{z} \end{pmatrix} \right\| \leq r < 1, \quad (2.10)$$

where $T(v) = v - DF\begin{pmatrix} x_r \\ z_r \end{pmatrix}^{-1} F(v)$.

In (2.9) one can see that there exists r_0, h_0 such that T is a strict contraction in $\left\{ \begin{pmatrix} x \\ z \end{pmatrix} : \left\| \begin{pmatrix} x \\ z \end{pmatrix} - \begin{pmatrix} x_r \\ z_r \end{pmatrix} \right\| \leq r_0 \right\}$, $0 < h \leq h_0$. We remark that r_0, h_0 depend on (t_r, x_r). As a simple observation from (2.9) we have the following proposition.

Proposition 2.1. If $F: R^n \to R^n$ is C^1 and $DF(x_0)^{-1}$ exists, then for every $x_1, x_2 \in R^n$, satisfying

$$\max_{0 \leq s \leq 1} \|DF(x_s) - DF(x_0)\| \|DF(x_0)^{-1}\| \leq \frac{1}{2},$$

$$x_s = x_2 + s(x_1 - x_2),$$

we have

$$\|Tx_1 - Tx_2\| < \frac{1}{2},$$

where $Tx = x - DF(x_0)^{-1}F(x)$.

We notice that at the given point (t_r, x_r) we have to find h to fulfill the condition (2.9) and then form the iterates (2.10). That is, we can get a numerical approximation of the solution of the random integro-differential equation for arbitrary large values of t. Since the right hand side of (2.9) is $r(1+0(h))$, if K_x, g_x are bounded, it is clear from (2.9) that we can choose one value of h to work at any (t_r, x_r).

3. Example

In this section we illustrate one method by an example. We consider

$$x'(t) = x(t) - te^{-t^2} - 2 \int_0^t ste^{-x^2(s)} ds + R_1(w), \quad x(t_0, w) = x_0(w) \quad (3.1)$$

where $R_1(w) \in N(1,\sigma)$, $R_2(w) \in N(0,\sigma)$, and $N(m,\sigma)$ is the normal distribution with mean m and standard deviation σ. We applied the method in Section 2 to (3.1). We have presented selected simulated values in Table 1. These results are based on a sample size of 185. By an application of Kolmogorov-Smirnov test, we tried to fit the distribution of x. For the sample size of 185, the D-values tested for normal distribution are also included in Table 1. It is explicit that the solution property of (3.1) is normal and D-values also justify the same. The frequency distribution of x at selected values are given in Figures 1-4.

209

Table I

| \multicolumn{4}{c}{$\sigma = 0.1$} |
t	E(x)	Var(x)	D
0.2	0.19966	0.00592	0.04793
0.4	0.39998	0.00725	0.04797
0.7	0.70049	0.01007	0.04910
1.0	1.00126	0.01509	0.04922

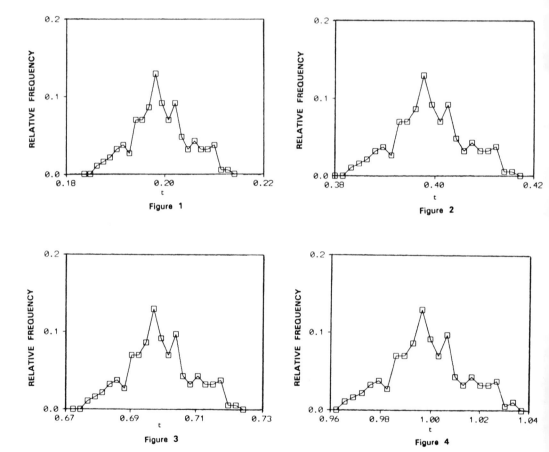

Figure 1

Figure 2

Figure 3

Figure 4

Acknowledgement: Research supported by NSF-R11-8905084, ONR 00014-87-K-0276 and NSF-R11 9014051.

References

[1] Delves, L. M. and Mohamed, J. L., Computational methods for integral equations. Cambridge University Press, 1985.

[2] Lax, M. D., Approximate solution of random differential and integral equations, Applied Stochastic Processes, (Ed. G. Adomian) Academic Press, 1980.

[3] Medhin, N. and Sambandham, M., Numerical solution of random Volterra integral equations I: Successive approximation method. Comp. Math. Appl. 15 (1988) 97-105.

[4] Medhin, N., Sambandham, M. and Zoltari, C., Numerical solution of random Volterra integral equations II: Newton's method. Stoch. Analy. Appl. 8 (1990) 105-125.

[5] Ripley, B. D. Stochastic Simulation, John Wiley, 1987.

R SAXTON
Finite time boundary blowup for a degenerate, quasilinear Cauchy problem

In this paper we examine a class of quasilinear, hyperbolically degenerate, partial differential equations and certain blow-up properties of their solutions. The study is motivated in part by an example related to liquid crystals, which has been discussed previously in a nondegenerate situation.

I. Introduction.

In recent work ([3]) on an *ersatz* equation incorporating director field effects -a physical idealization of the type used in liquid crystals, leading to consideration of vector fields constrained to be of unit length (see [2] for an introduction to mathematical and physical aspects of the static problem)-a nonlinear geometric optics approach was adopted to infer that finite time breakdown of solutions would occur.

The results of [3] were derived assuming strict hyperbolicity of the problem, which is equivalent to certain "Frank" constants being positive. We relax this requirement here by setting one constant to zero, which makes the problem hyperbolically degenerate. The nonconservative nature of the equations is such that, in both cases, breakdown does not lead to shock formation.

In Section 2, we consider an initial boundary-value problem for which degeneracy leads to blow-up in a general setting. This is related to recent work of Sakamoto ([5]) on local existence, established for a class of "degenerate-structure generating" data (see also [1]). An interesting aspect of the problem is the possibility of obtaining a maximal time of existence for smooth solutions by considering pointwise boundary behaviour, then invoking Sobolev's lemma to establish blowup of high derivative integral norms defined over the whole domain. There, we consider the problem

$$u_{tt} - f(u)\Delta u = g(u)|\nabla u|^2, \quad \mathbf{x} \in \Omega \subset \mathbb{R}^n, \quad t > 0, \tag{1.1}$$

$$u(\mathbf{x},t) = 0, \quad \mathbf{x} \in \partial\Omega, \quad t \geq 0 \tag{1.2}$$

$$u(\mathbf{x},0) = u_0(\mathbf{x}), \quad u_t(\mathbf{x},0) = u_1(\mathbf{x}), \quad \mathbf{x} \in \overline{\Omega}, \tag{1.3}$$

with the condition that $f, g \in C^1(\mathbb{R})$,

$$f(0) = g(0) = 0 = f'(0), \tag{1.4}$$

and

$$g'(0) = \lambda \in \mathbb{R}. \tag{1.5}$$

As a consequence of (1.2) and (1.4), the equation becomes degenerate for $\mathbf{x} \in \partial\Omega$, and the set $\partial\Omega$ becomes characteristic.

Section 3 is concerned with applying the results of the previous section to the liquid crystal situation. We also show that, formally, small amplitude periodic C^0-solutions may exist for all time.

2. Finite Time Blow-Up in $\mathbb{R}^n$.

Here we consider the behaviour of solutions $u(\mathbf{x},t) \in C^1([0,T);C^1(\overline{\Omega}))$ to (1.1)-(1.5), with $T > 0$ giving some maximal time of existence, for a bounded domain $\Omega \subset \mathbb{R}^n$. Given $\lambda \in \mathbb{R}, c(.) \in C(\partial\Omega; \mathbb{R})$ and $d(.) \in C(\partial\Omega;(-\infty,0])$, let

$$t^* = \inf_{\mathbf{x}\in\partial\Omega} 2^{-\frac{1}{2}} \int_{|\nabla u_0(\mathbf{x})|^2}^{\infty} (\lambda y^3 + c(\mathbf{x})y + d(\mathbf{x}))^{-\frac{1}{2}} dy \quad , \tag{2.1}$$

wherever the integral is defined for some $\mathbf{x} \in \partial\Omega$. Otherwise set $t^* = \infty$. It will be assumed here that the initial data (1.3) satisfy

$$u_0(\cdot), u_1(\cdot) \in C^1(\overline{\Omega}) \cap C_0(\Omega).$$

Theorem. Let $u \in C^1([0,T);C^1(\overline{\Omega}))$ be a solution to (1.1) - (1.5) for some $T \leq \infty$. Suppose additionally that $\lambda > 0$. Then $T \leq t^*$, where t^* is given by (2.1). In particular, $t^* < \infty$ whenever $\|\nabla u_0 . \nabla u_1\|_{\infty;\partial\Omega} \neq 0$. Further, if $T = t^*$, then $\|\frac{\partial u}{\partial n}\|_{\infty;\partial\Omega}(t)$ and $\|\frac{\partial u_t}{\partial n}\|_{\infty;\partial\Omega}(t) \to \infty$ as $t \to t^*$.

Proof: We compute the evolution of $\nabla u(\mathbf{x},t)$ for $\mathbf{x} \in \partial\Omega$ and show that for $\nabla u_0 . \nabla u_1|_{\partial\Omega}$ not identically zero, then with $\lambda > 0, \|u\|_{1,\infty;\partial\Omega}(t)$ blows up in finite time. As a consequence, since $u|_{\partial\Omega} = 0 \ \forall t$, it follows that the normal derivative of u will become infinite on the part of the boundary given by $X^*(\partial\Omega) = \{\mathbf{x} \in \partial\Omega : |\nabla u(\mathbf{x},t^*)| = \infty\}$.

From (1.1), we have

$$\nabla u_{tt} = f'(u)\nabla u \Delta u + f(u)\nabla\Delta u + g'(u)\nabla u|\nabla u|^2 + 2g(u)\nabla u.\nabla\nabla u . \tag{2.2}$$

Therefore, on $\partial\Omega$, (2.2) reduces to

$$\nabla u_{tt} = \lambda \nabla u |\nabla u|^2 \tag{2.3}$$

using (1.2), (1.4) and (1.5). Taking the scalar product of (2.3) with ∇u_t and ∇u gives

$$|\nabla u_t|^2 = \frac{1}{2}(\lambda|\nabla u|^4 + c) \tag{2.4}$$

and

$$(\frac{1}{2}|\nabla u|^2)_{tt} = |\nabla u_t|^2 + \lambda|\nabla u|^4 \tag{2.5}$$

respectively, where

$$c = c(\mathbf{x}) = 2|\nabla u_1(\mathbf{x})|^2 - \lambda|\nabla u_0(\mathbf{x})|^4. \tag{2.6}$$

Combining (2.4) with (2.5) then shows

$$(|\nabla u|^2)_{tt} = 3\lambda|\nabla u|^4 + c \tag{2.7}$$

from which, multiplying by $(|\nabla u|^2)_t$ and integrating with respect to time,

$$\frac{1}{2}((|\nabla u|^2)_t)^2 = \lambda|\nabla u|^6 + c|\nabla u|^2 + d, \tag{2.8}$$

where

$$d = d(\mathbf{x}) = 2(|\nabla u_0(\mathbf{x}).\nabla u_1(\mathbf{x})|^2 - |\nabla u_0(\mathbf{x})|^2|\nabla u_1(\mathbf{x})|^2). \tag{2.9}$$

Equation (2.8) is an ordinary differential equation in $|\nabla u|^2$. Taking λ to be positive leads to the relation

$$t(\mathbf{x}) = \pm 2^{-\frac{1}{2}} \int_{|\nabla u_0(\mathbf{x})|^2}^{|\nabla u(\mathbf{x},t(\mathbf{x}))|^2} (\lambda y^3 + c(\mathbf{x})y + d(\mathbf{x}))^{-\frac{1}{2}} dy \quad , \tag{2.10}$$

between ∇u and t, evaluated at $t = t(\mathbf{x})$ for each $\mathbf{x} \in \partial\Omega$. Provided the integral in (2.10) converges as the upper limit tends to infinity, finite time breakdown of $\nabla u(.,t)$ occurs in $\overline{\Omega}$ as $t \to T \leq t^*, t^*$ given by (2.1). Returning to (2.10) and setting

$$h(y) = \lambda y^3 + cy + d, \lambda > 0 \tag{2.11}$$

with

$$y = s|\nabla u_0(\mathbf{x})|^2, s \geq 1, \tag{2.12}$$

it follows, suppressing the $\mathbf{x}$ dependence, that

$$h(|\nabla u_0|^2) = 2|\nabla u_0.\nabla u_1|^2 \geq 0 \tag{2.13}$$

and

$$\frac{d}{ds}h(s|\nabla u_0|^2) = \lambda|\nabla u_0|^6(3s^2 - 1) + 2|\nabla u_0|^2|\nabla u_1|^2 \geq 0 \quad \text{for} \quad s \geq 1. \tag{2.14}$$

So, provided $\nabla u_0.\nabla u_1 \neq 0$ at some $\mathbf{x} \in \partial\Omega$, then $h(y) > 0 \; \forall y \in [|\nabla u_0(\mathbf{x})|^2, \infty)$. Thus $h(y)^{-\frac{1}{2}} \in L^1(|\nabla u_0(\mathbf{x})|^2, \infty)$ by an application of the dominated convergence theorem, and the result follows from (2.10), with $|\nabla u_t(\mathbf{x})| \to \infty$ as $t \to t^*$ by (2.4).

Remark 1. It is an immediate consequence of the imbedding $W_0^{2,p}(\Omega) \subset C^1(\overline{\Omega})$, ([4]), that $||u||_{2,p;\Omega}(t) \to \infty$ as $t \to t^*$, for any $p > n$.

Remark 2. If $\lambda < 0$, (2.8) shows instead that blowup of $|\nabla u|$ cannot occur on $\partial\Omega$. Also, by (2.7), only linear growth of $|\nabla u|$ takes place there as $t \to \infty$ for $\lambda = 0$.

3. The Liquid Crystal Example and Small Amplitude C^0-Solutions.

Here we examine the S^1-director model from [3], described by the equation

$$u_{tt} - (k_1\sin^2 u + k_3\cos^2 u)u_{xx} = (k_1 - k_3)\sin u \cos u \cdot u_x^2 \tag{3.1}$$

where $k_1, k_3 \geq 0$ are constitutive constants. In [3], both k_1 and k_3 were taken to be strictly positive; instead, here we allow either k_1 or k_3 to be zero. Without loss, let

$k_1 = 1$ and $k_3 = 0$. (The following arguments also apply to $k_1 = 0$ and $k_3 = 1$ with Dirichlet boundary conditions for u replaced, for example, by $u = \pi/2$ at the endpoints.) Set $\Omega = (0,1)$ and consider the problem

$$u_{tt} - \sin^2 u \cdot u_{xx} = \sin u \cos u \cdot u_x^2, \quad x \in (0,1), \, t > 0 \tag{3.2}$$

$$u(0,t) = u(1,t) = 0, \, t \geq 0 \tag{3.3}$$

$$u(x,0) = u_0(x), \, u_t(x,0) = u_1(x), \, t \geq 0. \tag{3.4}$$

Then, taking

$$f(u) = \sin^2 u, \text{ and } g(u) = \sin u \cos u \tag{3.5}$$

will satisfy (1.4) and (1.5), with $\lambda = 1$. Consequently, the Theorem of Section 2 applies, and we obtain the following

Corollary. Let $u \in C^1([0,T]; C^1[0,1])$ satisfy (3.2)-(3.5). Then $T \leq t^*$ where

$$t^* = \min_{i=1,2} 2^{-\frac{1}{2}} \int_{|u_0'(i)|^2}^{\infty} (y^3 + c(i)y)^{-\frac{1}{2}} dy , \tag{3.6}$$

with

$$c(i) = 2|u_1'(i)|^2 - |u_0'(i)|^4 \quad , i = 0, 1. \tag{3.7}$$

In particular, $t^* < \infty$ provided $u_0'(i) u_1'(i) \neq 0$ for either $i=0$ or 1.

Proof: By (2.9), $d(i) = 0$ for $i = 0$ or 1. (2.1) then leads to the result, with $|u_x(i,t)| \to \infty$ for $i = 0$ or 1 as $t \to t^*$, when $T = t^*$.

We remark, finally, that small amplitude, global C^0-solutions can be found formally, by approximating $\sin^2 u$ with u^2 and $\sin u \cos u$ with u in (3.2). This gives the equation

$$u_{tt} - u^2 u_{xx} = u u_x^2 \quad , \quad |u| \ll 1 \tag{3.8}$$

associated with (3.3) and (3.4). (3.8) is seen to possess solutions which are periodic in time with arbitrary period $\tau > 0$, ([5]), obtained by separation of variables. Setting $u(x,t) = X(x)T(t)$ shows $X(x) = \mu x^{\frac{1}{2}}(1-x)^{\frac{1}{2}}, \mu \in \mathbb{R}$. $T(t)$ satisfies $\dot{T}^2 + \frac{\mu}{2}T^4 = \nu$, $\nu \in \mathbb{R}$. For $\mu, \nu > 0$, phase plane arguments lead easily to T, hence u, having the given properties.

References.

[1] Craig,W., Nonstrictly hyperbolic nonlinear systems. Math. Ann. 277 (1987), pp.213-232.

[2] Ericksen, J., L.,"Introduction to the Thermodynamics of Solids". Applied Mathematics and Mathematical Computation, 1. Chapman and Hall, 1991.

[3] Hunter, J. and Saxton, R., S.I.A.M. J. Appl. Math. 51 (1991), 6.

[4] Gilbarg, D. and Trudinger, N.S.,"Elliptic Partial Differential Equations of Second Order", 2nd edition. Grundlehren der Mathematischen Wissenschaften, 224. Springer-Verlag, 1983.

[5] Sakamoto, R., Local existence theorems and blow-up of solutions for quasi-linear hyperbolic problems in a domain with characteristic boundary generated by initial data. Publ. RIMS, Kyoto Univ. 25 (1989), pp.381-405.

S SHAO AND H K CHENG
The Maxwell transfer equations: modification of Grad's system, I

By Modifying Grad's thirteen-moment theory, we provide a thorough formulation of Maxwell transfer equations for fourteen-moment system which based on a modified form of velocity-distribution function used in Grad's original work. It is found that the solution of our transfer equations is valid for arbitrary large shock Mach numbers in the upstream of a plane normal shock and Grad's system fails at shock Mach number exceeding 1.65 .

1. The Transfer Equations of 14-Moment System.

In many gas dynamic problems, the Navier-Stokes (NS) equations cannot describe the flow physics correctly in the interior structure of a shock wave if the mean-free path in the flow regions becomes comparable to the length scale of the field gradient. "Direct Simulation Monte Carlo" (DSMC) method and some other particle simulation techniques are powerful numerically tools in study such types of highly rarefied flows but are costly. Therefore, people in this field try to extend and improve the validity range of the NS equations via either the Burnett expansion of the Boltzmann equation or Maxwell transfer equations. Grad's thirteen (13)-moment theory [6,7] is applicable to highly nonequilibrium flows behind strong bow shocks and his moment approach could describe the transition from the continuum to the free-molecule limits rather well. However, the 13-moment system fails in the plane-shock problem at shock Mach number exceeding 1.65 [6, 7].

The reason why Grad's 13-moment theory cannot apply to the plane-shock structure problem when shock Mach number exceed 1.65 is believed to result from the lack of fidelity in his stipulated form of the distribution function. Meanwhile, both experimentally inferred data and the corresponding DSMC calculations made recently reveals a distinctly spike-like feature resembling a slightly distorted Gaussian form of the f for a high Mach-number shock. The spike like feature signifies a strong *persistence* of the upstream influence. Recalling Mott-Smith's rather successful bi-modal model, we constructed a modified form of velocity-distribution function (1.1) by adding of two Gaussian distributions symmetric with respect to the upstream state and downstream state respectively, which allow the additional terms in Grad's original velocity distribution function to reflect persistence of the upstream state and the smaller contributions from the downstream.

$$f = \alpha' e^{-\beta'(c')^2} + \exp[-\sum_k \beta_k c_k^2][1+ \sum_{ji} A_{ij}\, c_i c_j + B_i c_i + C_i c^2 c_i\,)] + \alpha'' e^{-\beta''(c'')^2} \qquad (1.1)$$

where α' and α'' are unknown functions of the space and time, u' and u'' be the upstream and downstream velocity components respectively, and β', β'', β_k, $(c'')^2$, c_k^2 and $(c')^2$ are defined by

$$\beta' \equiv \frac{1}{2RT'},\ \beta'' \equiv \frac{1}{2RT''},\ (c')2 = \sum_k (\xi_k - u'_k)^2,\ (c'')2 = \sum_k (\xi_k - u''_k)^2 \qquad (1.2)$$

and $\beta_i = \frac{1}{2RT_i}$, with T_i being the "temperature" associated with the i-th thermal-velocity component. The use of $\beta_i \neq \beta$ may better the characterization of the velocity distributions in the parallel and transverse directions. The system based on (1.1) with α', α'' and β_i's chosen requires the addition of one transfer equations to Grad's system, namely, 14-moment system. The coefficients α', α'', A_{ij}'s, B_i's and C_i's in (1.1) may be expressed in terms of 14-moments.

Among fourteen PDE's, thirteen are the transfer equations governing the moments ρ (density), u_i's (three macroscopic velocity components), P_{ij}'s (six pressure tensor components) and S_i (three heat-flux components as in Grad's 13-moment system corresponding to the ξ^0; ξ_i, $\xi_i \xi_j$, and $\xi_i \sum_j \xi_j^2$ moments of the Boltzmann equation. The rest one equations can be generated from the three $\sum_i \xi_i^3$-moment of the Boltzmann equation with $\sum_k S_{kkk}$ as new unknown moment. The collisional contribution on the right hand side of the corresponding equation are evaluated by Quarature, irrespective of the distribution function (Approximate evaluation of the collisional terms can be made for molecules with other spherically symmetric law) (ref.[6]). Then the transfer equations governing the moments ρ, u_i's, P_{ij}'s, S_i, $\sum_i S_{iii}$, in 14-moment system may be written for a steady flow as

$$\frac{\partial}{\partial x_k}(\rho u_k) = 0, \qquad (1.3)$$

$$\frac{\partial}{\partial x_k}(\rho u_k u_i + P_{ik}) = 0 \qquad (1.4)$$

$$\frac{\partial}{\partial x_k}(\rho u_k \sum_i u_i^2 + 2 \sum_i u_i P_{ik} + 3 u_k p - S_k) = 0 \qquad (1.5)$$

$$\frac{\partial}{\partial x_k}(\rho u_i u_j u_k + u_i P_{jk} + u_j P_{ik} + u_k P_{ij} + S_{ijk}) = -\frac{p}{\mu} P_{ij} \qquad (1.6)$$

$$\frac{\partial}{\partial x_k}[(\sum_j u_j^2)\rho u_k u_i + P_{ik}) + 2\sum_{ij} u_j P_{ji}\, u_k + 2 \sum_i u_j P_{jk}\, u_i + 2 \sum_i u_j S_{jik}$$

$$+ 3u_i u_k p + u_k S_i + u_k S_k] + \frac{\partial}{\partial x_k} \sum_j Q_{ikjj} = -\frac{p}{\mu}[\frac{2}{3} S_i + 2u_k P_{ik}] \qquad (1.7)$$

$$\frac{\partial}{\partial x_k}[\rho u_k \sum_i u_i^3 + 3u_k \sum_i u_i P_{ii} + 3 \sum_i u_i^2 P_{ik} + u_k \sum_i S_{iii} + 3 \sum_i u_i S_{kii}]$$

217

$$+ \frac{\partial}{\partial x_k} \sum_i Q_{iiik} = \frac{p}{\mu} \left(\frac{1}{2} \left(\sum_i S_i - 3 \sum_i S_{iii} \right) - 3 \sum_i u_i p_{ii} \right) \qquad (1.8)$$

where $S_i = \sum_k S_{ikk} = \iiint c_i c_k^2 f \, d\bar{c}$, $S_{ijk} = \iiint c_i c_j c_k f \, d\bar{c}$, $Q_{ijkh} = \iiint c_i c_j c_k c_h f \, d\bar{c}$,

$u_i = \frac{1}{\rho} \iiint \xi_i f \, d\bar{c}$, $P_{ij} = \iiint c_i c_j f \, d\bar{c}$, $d\bar{c} = dc_1 \, dc_2 \, dc_3$,

$$P_{ij} = p_{ij} + p \, \delta_{ij}, \quad p = \frac{P_{kk}}{3}, \quad \sum_\kappa p_{kk} = 0.$$

We also have (1.9) $\left(\frac{\pi^3}{\beta_1 \beta_2 \beta_3} \right)^{1/2} = \rho \left(1 + \theta_1 + \theta_2 + \frac{1}{2} \sum_k \frac{A_{kk}}{\beta_k} \right)^{-1}$ by $\rho = \iiint f \, d\bar{c}$,

here $\theta_1 = \alpha' \left(\frac{\beta_1 \beta_2 \beta_3}{\beta'^3} \right)^{1/2} = \left(\frac{T'^3}{T_1 T_2 T_3} \right)^{1/2} \alpha'$, $\theta_2 = \alpha'' \left(\frac{\beta_1 \beta_2 \beta_3}{\beta'^3} \right)^{1/2} = \left(\frac{T'^3}{T_1 T_2 T_3} \right)^{1/2} \alpha''$.

2. Structure of Plane Shocks at 14-Moment Level.

The value in developing an improved version of Grad's theory beyond the 13-moment level will hinge on success in analyzing the shock-wave structure. In the following, we examine the problems of applying the moment formulation of Sec.1 to a plane normal shock.

For a stationary plane normal shock, the moments u_i's, S_i's, P_{ij}'s, $\sum_i S_{iii}$ and their derivatives vanish except ρ, u_1, S_1, P_{11}, p, S_{111} and $\frac{\partial}{\partial x}$. We first normalize ρ, u, P_{11}, p, S_1, S_{111}, Q_{11kk} and Q_{1111} by the upstream values of ρ, u, ρu, ρu^2, ρu^3 and ρu^4 respectively; also we define η be a transformed spatial variable $\eta = \int \frac{p}{\mu} d\xi$, $\mu = \frac{\mu}{\mu_{-\infty}}$. Then system (1.3)-(1.8) with $P = \rho RT$ can be written as

$$\rho u = 1, \qquad (2.1)$$
$$u + P = 1 + p_1, \qquad (2.2)$$
$$u^2 + 2uP + 3up + S_1 = 1 + 5p_1 \qquad (2.3)$$
$$\frac{d}{d\eta} \left[3(1+p_1)u - 2u^2 + S_{111} \right] = -(P_{11} - p) \qquad (2.4)$$
$$\frac{d}{d\eta} \left[(1 + 5p_1)u + 3(1+p_1)u^2 - 3u^3 + 2uS_{111} + uS_1 + Q_{11kk} \right] = -\frac{2}{3} S_1 - 2u(P_{11} - p) \qquad (2.5)$$
$$\frac{d}{d\eta} \left[2(1+p_1)u^2 - \frac{3}{5}u^3 + \frac{4}{3}uS_{111} + uS_1 + \frac{1}{3}Q_{1111} \right] = -\frac{1}{2} S_{111} + \frac{1}{6} S_1 - u(P_{11} - p) \qquad (2.6)$$

unknowns: $u, S_{111}, \vartheta_1, \vartheta_2$, where $p_1 = \frac{1}{\gamma M_a^2}$ is the upstream value of the normalized p, M_a is the shock Mach number and $\gamma = 1.4$ is the specific-heat, T_i's are local temperatures, T'' is the downstream temperature and u'' is the downstream velocity. Then we obtained simply three ordinary differential equations (2.4)-(2.5) and one algebraic equation (cf. Sect. 1(1.8))

$$\mu\left(8T_1T_2^2\pi^3\right)^{1/2} \equiv \left(1-\vartheta_1|\frac{5}{2}-\frac{p_1}{2}(\frac{1}{T_1}+\frac{2}{T_2})|-\vartheta_2|\frac{5}{2}-\frac{T''}{2}(\frac{1}{T_1}+\frac{2}{T_2})-\frac{(u''-1)^2}{2T_1})|\right)^{-1} \quad (2.7)$$

Critical to the shock structure problem is the solution behavior on the upstream side ($\eta \to \infty$). The behavior in the question may be inferred from a limiting form near asymptotic limits. Since $u-1 = S_1 = S_{111} = \vartheta_2 = 0$, $\vartheta_1 = \vartheta_{-\infty} = 1 - \left(\frac{1}{2p_1\pi}\right)^{3/2}$. Therefore, we introduce small perturbations u^*, S^*, ϑ^*_1 and ϑ^*_2 near the upstream asymptotic limits

$$u = 1 + u^* + O(\varepsilon^2), \; S_{111} = S^* + O(\varepsilon^2), \; \vartheta_1 = \vartheta_{-\infty} + \vartheta_1^* + (\varepsilon^2), \; \vartheta_2 = \vartheta^*_2 + (\varepsilon^2)$$

where $u^* = S^* = \vartheta_1^* = \vartheta_2^* = O(\varepsilon)$, and ε is a small parameter, $S_1^* = \frac{5}{3}S^* + p_2\vartheta^*_2$. The full details of the ODE system governing u^*, S^*, ϑ_1^* and ϑ^*_2 in the upstream limit of leading order are as follows

$$(3p_1-1)\frac{du^*}{d\eta} + \frac{dS^*}{d\eta} = \left(1-\frac{5}{3}p_1\right)u^* - \frac{5S^*}{9} - \frac{p_2\vartheta^*_2}{3} \quad (3.16)$$

$$|-2 + 4p_1|\frac{du^*}{d\eta} + \left(\frac{11}{3} - \frac{5}{3}p_1\right)\frac{dS^*}{d\eta} + (-p_1p_2+p_2+p_3)\frac{d\vartheta^*_2}{d\eta}$$
$$= 2\left(1-\frac{5}{3}p_1\right)u^* - \frac{20S^*}{9} - \frac{4}{3}p_2\vartheta^*_2 \quad (3.17)$$

$$(-1 + 2p_1 + p_1^2)\frac{du^*}{d\eta} + \frac{4}{3}\frac{dS^*}{d\eta} + p_4\frac{d\vartheta^*_2}{d\eta}$$
$$= \left(1-\frac{5}{3}p_1\right)u^* - \frac{7S^*}{9} - \frac{1}{6}p_2\vartheta^*_2 \quad (3.18)$$

$$\mu\left(8T_1T_2^2\pi^3\right)^{1/2} \equiv \left(1-\vartheta_{-\infty}-\vartheta_1^*|\frac{5}{2}-\frac{p_1}{2}(\frac{1}{T_1}+\frac{2}{T_2})|-\vartheta^*_2|\frac{5}{2}-\frac{T''}{2}(\frac{1}{T_1}+\frac{2}{T_2})-\frac{(u''-1)^2}{2T_1})|\right)^{-1} \quad (3.19)$$

here u^*, S^*, ϑ_1^* and ϑ_2^* are the four unknowns, p_3 and p_4 are constants which can be represented by u^{**}, T^{**} and p_1. The system (3.16)-(3.18) is a simple first-order homogeneous system of ordinary differential equations and (3.19) is an algebraic equation and the system (3.16)-(3.19) is exact. Therefore, the upstream limiting solution of (3.17)-(3.18) has the form

$$\begin{pmatrix} u^* \\ S^* \\ \vartheta^*_2 \end{pmatrix} = \begin{pmatrix} A_1 \\ B_1 \\ C_1 \end{pmatrix} e^{\lambda\eta}, \text{ where } A_1, B_1, C_1 \text{ and } \lambda \text{ are constants to be determined.}$$

We note that the ODE system (3.16)-(3.18) give three roots of λ for all Ma >1. The solution is stable if λ has positive value as $\eta \to -\infty$. We summarize in Fig.1 the relation of Ma and the eigenvalues λ. We found that λ_1 are positive values for all Ma >1.

Therefore, $\begin{pmatrix} A_1 \\ B_1 \\ C_1 \end{pmatrix} e^{\lambda_1\eta}$ is the appropriate solution in the upstream.

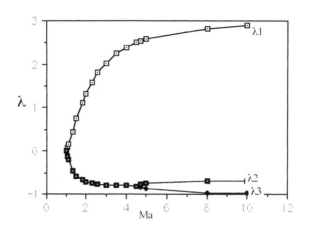

Fig. 1. Three solutions corresponding to λ_1, λ_2 and λ_3.

Conclusion. By modifications of Grad's system, we obtained a thorough formulation of transfer equations for a 14-moment system, which improves the fidelity of Grad's 13-moment system at the shock transition zone for a normal plane-shock. We can show that the steady-state one-dimensional solution is valid for arbitrary larger (Ma $\to \infty$) shock Mach numbers in the upstream. It is not an ending but a starting point. The dwonstream behavior we shall discuss in some where else and may develop an asymptotic theory which allows matching of upstream solution with downstream solution on the basis of 14-moment system inside the shock transition zones to provide a much better solution describing a plane-shock and oblique shock.

REFERENCES

[1] D. BURNETT, Proc. London Math. Soc., Ser 2, 40, 382 (1936).
[2] S. CHAMPMEN AND T. G. COWLING, *The Mathematical Theory of Nonuniform Gases*, Cambridge Univ. Press, 1953.
[3] H. K. CHENG, C. J. LEE, E. Y. WONG & H. T. YANG, AIAA paper 89-166(1989).
[4] V. K. DOGRA , J. N. MOSS AND J. M. PRICE, *Rarefied Gas Dynamics*, ed. E. P. Muntz, et al., AIAA Conf. Series, Acad. Press, 1989.
[5] H. GRAD, *On the Kinetic of Rarefied Gases*, Comm. Pure Appl. Math., 2, no.4, (1949), pp.331-407.
[6] H. GRAD, *The Profile of a Steady Plane Shock Wave*, Comm. Pure Appl. Math., 5, no. 3, (1952), pp.257-300.
[7] M. N. KOGAN, *Rarefied Gas Dynamics*, trans. ed. L. Trilling, Plenum Press, 1969.
[8] H. M. MOTT-SMITH, *The Solution of the Boltzmann equation for a shock Wave*, Phys. Rev., 82, (1951), pp.855-892.

DEPT. OF MATHEMATICS, CLEVELAND STATE UNIV, CLEVELAND, OH 44115.

DEPT. OF AEROPSPACE ENGINEERING, UNIV. OF SOUTHERN CALIF., LOS ANGELES, CA 90089-1191.

P X SHENG
Fixed point theorems and topological degrees

In the literature of fixed point theorems and topological degrees, many powerful theorems for completely continuous operators (compact operators on bounded subsets of a Banach space) have been obtained ([1], [5], [7], [11]). Some weak operators such as k–set contractions, condensing mappings, etc, and degrees for them are created by Nussbaum, Fenske, Browder, Petryshyn, Gupta and others. It turns out that these results are powerful in applications of nonlinear problems on both ODE and PDE. For continuous operators, it seems that a few results appeared. In this paper, we mainly concentrate on non–degenerate (with infinite dimensional range) completely continuous operators defined on an infinite dimensional Banach space from application point of view. The idea comes from the analysis of inherent difficulty of nonisolation of solutions. We find that topological degrees defined on a finite dimensional Banach space and defined on an infinite dimensional Banach space have a significant difference. More specifically, the Leray–Schauder degree defined on an infinite dimensional Banach space has a significant property while the Brouwer degree defined on a finite dimensional Banach space has no such property although the Leray–Schauder degree is approximated by the Brouwer degree in some sense.

Now we are going to consider any non–degenerate (with infinite dimensional range) completely continuous operator T defined on an infinite dimensional Banach space X. The following definitions are used.

<u>Definition 1:</u> If T has a fixed point at x_0, and $T+\epsilon T$ has a fixed point for ϵ sufficiently small, then T is said to have an ϵ–extension of fixed point.

<u>Definition 2:</u> If T has no other fixed point in any small neighborhood of x_0, then it is said to have an isolated fixed point at x_0. If any small perturbation operator $T+\epsilon$ of T has a fixed point near x_0, then T is said to have a non–degenerate fixed point at x_0.

(a) If T has a fixed point, can an approximated operator T_ϵ of T with finite rank have at least one fixed point ?

(b) In order to conclude that T has an isolated fixed point at x_0, what conditions do we need for T ?

(c) To claim that there exists some open bounded subset D containing x_0 such that the Leray–Schauder degree deg(I–T, D, 0)$\neq$0, what conditions do we need for T ?

(d) For what conditions, T must have an ϵ-extension of fixed point ?

(e) For what conditions, T must have a non–degenerate fixed point ?

<u>Theorem 1</u>: Suppose that X is a real Banach space, M is a bounded subset of X and T:M $\to$ X is compact, T has a fixed point at x_0. Given $\epsilon>0$ there is a continuous mapping T_ϵ: M $\to$ X whose range $T_\epsilon(M)$ is finite dimensional such that $\|T(x)-T_\epsilon(x)\|<\epsilon$ ($x\in M$) and $T_\epsilon(x_0)=x_0$.

The proof is trivial by using the Leray–Schauder projection map.

<u>Theorem 2</u>: Let T have a fixed point at x_0 and T be Frechet differentiable (or Gateaux differentiable). If $DT(x_0)h=h$ (or $DT(x_0, h)=h$) if and only if h=0, then T has an isolated fixed point at x_0.

<u>Theorem 3</u>: If a non–degenerate completely continuous operator T defined on an infinite dimensional Banach space X has an isolated fixed point at x_0, then there exists some open bounded subset D containing x_0 such that the Leray–Schauder degree deg(I–T, D, 0)$\neq$0.

Proof: Since T has an isolated fixed point at x_0, we consider some bounded open subset U containing x_0 such that T has no other fixed point in U. Since $\overline{T(U)}$ is compact, and so is covered by a finite number of balls of radius ϵ; let these be $B(v_i,\epsilon)$ (i=0, 1, 2, $\cdots$, N) where $v_0=x_0$ and $v_i \in \overline{T(U)}$. Define $m_i(x,\epsilon)=\max\{0, \epsilon-\|T(x)-v_i\|\}$ ($x\in U$) and $\varphi_i(x,\epsilon)=m_i(x,\epsilon)/[\sum_{k=0}^{N} m_k(x,\epsilon)]$. Now define $T_\epsilon(x)= \sum_{i=0}^{N} \varphi_i(x,\epsilon)v_i$, where T_ϵ: U $\to$ X is continuous, $T_\epsilon(U)$ is contained in the finite dimensional linear space spanned by the vectors v_0, v_1, $\cdots$, v_N. Let D= U$\cap$B(x_0,ϵ/K) for some large K>0 (an open subset in X), we restrict T_ϵ on D, for an appropriate K, $T_\epsilon(D)$ is contained in one dimensional linear space spanned by x_0, say V=$\{cx_0, c\in R^1\}$. Since any one dimensional normed space is topologically equivalent to R^1, and T_ϵ restricted on D$\cap$V behaves like a linear function on R^1, we can construct a C^1 function η defined on an open set O$\subset R^1$ with only one isolated and non–degenerate fixed point in O such that η operates on O is topologically equivalent to that I–T_ϵ operates on D$\cap$V. Hence by the definition of the Brouwer degree we know that $\deg(\eta,O,0)\neq 0$, hence $\deg(I-T_\epsilon, D\cap V, 0)\neq 0$. We conclude that $\deg(I-T,D,0)\neq 0$.

An alternative proof: As above notations, $T_\epsilon: U \to X$ is continuous and $T_\epsilon(U)$ is contained in the finite dimensional linear space V spanned by the vectors $v_0, v_1, \cdots, v_N$. Let $r=\inf\{\|(I-T)(x)\|: x\in\partial U\}$, then $r>0$ since T has no fixed point on ∂U. Choose $\epsilon>0$ sufficiently small such that $\epsilon+\epsilon/K<r$. Define another operator $\hat{T}_\epsilon = \sum_{i=0}^{N} \varphi_i(x,\epsilon)v_i+\epsilon/K$, then $\|\hat{T}_\epsilon(x)-T(x)\|\leq \epsilon+\epsilon/K<r$ for $x\in U$. We can choose K sufficiently large such that $x_0+\epsilon/K\in B(x_0,\epsilon)$ and $x_0+\epsilon/K\notin B(v_i,\epsilon)$ for $i=1, 2, \cdots, N$. Hence $x_0+\epsilon/K$ is a fixed point of $\hat{T}_\epsilon$. It implies that T_ϵ has a non-degenerate fixed point at x_0. It is easy to see that for any $x\in B(x_0,\epsilon/K)$, $T_\epsilon(x)=x_0$. Hence T_ϵ has an isolated fixed point at x_0. By the property of the Brouwer degree, we know that $\deg(I-T_\epsilon, U\cap V, 0)\neq 0$, hence $\deg(I-T,U,0)\neq 0$. Let $D=U$, the proof is complete.

Note: The theorem is not true if T is defined on a finite dimensional Banach space or if T defined on an infinite dimensional Banach space is degenerated (with finite dimensional range). If T has no fixed points in a bounded open set U, it is not true in general that there exists an open bounded subset D of U such that $\deg(I-T, D,0)=0$.

<u>Theorem 4</u>: If λT has (isolated) fixed points for $\lambda\in \Lambda\subset(0,\epsilon)$, where Λ is a dense subset of $(0,\epsilon)$ for $\epsilon>0$ sufficiently small, then λT has a fixed point for $\lambda\in(0,\epsilon)$.

The proof follows from the property of the Leray–Schauder degree and theorem 3.

<u>Theorem 5</u>: If a non-degenerate completely continuous operator T defined on an infinite dimensional Banach space X has an isolated fixed point. It must have an ϵ-extension of fixed point.

Proof: We only need to show that $(1+\epsilon)T$ has a fixed point for $\epsilon>0$ sufficiently small by the definition of the ϵ-extension of fixed point. Define $T_\lambda=\lambda T$, then $T_1=T$ has an isolated fixed point. By the theorem 4, we only need to consider that for $\lambda\in(1,1+\epsilon)$, T has no fixed points. Define $F_t=I-T_{(1-t)+t\lambda}$ for $0\leq t\leq 1$. Since F_0 has an isolated solution, by the theorem 3, there exists a bounded open subset $U\subset X$ such that $F_0(x)\neq 0$ for any $x\in\partial U$ and $\deg(F_0,U,0)\neq 0$. Since F_t has no solutions for $0<t<1$, if F_1 has a solution on ∂U, then $T_\lambda=\lambda T$ has a fixed point, if F_1 has no solution on ∂U, then F_1 is fp-homotopic to F_0 and the Leray–Schauder degree for F_t is well defined. By homotopy invariance we have $\deg(F_1,U,0)=\deg(F_0,U,0)\neq 0$, so F_1 has a solution, namely T_λ has a fixed point. Hence T has an ϵ-extension of fixed point.

<u>Theorem 6</u>: If a non-degenerate completely continuous operator T defined on an infinite dimensional Banach space X has an isolated fixed point, then $T+\epsilon$ must have a fixed point for ϵ sufficiently small.

Proof: Since T has an isolated fixed point x_0, by the theorem 3, there exists an open subset D such that $\deg(I-T,D,0) \neq 0$. Let $F=I-T$, ρ the distance induced by the norm on X. Since x_0 is an isolated fixed point of T, $0 \notin F(\partial D)$. Let $r=\rho(0,F(\partial D))=\inf\{\|F(x)\| \mid x \in \partial D\}$, it is obvious that $r>0$. We construct a mapping $T_\epsilon: \overline{D} \to X$ with finite dimensional range such that $\|T(x)-T_\epsilon(x)\|<\epsilon$ for $x \in D$. By the property of the Leray-Schauder degree that $\deg(F_\epsilon,D_\epsilon,0)$ is independent of ϵ for $0<\epsilon<r$, we know that T_ϵ can also be the approximate operator of $T+\epsilon$ for $2\epsilon<r$. In fact, if $T+\epsilon$ has a fixed point on ∂D, then there is nothing to prove, if $T+\epsilon$ has no fixed point on ∂D, then $\|T+\epsilon-T_\epsilon\|<2\epsilon<r$. Hence $\deg(I-(T+\epsilon),D,0)=\deg(I-T,D,0) \neq 0$. It shows that $T+\epsilon$ has a fixed point for ϵ sufficiently small.

Note: We only conclude that $(1+\epsilon)T$ and $T+\epsilon$ have a fixed point, not an isolated fixed point. The theorem 5 and 6 are useful in showing global or structural stability.

<u>Theorem 7</u>: If a non-degenerate completely continuous operator T defined on an infinite dimensional Banach space has no fixed point, then λT has at least one fixed point for $0 \leq \lambda < \epsilon$ with ϵ sufficiently small.

<u>Theorem 8</u>: If a Gateaux differentiable and non-degenerate completely continuous operator T defined on an infinite dimensional Banach space X has only isolated fixed points with $DT(x,h)=h$ if and only if $h=0$, then λT for any given $0<\lambda<1$ only has isolated fixed points.

<u>Definition 3</u>: If $x_n=\lambda_n T(x_n)$, $\|x_n\| \to \infty$ and $\lambda_n \to \lambda_0$, then we say that $\lambda_0 T$ has a fixed point at infinity in generalized sense. Moreover, if $\|x_n\|<\|x_{n+1}\|$ for n sufficiently large, and $c_n x_n + \delta_n \epsilon B(x_n,\epsilon_n)$ with $\|c_n x_n + \delta_n\| \leq \|c_{n+1} x_{n+1} + \delta_{n+1}\|$ for n sufficiently large, $\|\lambda_n T(c_n x_n + \delta_n)\| \sim \|c_n x_n + \delta_n\|$ as $\|c_n x_n + \delta_n\| \to \infty$, then we say that $\lambda_0 T$ has a non-isolated fixed point at infinity in generalized sense.

<u>Theorem 9</u>: If a non-degenerate completely continuous operator T and λT for $0<\lambda<1$ defined on an infinite dimensional Banach space X can only have isolated fixed point at infinity in generalized sense, then T must have a fixed point.

In applications, for a given integral operator, questions leave to how to check the isolation of fixed point at infinity in generalized sense.

References

[1] M.S. Berger, Nonlinearity and functional analysis, Lectures on nonlinear problems in mathematical analysis (1977).

[2] S.R. Bernfeld & V. Lakshmikantham, An introduction to nonlinear boundary value problems (1974).

[3] F.X. Browder, Nonlinear operators and nonlinear equations of evolution in Banach space, being nonlinear functional analysis, Vol 2, Amer. Math. Soc. 72, 571–5.

[4] J. Cronin, Fixed points and topological degree in nonlinear analysis, Amer. Math. Soc. 1964.

[5] James Dugundji & Andrzej Granas, Fixed point theory, Vol 1, 1982.

[6] Jack K. Hale (editor), Studies in ODEs, MAA studies in mathematics, Vol 14, 1977.

[7] N.G. Lloyd, Degree theory, Cambridge University Press, Cambridge, 1978.

[8] R.D. Nussbaum (1969), The fixed point index and asymptotic fixed point theorems for k–set contractions, Bull. Amer. Math. Soc. 75, 490–5.

[9] R.D. Nussbaum (1972), Degree theory for local condensing maps, J. Math. Anal. Appl. 37, 741–66.

[10] W.V. Petryshyn (1973), Fixed point theorems for various classes of 1–set contractive and 1–ball contractive mappings in Banach spaces, Trans. Amer. Math. Soc. 182, 323–52.

[11] D.R. Smart, Fixed point theorems (1974).

[12] I. Stakgold, Boundary value problems of mathematical physics, V1 & V2, 1967.

M W SMILEY[*]

On the asymptotic behavior of nonautonomous dissipative abstract differential equations

1. Problem statement and *a priori* estimates

In this article we discuss two aspects of asymptotic behavior of solutions of the equation $u' + Au + G(u) = f(t)$, in which A is an unbounded linear operator in a Hilbert space H and G is a Lipschitz nonlinear operator in H. The two aspects involve the concepts of attractors and approximate inertial manifolds. Monotonicity plays a role in our work, but we do not assume $A+G$ is monotone. In general, f is assumed to be Hölder continuous and bounded on $\mathbb{R}^+ = [0, +\infty)$. The operator A is assumed to be closed and symmetric, and to be the generator of an analytic semigroup on H which is uniformly bounded for $t \in \mathbb{R}^+$. For convenience, we assume A is positive definite so that $\mathcal{D}(A^{1/2})$ is a Hilbert subspace of H, with the graph norm $\|u\| = |A^{1/2}u|$.

Under these assumptions, it is well-known (cf. [3]) that the initial-value problem

$$(1) \qquad u' + Au + G(u) = f(t), \quad u(0) = u_0 \in H,$$

has a unique (classical) solution $u \in C((0, +\infty), \mathcal{D}(A))$, for every $u_0 \in H$. It is also known that u, u', Au are all Hölder continuous on $[t_0, T]$, for any $0 < t_0 < T < +\infty$, with exponent $\nu = \min(\beta, \theta)$ where θ is the Hölder exponent of f and $\beta \in (0,1)$ is arbitrary. Moreover it is not difficult to show that the solution $u = u(t, u_0, f)$ depends continuously on the initial data and the forcing term $f \in L^\infty(\mathbb{R}^+, H)$. This continuous dependence on f remains valid if the topology of uniform convergence on compact subsets of $\mathbb{R}^+$ is used on the forcing terms.

The forcing functions $f(t)$ will subsequently be required to be uniformly locally Hölder continuous on $\mathbb{R}^+$. This property is characterized in terms of the Hölder norm $[f]_\theta = \sup\{[f]_{\theta,J}: J \subset \mathbb{R}^+, |J| \leq 1\}$; $[f]_{\theta,J} = \sup\{|f(t)-f(s)|/|t-s|^\theta: t, s \in J, t \neq s\}$ where $\theta \in (0,1]$ is a given constant. If $[f]_\theta < +\infty$, then $|f(t) - f(s)| \leq K|t-s|^\theta$, where K depends only on the length of the interval containing t, s.

Solutions of (1) will be uniformly bounded on $\mathbb{R}^+$ under the following abstract dissipativity condition. Let $\lambda_* = \inf\{|A^{1/2}u|^2/|u|^2: u \neq 0\}$. The equation will be dissipative if there are constants $G_0 > 0$ and $m_* < \lambda_*$ such that

$$(2) \qquad (G(u), u) > -G_0 - m_*|u|^2, \quad \text{for all } u \in H.$$

Notice that (2) follows from the asymptotic condition: $\liminf (G(u), u)/|u|^2 > -\lambda_*$, as $|u| \to \infty$.

Lemma 1. *Suppose that $G \in \text{Lip}(H,H)$ satisfies (2), and that $f \in L^\infty(\mathbb{R}^+, H)$ is Hölder continuous. Then the solution u of (1) belongs to $L^\infty(\mathbb{R}^+, H)$ and satisfies*

$$(3) \qquad |u(t)|^2 \leq R^2 e^{-(\lambda_* - m_*)t} + C_0(G_0, B, \lambda_* - m_*), \quad t \geq 0,$$

whenever $|u_0| \leq R$ and $\sup\{|f(t)|: t \in \mathbb{R}^+\} \leq B$. (Here and subsequently $C_0(\ldots)$ denotes a constant depending only on the indicated quantitites.) In addition $A^{1/2}u \in L^\infty([t_0, +\infty), H)$, for any $t_0 > 0$, and satisfies

$$(4) \qquad \|u(t)\| \leq C_1(t_0, R, G_0, G_1, B, \lambda_* - m_*), \quad t \geq t_0 > 0$$

---
[*]Iowa State University, Department of Mathematics, Ames, Iowa 50011

where G_1 is the Lipschitz constant for $G(u)$.

Proof. To obtain (3), multiply (1) by u and use (2) to arrive at

$$(5) \quad \left(\frac{1}{2}|u|^2\right)' + |A^{1/2}u|^2 \leq F_0 + m_*|u|^2 + (f, u), \quad t > 0.$$

Using the definition of λ_* to bound the left side from below, and Cauchy's inequality with $\epsilon = (\lambda_* - m_*)/2$ to bound the last term on the right from above yields a differential inequality that directly integrates to (3). To obtain (4), one first observes that (5) and (3) imply that for any $t_0 > 0$, the mean value of $A^{1/2}u(t)$ on $(0, t_0)$ is bounded by a constant C of the type indicated in (4). Hence, given $t_0 > 0$, there is a $t_1 \in (0, t_0)$ such that $|A^{1/2}u(t_1)| \leq C$. Next, from the differential equation in (1), it follows that $(|A^{1/2}u|^2)' = 2(Au, f - Au - G(u)) \leq -\lambda_*|A^{1/2}u|^2 + \|f - G(u)\|_{L^\infty(\mathbb{R}^+, H)}$. Integration of this inequality twice, first over (t_1, t_0) and then over (t_0, t) leads immediately to (4) and completes the proof.

A consequence of the boundedness of solutions is strengthen regularity. By a slight modification of a proof in [3] (see [5]) one can show that u is uniformly locally Hölder continuous, with any exponent $\nu \in (0, 1)$. The uniformity of this property becomes important in showing the existence of approximate inertial manifolds.

2. Global attractors in the skew-product flow

For given constants $\theta \in (0, 1]$ and $B, K > 0$, we define the set of functions $\mathcal{F} = \mathcal{F}(\theta, B, K) = \{f \in C(\mathbb{R}^+, H): \sup |f(t)| \leq B, [f]_\theta \leq K\}$. Equipped with the topology of uniform convergence on compact subsets of $\mathbb{R}^+$, $\mathcal{F}$ is a complete metric space. For $\tau > 0$, set $f_\tau(t) = f(t + \tau)$. This is a continuous mapping of $\mathcal{F}$ into itself. In fact, it can be easily verified that $(\tau, f) \to f_\tau$ is a continuous semi-flow on the space $\mathcal{F}$. The skew-product (semi-)flow associated to (1) employs this shift mapping. Define a map $S: \mathbb{R}^+ \times H \times \mathcal{F} \to H \times \mathcal{F}$ by $S(t, u_0, f) = (u(t, u_0, f), f_t)$, where as above $u(t, u_0, f)$ denotes the solution of (1). By uniqueness of solutions we have $S(t+\tau, u, f) = (u(t, u(\tau, u_0, f), f_\tau), f_{t+\tau}) = S(t, u(\tau, u_0, f), f_\tau) = S(t, S(\tau, u_0, f))$. Thus $S(0, \cdot)$ is the identity on $H \times \mathcal{F}$ and $S(t + \tau, \cdot) = S(t, S(\tau, \cdot))$. Since u is a continuous function of its arguments, this shows that $S: \mathbb{R}^+ \times H \times \mathcal{F} \to H \times \mathcal{F}$ is a dynamical system. This is the skew-product flow associated with (1) (cf. [2],[4]).

In the finite-dimensional setting, uniform ultimate boundedness of a semi-flow S would be sufficient to insure the existence of an attractor. However, in the present infinite-dimensional situation compactness must also be shown. The *a priori* estimates (3)-(4) verify the appropriate compactness property in the first component of S, when $\mathcal{D}(A^{1/2})$ is compact in H. For compactness in the second component, the property of asymptotically almost periodicity (a.a.p.) is required, and in fact is equivalent to compactness of the shift flow (cf. [7]). A function $f(t)$ is said to be a.a.p. if $f(t) = g(t) + h(t)$, where $g(t)$ is almost periodic (a.p.) and $h(t)$ converges to zero as $t \to +\infty$.

Theorem 1. *Suppose that $\mathcal{D}(A^{1/2})$ is compactly contained in H, and that $G \in \text{Lip}(H, H)$ satisfies (2). If $f \in \mathcal{F}$ is a.a.p. then there is a compact global attractor $\mathcal{A} \subset H \times \mathcal{F}$ for the semi-flow S.*

Proof. Let $\mathcal{H}(f)$ be the closure of the orbit (i.e. the hull) of f under the shift flow $(\tau, f) \to f_\tau$. Define a complete metric space $\Sigma = H \times \mathcal{H}(f) \subset H \times \mathcal{F}$. Then the restriction of S to Σ defines a continuous semi-flow. By the *a priori* estimate (4) and the compactness of the shift flow, it follows that $S(t,\cdot)\colon \Sigma \to \Sigma$ is uniformly compact for $t \geq t_0 > 0$, for any $t_0 > 0$. Let $\rho > C_0$, where C_0 is the constant appearing in (3). Then $\mathcal{B}(\rho) = \{(v,\psi)\colon |v| \leq \rho,\ \psi \in \mathcal{H}(f)\}$ is an absorbing ball in Σ. Hence (cf. [2],[6]), there is a compact global attractor $\mathcal{A} = \Omega(\mathcal{B}(\rho))$ for S in Σ.

Remark. Although the attractor $\mathcal{A}$ is invariant for the flow S, its projection $\pi_H(\mathcal{A}) \subset H$ onto the first component is not necessarily invariant. It is however quasi-invariant (cf. [4]) in the sense that for every $v_0 \in \pi_H(\mathcal{A})$, there is a function $f_\infty \in \Omega(f)$ such that the solution $u(t, v_0, f_\infty)$ of the limiting equation $u' + Au + G(u) = f_\infty$, satisfying $u(0) = v_0$, remains in $\pi_H(\mathcal{A})$ for all $t \geq 0$. If, for example, $f(t) = g + h(t)$, where $g \in H$ and $h(t) \to 0$ as $t \to +\infty$, then $\pi_H(\mathcal{A})$ must be contained in the attractor of the corresponding autonomous problem with right hand side g.

3. Finite-dimensionality of steady-states

When the operator $A+G$ is strongly monotone, and the forcing term f is independent of t, there is a unique equilibrium solution of (1), which is (globally) asymptotically stable. When $A+G$ is not monotone, multiple steady-states may exist. However, when $A + G + mI$ is monotone and coercive, for some $m > 0$, all steady-states are generally found to lie on a finite-dimensional manifold.

To see this assume that there is an orthogonal decomposition of H into subspaces X and Y, $H = X+Y$, with both X and Y invariant with respect to A and $\dim(X) < +\infty$. Let P and Q be the corresponding projections, $P\colon H \to X$ and $Q\colon H \to Y$. Then the problem $Au + G(u) = f$ (independent of t) can be written as an equivalent system

(6) $$Ax + PG(x+y) = g = Pf,$$
(7) $$Ay + QG(x+y) = h = QF.$$

We suppose further that $|A^{1/2}y|^2 \geq \mu|y|^2$ for all $y \in Y^1 = Y \cap \mathcal{D}(A^{1/2})$. Clearly this would be the case if for example there is a complete orthonormal set of eigenvectors $\{\varphi_i\}$ for A and $Y = c\ell\,(\mathrm{span}\{\varphi_i\colon \lambda_i \geq \mu\})$. In this case one can in fact choose $\mu > 0$, and generate the subspaces X, Y.

The following condition on G will be used. Assume that, for a constant m,

(8) $$(G(u) - G(v), u - v) \geq -m|u-v|^2, \quad \text{for all } u, v \in H.$$

Actually this is not an additional assumption. We may choose $-m$ as the infimum, of the form $(G(u) - G(v), u - v)/|u-v|^2$, over all $u, v \in H$. Since G is Lipschitz this form is bounded from below.

Lemma 2. *Under the notations and assumptions above, if $\mu > m$ then there is a mapping $\sigma\colon X \times Y \to Y^1$ such that $y = \sigma(x,h)$ is the unique solution of (7), with x and h held fixed. The map σ is Lipschitz continuous into both Y and Y^1 with Lipschitz constants $L_0 = \max\{1, G_1\}/(\mu - m)$ and $L_1 = \max\{1, G_1\}/\delta(\mu - m)$, where δ is any number in the interval $(0, (\mu - m)/\mu)$.*

Proof. With $x \in X$ fixed, define an operator $T(y) = Ay + QG(x+y)$. This is a continuous monotone and coercive operator from Y^1 to its dual. Therefore there is a unique solution $y = \sigma(x,h)$ for every $h \in Y$. By exploiting the monotonicity property of T and the Lipschitz property of G one verifies (cf. [5]) that σ is Lipschitz into both Y and Y^1, with the constants given above.

4. Approximate inertial manifolds

It is natural to exploit the existence of a manifold containing all steady-states in characterizing the asymptotic behavior of (1). Ideally, in the autonomous case, one would like to obtain an invariant manifold that exponentially attracts all solutions of the problem and is of low dimension. Center manifolds have these properties (locally) and have been used extensively to investigate the behavior of solutions near non-hyperbolic rest points. Inertial manifolds (cf. [6]) are also manifolds of this type. However, in general, the existence of inertial manifolds is difficult to verify.

Approximate inertial manifolds are more readily obtained, but do not have all of the desirable properties suggested above. By an approximate inertial manifold we mean (cf. [1]) a finite-dimensional Lipschitz manifold which has a neighborhood that is exponentially attracting under the flow of the system. Thus it need not be invariant nor attracting in its own right. However, if the attracting (an necessarily invariant) neighborhood is very thin, these properties are almost possessed by such a manifold. In the present situation, the equations are nonautonomous and consequently the definitions above have to be modified to account for the time dependence in the problem.

We consider the manifold generated by the map σ of the previous section as a candidate for an approximate inertial manifold. We must show that solutions are drawn towards this manifold. Consider the evolution equation $z' + Az + QG(x + z) = h$, on an interval $J = [a, b]$, where $x \in X$ and $h \in Y$ are independent of t. Here Q denotes the projection described in the previous section. Throughout we assume that the hypotheses of the last section are valid. By definition of σ and since $A + QG(x + \cdot)$ is monotone it is easy to see that any solution $z(t)$ satisfies

(9) $\qquad |z(t) - \sigma(x, h)| \leq |z(a) - \sigma(x, h)| e^{-(\mu - m)(t-a)}, \quad$ for all $t \in J$.

A similar estimate holds when the norm $|\cdot|$ is replace by the norm $\|\cdot\|$. Estimates of this type do not however completely resolved the problem.

Let $u(t)$ be the solution of (1). Decomposing (1), and $u(t)$, according to the decomposition of $H = X + Y$, results in the system

(10) $\qquad x' + Ax + PG(x + y) = g(t), \quad y' + Ay + QG(x + y) = h(t),$

where $x(t) = Pu(t)$, $y(t) = Qu(t)$, $g(t) = Pf(t)$, $h(t) = Qf(t)$. The second equation is analogous to the one considered in deriving (9), but has both x and h depending on t. This suggests working with $\sigma(x(t), h(t))$. Fortunately continuous dependence of solutions on parameters can be used to bridge the gap between these two cases. Consider the pair of equations $y' + Ay + QG(x + y) = h$, $z' + Az + QG(\bar{x} + z) = \bar{h}$, in which $\bar{x}$ and $\bar{h}$ are constants. Using the uniform local Hölder continuity of x and h and the monotonicity of $A + QG$ one is able to show that the solutions of these equations, determined by the initial data $y(\bar{t}) = z(\bar{t}) = \bar{y}$, satisfy

(11) $\qquad |y(t) - z(t)| \leq C(\mu - m)^{-1} \Delta t^\theta, \quad \bar{t} \leq t \leq \bar{t} + \Delta t$

where C is a constant independent of both $\bar{t}$ and t. This constant may depend on Δt.

Theorem 2. *Under the hypotheses of Lemma 2, any solution of (1) has the form $u(t) = x(t) + y(t)$, where $y(t)$ satisfies*

(12) $\qquad |y(t) - \sigma(x(t), h(t))| \leq C(\mu - m)^{-(1+\theta)} + d_0 e^{-(\mu - m)(t - t_0)}, \quad 0 \leq t_0 \leq t,$

where $d_0 = |y(t_0) - \sigma(x(t_0), h(t_0))|$. A similar estimate on $\|y(t) - \sigma(x(t), h(t))\|$ holds, with the exponent $(1+\theta)$ replaced by $\left(\frac{1}{2} + \theta\right)$ and with the restriction $t_0 > 0$.

Proof. Fix $\Delta t > 0$, and define sequences $\{t_n\}$, $\{x_n\}$, $\{h_n\}$, $\{d_n\}$ by setting $t_n = t_0 + n\Delta t$, $x_n = x(t_n) = Pu(t_n)$, $h_n = h(t_n) = Qf(t_n)$, $d_n = |y(t_n) - \sigma(x_n, h_n)|$ where $y(t) = Qu(t)$. Then (9) shows that $|z_n(t) - \sigma(x_n, h_n)| \le d_n \exp[-(\mu - m)(t - t_n)]$, $t_n \le t \le t_{n+1}$, where z_n is the solution of the equation with $\bar{x} = x_n$, $\bar{h} = h_n$ and $z_n(t_n) = y(t_n)$. From this and (11) it then follows that $d_{n+1} \le C(\mu - m)^{-1} \Delta t^\theta + d_n \exp[-(\mu - m)\Delta t]$. The constant C is independent of n by the uniform nature of the Hölder continuity. Iterating this inequality gives a discrete version of (12). The discrete version then extends to the continuous version, since σ is Lipschitz continuous. The proof of the analogous estimate in the $\|\cdot\|$ norm is similar.

Let $\varphi \colon \mathbb{R}^+ \times X \to Y^1$ be defined by $\varphi(t, x) = \sigma(x, Qf(t))$. An immediate consequence of Theorem 2 is that trajectories $(t, u(t))$ of solutions converge to a neighborhood of the manifold $\mathcal{M} = \{(t, u) \in \mathbb{R}^+ \times H \colon u = x + \varphi(t, x)\}$. This is a nonautonomous version of an approximate inertial manifold.

References

[1] C. Foias, O. Manley, and R. Temam, *Sur l'interaction des petits et grands tourbillons dans des écoulements turbulents*, C. R. Acad. Sci. Paris Sér. I. Math. **308** (1987), 497–500.

[2] J. Hale, *Asymptotic Behavior of Dissipative Systems*, Math. Surveys and Monographs, vol. 25, AMS, Providence, RI, 1988.

[3] A. Pazy, *Semigroups of Linear Operators and Applications to Partial Differential Equations*, Appl. Math. Sci., vol. 44, Springer-Verlag, New York, 1983.

[4] G. Sell, *Topological Dynamics and Ordinary Differential Equations*, Van Nostrand Reinhold, London, 1971.

[5] M. Smiley, *Global Attractors and Approximate Inertial Manifolds for Nonautonomous Dissipative Equations*, J. Diff. Eqns., submitted.

[6] R. Temam, *Infinite-dimensional Dynamical Systems in Mechanics and Physics*, Appl. Math. Sci., vol. 68, Springer-Verlag, New York, 1988.

[7] S. Zaidman, *Almost-Periodic Functions in Abstract Spaces*, Research Notes in Math., Pitman, Boston, MA, 1985.

A numerical method for control of a distributed parameter equation

The following equation describes the evolution of a controlled quantum particle-in-a-box system with one degree of freedom, whose initial state ψ^0, is zero :

$$i\psi_t = \psi_{xx} + u$$
$$\psi(0,t) = \psi(1,t) = 0$$
$$\psi(x,0) = \psi^0(x) = 0$$

The control problem is :

Given $T > 0$ and $\phi \in L^2(0,1;\mathbf{C})$, find $u \in L^2([x_l, x_r] \times [0,T]; \mathbf{C})$ so that the solution to equation (1), satisfies $\psi(x,T) = \phi(x)$.

We will show that a solution exists by a constructive method which lends itself to computation. The parameters of interest here are those that describe the interval $[x_l, x_r]$, the interval of support for the control, and we wish to show the effect on the computed solution as these parameters are varied.

Controllability

Here we show that the control problem has a solution, and indicate its construction. Let

$$y : [x_l, x_r] \times [0,T] \longrightarrow \mathbf{C}$$

be any function that satisfies

$$\int_0^T \int_{x_l}^{x_r} y\,\overline{w}\,dx\,dt = \int_0^1 \phi\,\overline{\eta}\,dx \tag{1}$$

Where $\eta \in L^2(0,1)$ is arbitrary and w satisfies

$$iw_t = w_{xx}, \qquad w(x,T) = \eta$$

If $\tilde{y}$ denotes extension by zero to the rest of Q_T, then the function $u = i\tilde{y}$ solves the control problem. To see this we effect the following substitution

$$\int_{x_l}^{x_r} y\overline{w}dx = \int_0^1 -iu\overline{w}dx = \int_0^1 (\psi_t + i\psi_{xx})\overline{w}dx$$

The equation (1) , via integration by parts becomes

$$\int_0^T \int_0^1 (-\psi\overline{w}_t + i\psi_{xx})dx\,dt + \int_0^1 (\psi(x,T)\overline{w}(x,T) - \psi(x,0)\overline{w}(x,0))\,dx = \int_0^1 \phi\overline{\eta}\,dx$$

or,

$$\int_0^1 (\psi(T) - \phi)\bar{\eta}\, dx = 0.$$

Since η is arbitrary, $\psi(x,T) = \phi(x)$. If we restrict the search to function y that satisfies $iy_t = y_{xx}$, then we get a constructive method for finding u. To this end we repose the control problem as a variational problem :
Find $v \in L^2(0,1;\mathbf{C})$ such that

$$\int_0^T \int_{x_l}^{x_r} y\bar{w}\,dx dt = \int_0^1 \phi\bar{\eta}dx, \qquad \forall \eta \in L^2(0,1;\mathbf{C}) \qquad (2)$$

where

$$iy_t = y_{xx}, \qquad y(x,T) = v$$

and

$$iw_t = w_{xx}, \qquad w(x,T) = \eta$$

The left hand side of equation (2) is a continuous sesquilinear form on $L^2(0,1;\mathbf{C})$ and the solution to this problem clearly hinges on a positivity property :

$$\int_0^T \int_{x_l}^{x_r} | S_{-t}v |^2 dx dt \geq \alpha |v|_{L^2}^2,$$

where S_{-t} denotes the general solution to the backward Schrödinger equation. It is easily seen that the inequality holds with $\alpha = \frac{1}{T}$ in the case where $x_l = 0$, $x_r = 1$, and that α, if positive, must decrease with respect to both T and $\delta = \frac{x_r - x_l}{2}$. By a very elementary method we will show that this inequality holds for $\alpha(T) > 0$ for T large enough. A more sophisticated analysis would show that it holds for arbitrary $T > 0$.
Let v be a trigonometric polynomial over $\mathbf{C}$; i.e.,

$$v(x) = \sum_{k=1}^N a_k \sin k\pi x, \qquad x \in [0,1], a_k \in \mathbf{C}.$$

For such a function ,

$$S_t v = \sum_{k=1}^N a_k\, e^{ik^2\pi^2 t}\, \sin k\pi x,$$

and so

$$\int_0^T \int_{x_l}^{x_r} (S_t v)\overline{(S_t v)}\, dx\, dt = \int_0^T \int_{x_l}^{x_r} \sum_{k,l} a_k\, \overline{a_l}\, e^{i(k^2-l^2)\pi^2 t}\, \sin k\pi x\, \sin l\pi x\, dx\, dt.$$

It is convenient to split this double series into diagonal and off-diagonal terms. If we define $x_c = \frac{x_l+x_r}{2}$, $\delta = \frac{x_r - x_l}{2}$, and

$$\Delta_k = \int_{x_l}^{x_r} (\sin k\pi x)^2 dx = \delta + \frac{1}{4k\pi}(\sin 2k\pi x_r - \sin 2k\pi x_l)$$

$$= \delta + \frac{\sin 2k\pi\delta \cos 2k\pi x_c}{2k\pi},$$

then it is easy to see that, for each δ, there is a lower bound to the Δ_k, independent of k. For example, let $k^* \in Z$ be such that $(4k^*)^{-1} < \delta < (4k^* - 4)^{-1}$. Then

$$\Delta_k \geq \frac{1}{4k^*} - \frac{1}{2k^*\pi} = \frac{1}{2k^*}(\frac{1}{2} - \frac{1}{\pi}) \geq \frac{\delta}{3}.$$

So,

$$\int_0^T \int_{x_l}^{x_r} \sum_{k=1}^N |a_k|^2 (\sin k\pi x)^2 dx dt \geq \frac{T\delta}{3} \sum_{k=1}^N |a_k|^2.$$

Now if we define the terms

$$\Omega_{kl} = \int_{x_l}^{x_r} \sin k\pi x \sin l\pi x dx = \int_{x_l}^{x_r} \cos(k-l)\pi x - \cos(k+l)\pi x dx$$

$$= \frac{1}{2\pi}[\frac{\cos(k-l)\pi x_c \sin(k-l)\pi\delta}{k-l} - \frac{\cos(k+l)\pi x_c \sin(k+l)\pi\delta}{k+l}],$$

then the off-diagonal terms are

$$2 \int_0^T \int_{x_l}^{x_r} \sum_{k>l} \Re(a_k \overline{a_l} e^{i(k^2-l^2)\pi^2 t}) \sin k\pi x \sin l\pi x \, dx \, dt$$

$$= 2 \sum_{k>l} \Omega_{kl} \int_0^T \Re(a_k \overline{a_l} e^{i(k^2-l^2)\pi^2 t}) \, dt.$$

These are bounded in modulus, by

$$\frac{2}{\pi^2} \sum_{k>l} |\Omega_{kl}| \, |\frac{\Re(a_k \overline{a_l})}{k^2 - l^2}| \, | -e^{i(k^2-l^2)\pi^2 T} | \leq \frac{2}{\pi^2} \sum_{k>l} |\Omega_{kl}| \frac{|a_k|^2 + |a_l|^2}{k^2 - l^2}$$

$$\leq \frac{2}{\pi^2} \frac{2}{3\pi} \sum \frac{|a_k|^2 + |a_l|^2}{k^2 - l^2} \leq \frac{1}{9\pi} \sum |a_k|^2,$$

where we have used the rough estimate for Ω_{kl},

$$|\Omega_{kl}| \leq \frac{2}{3\pi}, \quad k \neq l.$$

Computational Results

The variational problem just outlined lends itself to a straightforward Galarkin approximation. We will now describe this before we describe numerical methods. If $\{\epsilon_k^h\}$ is a basis for Σ_h which is a finite-dimensional subspace of L^2, we can expand the test function, η, and the unknown v^h in terms of this basis.

$$\eta = \sum_{}^{N} \eta_l \epsilon_l \qquad v^h = \sum_{}^{h} v_k \epsilon_k.$$

After assembling the data vector,

$$F_k = \int_0^1 \phi \, \epsilon_k \, dx,$$

and the Gram matrix,

$$G(\eta, v^h)_{kl} = \int_0^T \int_{x_l}^{x_r} (S_{-t}\epsilon_k)(S_{-t}\epsilon_l) \, dx \, dt,$$

we solve the matrix equation

$$GV_N = F.$$

Because of controllability, we know that G is Hermitian, positive definite and that there is a unique solution. To compute the solution to the control problem, we let

$$v^h = V_N \bigotimes \epsilon_k^h$$

and then compute

$$iy_t = y_{xx}, \qquad y(x,T) = v^h(x)$$

on the interval [0,T].
Then we store the result

$$u^h = iy|_{[x_l, x_r] \times [0,T]}.$$

The approximate solution can then be tested as follows : With u^h computed as above, we solve,

$$i\psi_t^h = \psi_{xx}^h + u^h$$
$$\psi^h(x,0) = 0$$

over the rectangle $[0,1] \times [0,T]$, and compare $\psi^h(x,T)$ with the known data $\phi(x)$.
The discrete form of the PDE was implemented by the Crank-Nicolson finite difference method. This is an implicit 2-step method with unitary step forward matrix ; the scheme is non-dissipative. In practice we only computed the forward equation; that is, rather than computing

$$iy_t = y_{xx}, \qquad y(x,T) = v(x)$$

the program computes the IVP

$$-iy_t = y_{xx}, \qquad y(x,0) = v(x)$$

If we store $iY|_{x_l, x_r}$, then the desired solution to the control problem is

$$u(x,t) = iY(x, T-t).$$

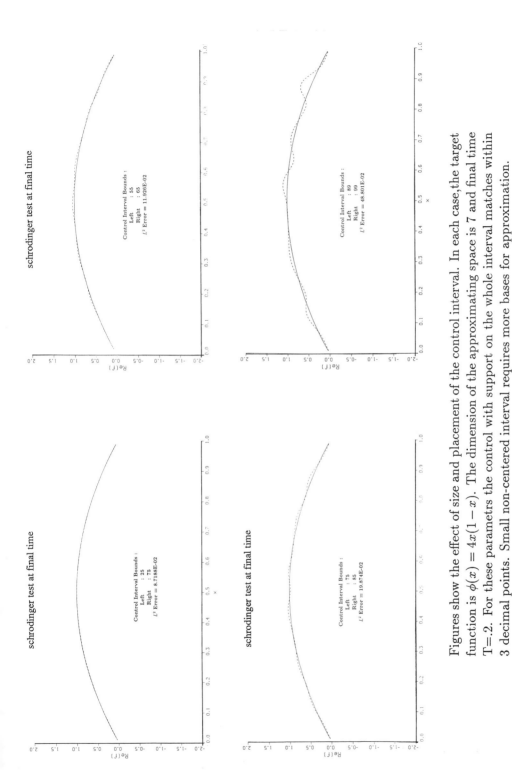

Figures show the effect of size and placement of the control interval. In each case, the target function is $\phi(x) = 4x(1-x)$. The dimension of the approximating space is 7 and final time T=.2. For these parametrs the control with support on the whole interval matches within 3 decimal points. Small non-centered interval requires more bases for approximation.

T J TABARA

On an existence problem for the boundary value problem of a second order linear differential equation with polynomial coefficients

0. INTRODUCTION

In [3], we considered the following differential equation

$$(\delta^2 - p\delta)y - (x^m + a_1 x^{m-1} + \ldots + a_{m-1}x)y = 0, \qquad \delta = x\frac{d}{dx} \tag{E}$$

where m and p are fixed integers with $m \geq 2$ and $1 \leq p \leq m-1$ and $a_1, \ldots, a_{m-1}$ are complex parameters under the assumption that $x = 0$ is an apparent singular point. To satisfy this assumption, a_{m-p} must be a certain polynomial in terms of $a_{m-(p-1)}, \ldots, a_{m-1}$ for each pair of (m, p) (See [1], [3] and [4]). This condition, called (SC) condition, will give us two linearly independent solutions that are entire in both x and the parameters. We note that (E) is the type Y. Sibuya considered previously when $p = 1$ ([2]).

We set

$a = (a_1, \ldots, a_{m-(p+1)}, a_{m-(p-1)}, \ldots, a_{m-1}) \in C^{m-2}$, and $M = \{0, 1, \ldots, m-1\}$

$S_k = \{ x \in C : \left|\arg x - \frac{2k\pi}{m}\right| < \frac{\pi}{m} \}$,

$R(x,a) = \dfrac{y_1(x, a)}{y_2(x, a)}$, where $y_1(x, a)$ and $y_2(x, a)$ are linearly independent solutions of (E).

The following existence and uniqueness problems were discussed:

i) *Existence:*

Given extended-valued complex numbers $c, c_0, \ldots, c_{m-1}$, is there a differential equation of the type (E) with linearly independent solutions $y_1(x, a)$ and $y_2(x, a)$ such that $R(0, a) = c$, and $R(x, a) \to c_k$ as $x \to \infty$ in S_k for all $k \in M$?

ii) *Uniqueness:*

If so, to what extent is the vector $a = (a_1, \ldots, a_{m-(p+1)}, a_{m-(p-1)}, \ldots, a_{m-1})$ produced by the numbers $c, c_0, c_1, \ldots, c_{m-1}$ unique?

For existence i), the following results are obtained as the necessary and sufficient conditions:

THEOREM 0.1 Let m and p be integers with $m \geq 4$ and $1 \leq p \leq m-1$, $a_1, \ldots, a_{m-1}$ be complex parameters. In addition if m is even, assume that $1 \leq p < (1/2)m$. Then there exist $m + 1$ extended-valued complex numbers $c, c_0, c_1, \ldots, c_{m-1}$ satisfying

(S1) $c_k \neq c_{k+1}$ for all $k \in M$, where $c_m = c_0$, and

(S2) the set $\{c_0, c_1, \ldots, c_{m-1}\}$ contains at least three distinct points

if and only if there are a vector $a = (a_1, ..., a_{m-(p+1)}, a_{m-(p-1)}, ..., a_{m-1})$ *under* (SC) *condition and linearly independent solutions* $y_1(x, a)$ *and* $y_2(x, a)$ *of the differential equation* (E) *such that* $R(0, a) = c$, *and* $R(x, a) \to c_k$ *as* $x \to \infty$ *in* S_k *for all* $k \in M$.

For uniqueness ii), global uniqueness can be easily ruled out by examining the case when $m = 4$ and $p = 1$ so that our main result is as follows ([3]) :

THEOREM 0.2 *Given complex numbers* $c, c_0, c_1, ..., c_{m-1}$ *satisfying* (S1) *and* (S2), *then there is a locally unique vector* $a = (a_1, a_2, ..., a_{m-1})$ *such that* $R(0, a) = c$, *and* $R(x, a) \to c_k$ *as* $x \to \infty$ *in* S_k.

Notice that when m is an even integer, the additional condition $1 \le p < (1/2)m$ has been imposed to guarantee that there is always a none zero Stokes multiplier in some sector ([3]). This note will explore a possible alternative to this condition that may still guarantee the existence problem. We shall examine the simplest cases of such, $m = 4$, $p = 2$, and $p = 3$. The main result of this note is :

THEOREM 0.3 i) *For* $m = 4$ *and* $p = 2$, *there exists a sufficiently small positive number* ρ *such that theorem* 0.1 *holds for* $0 < |a| \le \rho$.

ii) *For* $m = 4$ *and* $p = 3$, *there exists a transformation that reduces* (E) *to that of* $m = 4$ *and* $p = 1$.

To show the above, we assume that (S1) and (S2) hold for given numbers $c_0, c_1, ..., c_{m-1}$. Then there is the smallest integer $l \in M$ such that the three point c_{l-1}, c_l and c_{l+1} are distinct. Hence there is a Möbius transformation $T(z)$ that takes the ordered triple (c_{l-1}, c_l, c_{l+1}) into $(0, \infty, 1)$. Since the ratio of two linearly independent solutions is again expressed as the ratio of the two linear combination of these solutions under $T(z)$. We set $T(R(x, a)) = F(x, a)$ to begin with

$$F(0, a) = c \tag{0.1}$$

$$F(x,a) \to \frac{W_{12}(a) W_{0k}(a)}{W_{02}(a) W_{1k}(a)} \text{ for all } k \in M \text{ as } x \to \infty \text{ in } S_k, \tag{0.2}$$

where $W_{jk}(a) = \frac{1}{p} W_{jk}(x,a) x^{-(p-1)}$, and $W_{jk}(x,a)$ are the Wronskians of the solutions for (E).

In case of $m = 4$, c = arbitrary, $c_0 = 0$, $c_1 = \infty$, $c_2 = 1$, and $c_3 \ne 0, 1$ are assumed.

1. PROOF OF THEOREM 0.3

When $m = 4$ and $p = 3$, making the parameters simple we set $a_2 = a$ and $a_3 = 2b$. Then under (SC) the equation (E) becomes

$$(\delta^2 - 3\delta)y - (x^4 + 2b(a - b^2)x^3 + ax^2 + 2bx)y = 0. \tag{1.1}$$

There exists the following transformation

$$y \to [-b + (b^2 - a)x]y + y' \tag{1.2}$$

that reduces (1.1) to

237

$$(\delta^2 - \delta)y - (x^4 + 2b(a - b^2)x^3 + ax^2)y = 0,$$
$$y'' - (x^2 + a - b^2(a - b^2)^2)y = 0. \tag{1.3}$$

Hence the problem can be solved completely since this is of the Weber type ([2]).

When $m = 4$ and $p = 2$, letting $a_1 = 2a$ and $a_3 = b$, the equation (E) is given under (SC)
$$(\delta^2 - 2\delta)y - (x^4 + 2ax^3 + b^2x^2 + bx)y = 0. \tag{1.4}$$

Before we solve the problem, we shall list some of the known facts for (1.4):

FACT 3.1. There are two linearly independent solutions under (SC) for (1.4):

$$\begin{cases} \varphi_1(x, a, b) = 1 + \varphi_{11}(a, b)x + \sum_{h=3}^{+\infty} \varphi_{1h}(a, b)x^h \\ \varphi_2(x, a, b) = x^2 + \sum_{h=3}^{+\infty} \varphi_{2h}(a, b)x^h \end{cases}, \tag{1.5}$$

which are unique and entire in (x, a, b). Every solution of (1.4) is uniquely determined by the initial data $y(0)$ and $y''(0)$.

FACT 3.2. *There is a solution $\mathcal{Y}(x,a)$ of (1.4) with the following properties:*

i) $\mathcal{Y}(x, a, b)$ is an entire function of x and the parameters (a, b)

ii) $\mathcal{Y}(x, a\ b)$ satisfies the asymptotic representation

$$\mathcal{Y}(x, a, b) = x^{-\frac{1}{2}(b^2 - a^2)}(1 + O(x^{-1/2})) \exp\left[-(\tfrac{1}{2}x^2 + ax)\right] \tag{1.6}$$

uniformly on each compact set in the a-space as $x \to \infty$ in the sector

$$\bar{S}_3 \cup \bar{S}_0 \cup \bar{S}_1 = \{x \in C : |\arg x| < \tfrac{3\pi}{4}\}, \text{ where } \bar{S}_k \text{ means the closure of } S_k. \tag{1.7}$$

We note that condition (ii) determine uniquely the solution $\mathcal{Y}(x, a, b)$. Set $f_k(x, a, b) = \mathcal{Y}((-i)^k x, (-i)^k a, i^k b)$ for all k.

FACT 3.3. *For every k, the function $f_k(x, a, b)$ is a solution of (1.4) and*

$$f_k(x, a, b) = ((-i)^k x)^{-(-1)^k \frac{1}{2}(b^2 - a^2)} (1 + O(x^{-1/2})) \exp\left[-(-1)^k (\tfrac{1}{2}x^2 + ax)\right] \tag{1.8}$$

uniformly on each compact set in the a-space as $x \to \infty$ in the sector

$$\bar{S}_{k-1} \cup \bar{S}_k \cup \bar{S}_{k+1} = \{x \in C : \left|\arg x - \tfrac{2k\pi}{4}\right| < \tfrac{3\pi}{4}\}.$$

FACT 3.4. $f_k(x, a, b)$ *and* $f_{k+1}(x, a, b)$ *are linearly independent solutions of (1.4) for all k.*

FACT 3.5. (i) $f_k(x, a, b)$ *can be expressed by the central connection formula;*

$$f_k(x, a, b) = f_k(0, a, b)\varphi_1(x, a, b) + \tfrac{1}{2}f_k''(0, a, b)\varphi_2(x, a, b)$$

(ii) $f_k(x, a, b)$ *can also be expressed by the lateral connection formula;*

$$f_k(x, a, b) = C_k(a, b) f_{k+1}(x, a, b) + \tilde{C}_k(a, b) f_{k+2}(x, a, b)$$

where $C_k(a, b)$ and $\tilde{C}_k(a, b)$ are called the Stokes multipliers for $f_k(x, a, b)$ with respect to $f_{k+1}(x, a, b)$ and $f_{k+2}(x, a, b)$.

If $a = b = 0$, the equation (1.4) becomes $(\delta^2 - 2\delta)y - x^4 y = 0$ whose solutions $y_1(x,a) = \exp(-\frac{1}{2}x^2)$, and $y_2(x,a) = \exp(\frac{1}{2}x^2)$ can be computed directly. Then there are only two distinct c_k, i.e., $c_0 \neq c_1$ possible, and all of the Stokes multipliers are zero so that (S2) of the existence theorem 0.1 fails. In general, if $|a|+|b| \neq 0$, each of the Stokes multipliers
$C_0(a,b) = C(a,b)$, $C_1(a,b) = C(-ia,ib)$, $C_2(a,b) = C(-a,-b)$, and $C_3(a,b) = C(ia,-ib)$
can be computed directly from (1.4) (See [3]), and

$$C(a,b) = 2^{\frac{1}{2}}\sqrt{2\pi} \frac{e^{\pi i}}{(a-b)\Gamma(\frac{1}{2}(b^2-a^2))} e^{-\frac{1}{2}(b^2-a^2)\gamma} e^{g(a,b)}, \tag{1.9}$$

$$C(ia,-ib) = 2^{\frac{1}{2}}\sqrt{2\pi} \frac{e^{\pi i}}{i(a+b)\Gamma(-\frac{1}{2}(b^2-a^2))} e^{\frac{1}{2}(b^2-a^2)\gamma} e^{-g(a,b)}, \tag{1.10}$$

$$\tilde{C}(a,b) = e^{-\frac{\pi}{2}(b^2-a^2)i}, \tag{1.11}$$

where γ is the Euler's constant and $g(a,b)$ is an analytic function with $g(-ia,ib) = -g(a,b)$ and $g(0,0) = 0$.

We set $\alpha(a, b) = f_k(0, a, b) = y(0, (-i)^k a, i^k b)$ for all k. Then the existence problem can be restated as to whether or not the following equations are solvable for a and b: For c and c_3 fixed, by (0.1) and (0.2) we get the system of equations

$$\begin{cases} b^2 - a^2 = \frac{1}{\pi i} \log\left(\frac{c_3}{c_3 - 1}\right) \\ \frac{1}{C(a,b)} \frac{\alpha(a,b)}{\alpha(-ia,ib)} = c \end{cases}.$$

Since $c_3 \neq 0, \neq 1$, $b^2 - a^2$ takes a finite value from the first equation. Now our problem is whether or not to solve the second equation for a. It can be shown that $\alpha(0, 0) \neq 0$ for all k, $\partial C(0,0)/\partial a \neq 0$, $\partial C(0,0)/\partial b \neq 0$, $\partial \alpha(0,0)/\partial a \neq 0$ and $\partial \alpha(0,0)/\partial b \neq 0$ ([3]). Hence there exists a sufficiently small positive number ρ such that $\alpha(a,b) \neq 0$ for $0 < |a|+|b| \leq \rho$. By the direct computation, there is always a nonzero Stokes multiplier if $(a, b) \in \{0 < |a|+|b| \leq \rho\}$. The equation can be solved for a in this neighborhood. Consequently this system of equations can be solved in terms of c and c_3 by means of the complex implicit function theorem. This completes the proof.

2. REMARKS

We believe that in the case when m is even and $m \geq 6$, the similar method we employed above could be applicable to remove the condition $1 \leq p < (1/2)m$ should we calculate their Stokes multipliers explicitly. Unfortunately we have not succeeded to find them yet.

The author would like to thank Professor Y. Sibuya for stimulating conversations on the examples considered in this note during his two weeks' visit to the School of Mathematics at the University of Minnesota in May, 1991. This work was partially supported by NSF and by the Golden Gate University Faculty Development Fund.

References

[1] C. H. Lin and Y. Sibuya, Some applications of isomonodromic deformations to the study of Stokes multipliers, J. Fac. of Sci., Univ. of Tokyo, Sec. IA, 36 (1989), pp 649-663.

[2] Y. Sibuya, *Global Theory of a Second Order Linear Ordinary Differential Equations with a Polynomial Coefficient,* North-Holland Mathematics Studies 18, North-Holland/American Elsevier, 1975.

[3] T. J. Tabara, An asymptotic analysis and Stokes multipliers for a second order linear differential equation with polynomial coefficients, Ph.D. Dissertation, University of Minnesota, 1989.

[4] T. J. Tabara, A locally prescribed Stokes phenomenon, Funkcial. Ekvac., To appear.

[5] T. J. Tabara, Local uniqueness of a boundary value problem in the complex plane, Submitted.

Department of Mathematics, Golden Gate University, San Francisco, CA 94105-2968

L LIN[1], B TEMPLE[2] AND J WANG[3]

On the convergence of Glimm's method and Godunov's method when wave speeds coincide

Abstract

We discuss the convergence of Glimm's method and Godunov's method in the simplest setting where wave speeds coincide. For initial data of bounded variation, we use a sharp estimate to get estimates for the a rate of convergence of the residuals, and these are observed to agree when the problem is strictly hyperbolic, but an improvement is observed in the rate for the Godunov method when the problem is not strictly hyperbolic.

1 Introduction

We discuss the problem of the convergence of the 2×2 Glimm and Godunov methods in the presence of coinciding wave speeds. Details will be provided in the authors forthcoming paper. Our setting is the initial value problem for the 2×2 resonant nonlinear system of conservation laws

$$a_t = 0,$$
$$u_t + f(a, u)_x = 0, \tag{1}$$

---

[1]Department of Mathematics, Zhongshan University, Peoples Republic of China; Supported in part by National Science Foundation US-CHINA Cooperative Research Grant number DMS-8657319, and by the Institute of Theoretical Dynamics, UC-Davis, Spring 1990.

[2]Department of Mathematics, University of California, Davis, Davis CA 95616; Supported in part by National Science Foundation Grant Numbers DMS-8657319 and DMS-86-13450 (Principle Investigator), and by the Institute of Theoretical Dynamics, UC-Davis.

[3]Institute of Systems Science, Academia Sinica, Beijing, Peoples Republic of China; Supported in part by National Science Foundation US-CHINA Cooperative Research Grant number DMS-8657319, and by the Institute of Theoretical Dynamics, UC-Davis, Winter 1990.

$$a(x,0) = a_0(x) \equiv a(x),$$

$$u(x,0) = u_0(x), \tag{2}$$

where $u \in R$, $a \in R$, and we let $U = (a, u)$. This is the special case of an $n \times n$ resonant nonlinear system as introduced in [10,11]. System (1) is a system with the two wave speeds $\lambda_0(U) = 0$ and $\lambda_1(U) = \partial f/\partial u$, and we study solutions in a neighborhood of a state where $\lambda_1 = \lambda_2$. System (1) is a model for the resonant behavior that occurs in problems arising in elasticity, multi-phase flow, and for transonic duct flow [10,11,12]. Here we consider the problem of the convergence of the 2×2 Glimm and Godunov method for solutions of (1) generated from initial data of bounded variation, and taking values in a neighborhood of a state U_* where we assume (these assumptions were introduced in [10,11])

$$\lambda_1(U_*) = 0, \tag{3}$$

$$\frac{\partial}{\partial u}\lambda_1(U_*) \neq 0, \tag{4}$$

$$\frac{\partial}{\partial a}f(U_*) \neq 0. \tag{5}$$

Condition (3) states that the wave speeds λ_0 and λ_1 coincide at U_*, condition (4) states that the nonlinear family of waves is genuinely nonlinear at U_*, and condition (5) is a nondegeneracy condition introduced in [11] that determines a canonical structure to the nonlinear wave curves in a neighborhood of the state U_*. The initial value problem is more complicated when wave speeds coincide, and one can show that unlike the strictly hyperbolic problem, in the resonant problem (1), the total variation of the solution at time $t > 0$ (a measure of the strength of the nonlinear waves present) cannot be bounded by the total variation of the initial data in this setting, even for sufficiently weak solutions. Moreover, the process by which solutions decay to time asymptotic wave patterns is correspondingly more interesting and more complicated than in the strictly hyperbolic case [14,7]. Here we discuss forthcoming work of the authors in which we obtain time independent bounds on derivatives, prove convergence and establish a rate of convergence for the 2×2 Godunov numerical method as applied to solutions of (1), (2) in a neighborhood of such a state U_*. The fact that such time independent bounds exist at all is a purely nonlinear phenomenon and is surprising in that the resonant linearized system blows up. The analysis in [18] carries over directly to the problem (1), (2), thus providing a corresponding convergence theorem for Glimm's numerical method. Our methods are based on a sharpened estimate from [18], enabling us to obtain convergence rates for the residuals that reduce to the sharp rates when the problem is strictly hyperbolic.

More specifically, in the authors forthcoming paper we prove that the 2×2 Godonov method converges to a weak solution of (1), (2) (modulo extraction of a subsequence) by showing that the total variation of any approximate solution of (1), (2) at time $t > 0$ is bounded by the initial total variation when the total variation is measured under the singular transformation Ψ : $(a, u) \rightarrow (a, z)$ which was introduced in [18] (see also [8]). We sharpen the analysis in [18] that gives a rate of convergence for Glimm's method, and we use the improved estimates to conclude that the best rate of convergence of the 2 × 2 Godunov method implied by our analysis is

$$R(U_{\Delta x}, \phi) = O(\Delta x^{\frac{1}{1+p}}), \tag{6}$$

where $U_{\Delta x} \equiv (a_{\Delta x}, u_{\Delta x})$ denotes the Godunov approximate solution, p is the order of contact between the wave curves at the points of coinciding wave speeds, and R denotes the residual of the weak solution,

$$R(U, \phi) \equiv \int_0^\infty \int_{-\infty}^{+\infty} (U\phi_t + F(U)\phi_x) dx\, dt + \int_{-\infty}^{+\infty} U_0(x)\phi_0(x) dx, \tag{7}$$

where $\phi(x, t)$ is any smooth test function. We can compare this to the rate of convergence of Glimm's method in this problem, which by our methods (sharpening [18]) is given by

$$\int_{\Pi[0,1]_{ij}} R(U_{\alpha \Delta x}, \phi)^2 d\alpha = O(\Delta x^{\frac{1}{1+2p}}). \tag{8}$$

Here, $U_{\alpha \Delta x}$ denotes the approximate solution generated by the Glimm scheme, where $\alpha \in A \equiv \Pi[0, 1]_{ij}$ denotes the random sequence which determines the choice of sampling in Glimm's method. (The rate originally obtained in [18] was $\Delta x^{\frac{1}{3+2p}}$ which is not correct when $p = 0$. The improvement refines the estimate obtained in Proposition 6.2 of [18].) Note that when $p = 0$, both of the rates in (8) and (6) reduce to the known rate $O(\Delta x)$ for the convergence of these methods in the presence of a total variation bound in the conserved quantities. (In fact, the condition (4) forces p=2 at the state U_*, but we can consider the more general case $p \neq 2$ as well.) We attempt to make a meaningful comparison of these rates as follows: the error estimate in (8) is given in terms of the integral of $R(U_{\alpha\Delta x}, \phi)^2$ over the measure space of sampling sequences. Now estimate (8) can be interpreted as saying that the convergence rate of $R(U_{\alpha\Delta x}, \phi)^2$, averaged over all equidistributed sequences (a set of measure one in A), is $O(\Delta x)$, suggesting that the average rate of convergence of $R(U_{\alpha\Delta x}, \phi)$ is $O(\Delta x^{1/2})$. However, the best equidistributed sequences (a set of measure zero in A), give an improvement in convergence over the average by essentially a power of two (ignoring logarithms), thus we expect that in the best cases, the convergence of $R(U_{\alpha\Delta x}, \phi)$ in Glimm's method is $O(\Delta x)$. [1] Thus when $p = 0$, the comparison of the rates (7) and (8) should be valid, suggesting that (7) and (8) may provide a fair basis of comparison when $p > 0$. We believe that there is an improvement in the convergence of the Godonov method over the Glimm method for $p > 0$ due to the fact that the averaging in the Godunov method regularizes the oscillations that can occur in both methods at the states of coinciding wave speeds (see [18]). Specifically, the Glimm method is designed to give an accurate time evolution of elementary waves through an exact calculation of wave interactions. The problem here is that the coordinate system of wave curves is singular at the points of coinciding wave speeds, and thus the solution for smooth initial data which is transverse to the wave curves is not well approximated by elementary waves, and so the piecewise constant approximation scheme generates oscillatory waves that are exactly time evolved by the Glimm method. The averaging process in the Godunov method regularizes these oscillations at each time step.

We are only able to obtain the time independent bounds on derivatives for the 2×2 Glimm and Godunov methods, and have not succeeded in obtaining these bounds directly by any other 2×2 method (including the Lax-Friedrichs scheme). Moreover, because of the resonant behavior near the state U_*, the total variation of u at any time cannot be bounded by the total variation of the initial data when a is only of bounded variation, making it difficult to obtain estimates for the derivatives of u directly at each fixed time. Indeed,

---

[1] Proving this rigorously is an open problem. The conclusion is suggested by the convergence properties of a single shock. For a rigorous result in this direction, see Lemma 3.3 of A. Harten and P.D. Lax, SIAM J. Numer. Anal., Vol. 18, No. 2, 1981., where, in the strictly hyperbolic case, $R(U_{\alpha\Delta x}, \phi)$ is shown to be at least $O(\Delta x^{1/3})$ for all α except on a set of measure $\leq O(\Delta x^{1/3})$.

(3), (4) and (5) imply that the linearized equations for (1) at the state U_* are given by

$$a_t = 0,$$
$$u_t + f_a(U_*)a_x = 0,$$
(9)

and thus it is easy to see that the solution in the linearized case is $u(x,t) = -f_a(U_*)a'(x)t + u_0(x)$ Thus in the linear problem, u and all x-derivatives of u blow up at a linear rate. The results here and in [18] confirm that solutions of the nonlinear problem satisfy time independent bounds on derivatives when $u_0(\cdot)$ and $a(\cdot)$ are of bounded variation, and this shows that the nonlinear problem is more regular than the linearized problem. The results of Oleinik and Kruzkov [16,13], which apply to the inhomogeneous scalar conservation law $u_t + f(a(x), u)_x = 0$ when $a \in C^2$, give Gronwall type exponential growth bounds on the total variation in u. As far as we know, it has not been shown that the entropy solutions of Oleinik and Kruzkov satisfy the time independent bounds that we are obtaining via the 2×2 methods. It appears to be a subtle problem to verify the Kruzkov entropy condition for solutions contructed by 2×2 methods for arbitrary $a(\cdot)$ of bounded variation, because the argument of Kruzkov is easy to apply only when based on a total variation bound on u. The established total variation bound on z does not imply estimates on the total variation of u when $a(\cdot)$ is not smoother than C^1. Counterexamples show that there is no bound on the rate of blowup of u in the total variation norm that is based on the C^1-norm of a alone. The authors plan to address these issues in a subsequent paper. We believe that the averaging in Godunov's method can be used to advantage, and as far as we know, no one to date has obtained a total variation estimate for the conserved variable u based on the 2×2 Glimm's method. Anticipating future results here, we expect to obtain the result that for solutions of the 2×2 Godunov method, the total variation of u grows sublinear when a is sufficiently smooth. Such an estimate would establish the entropy conditions of Kruzkov and the time independent bounds on z-derivatives simultaneously in solutions generated by a 2×2 numerical method based on the 2×2 Riemann problem (Godunov's method). As far as we know, these results are new for scalar conservation laws as well. (See also [19] where exponential growth bounds on u are obtained under similar assumptions via an upwind scheme that applies to the polymer equations, (10) below, an Eulerian formulation of (1), (2)).)

It is important to note that (6) and (8) give rates of convergence which depend only on the total variation of the initial data $U_0(x) = (a_0(x), u_0(x))$ as measured under the singular transformation Ψ. There is no corresponding time independent convergence rate provided by the methods of Kruzkov and Oleinik [13], [16] since their methods do not provide time independent bounds on derivatives, and apply only when $a(x)$ is sufficiently smooth. A study of (1) when a is of bounded variation is equivalent to studying the time asymptotic wave patterns when a is smooth.

References

[1] Courant and Freidrichs, *Supersonic flow and shock waves*, John Wiley & Sons, New York, 1948.

[2] J. Glimm, *Solutions in the large for nonlinear hyperbolic systems of equations*, Comm. Pure Appl. Math. **18**(1965), pp. 697-715.

[3] S.K. Godunov, *A difference method for numerical calculations of discontinuous solutions of the equations of hydrodynamics*, Mat. Sb. **47**(1959), in Russion, pp. 271-306.

[4] S.K. Godunov, *Bounds on the discrepancy of approximate solutions constructed for the equations of gas dynamics*, J. Comput. Math. and Math. Phys., Vol. 1, 1961, pp. 623-637.

[5] E. Isaacson, *Global solution of a Riemann problem for a non-strictly hyperbolic system of conservation laws arising in enhanced oil recovery*, Rockefeller University preprint.

[6] E. Isaacson, D. Marchesin, B. Plohr, and B. Temple *The Riemann problem near a hyperbolic singularity: the csassification of solutions of quadratic Riemann problems I*, SIAM J. Appl. Math.,**48**(1988),pp.

[7] E. Isaacson, B. Temple, *The structure of asymptotic states in a singular system of conservation laws*, Adv. Appl. Math. (to appear).

[8] E. Isaacson, B. Temple, *Analysis of a singular system of conservation laws*, Jour. Diff. Equn., **65**(1965), pp.

[9] E. Isaacson, B. Temple, *Examples and classification of non-strictly hyperbolic systems of conservation laws*, Abstracts of AMS, January 1985.

[10] E. Isaacson, B. Temple, *Nonlinear resonance in inhomogeneous systems of conservation laws*, Contemporary Mathematics, Vol 108, 1990.

[11] E. Isaacson, B. Temple, *Nonlinear resonance in systems of conservation laws*, submitted, SIAM Jour. Appl. Anal.

[12] B. Keyfitz and H. Kranzer, *A system of non-strictly hyperbolic conservation laws arising in elasticity theory*, Arch. Rat. Mech. Anal., **72**(1980), pp.219-241.

[13] S.N. Kruzkov, *First order quasilinear equations with several space variables*, Math. USSR. Sb. 10 (1970), pp.217-243.

[14] T.P. Liu, *Resonance for a quasilinear hyperbolic equation*, J. Math. Phys. **28** (11), November 1987.

[15] D. Marchesin and P.J. Paes-Leme, *A Riemann problem in gas dynamics with bifurcation*, PUC Report MAT 02/84, 1984.

[16] O.A. Oleinik, *Discontinuous solutions of non-linear differential equations*, Uspekhi Mat. Nauk (N.S.), **12**(1957),no.3(75), pp. 3-73 (Am. Math. Soc. Trans., Ser. 2, **26**, pp. 195-172.)

[17] J. Smoller, *Shock waves and reaction diffusion equations*, Springer-Verlag, Berlin, New York, 1980.

[18] B. Temple, *Global solution of the Cauchy problem for a class of 2×2 nonstrictly hyperbolic conservation laws*, Adv. in Appl. Math., **3**(1982), pp. 335-375.

[19] A. Tveito and R. Winther, *Existence, uniqueness and continuous dependence for a system of hyperbolic conservation laws modelling polymer flooding*, Preprint, Department of Informatics, University of Oslo, Norway, January, 1990.

C VARGAS AND M ZULUAGA
On a nonlinear Dirichlet problem type at resonance and bifurcation

Introduction:

This paper deals with existence results for the nonlinear Dirichlet Problem

$$P(\lambda): \Delta u + \lambda u + g(u) = h(x) \quad \text{on } \Omega; \qquad u(x) = 0 \quad \text{on } \partial\Omega,$$

where $\Omega \subset \mathbf{R}^n$ is an open and bounded domain with smooth boundary and λ a parameter. We are mainly interested in the case $\lambda = \lambda_1$, the first eigenvalue of $-\Delta$ with zero Dirichlet condition on $\partial\Omega$.

Intensive research on this problems started after the 1970 paper of Landesman and Lazer [3], and several people have extended their results in various directions.

Our starting point has been the paper [1] of Anbrosetti and Mancini. They assumed $g \in C^1(\mathbf{R},\mathbf{R})$ and $h \in L^\infty(\Omega)$. Furthermore, they supposed that $\lambda_1 + g'(s) < \text{const} < \lambda_2$ and were able to prove the existence of an interval $[a,b] \subset \mathbf{R}$ such that, if $\int_\Omega h\varphi_1 \in [a,b]$, $P(\lambda_1)$ has a solution, otherwise $P(\lambda_1)$ has no solutions. They also obtained some multiplicity results for $P(\lambda_1)$, we refer the reader to their paper.

Here we assume some weaker conditions, we suppose that g has a Lipschitz constant K such that $K + \lambda_1 < \lambda_2$ and we are able to prove that there exists $(a,b) \subset \mathbf{R}$ such that $P(\lambda_1)$ has a solution if $\int_\Omega h\varphi_1 \in (a,b)$ and $P(\lambda_1)$ has no solutions if $\int_\Omega h\varphi_1 \notin [a.b]$. Besides of weakening the conditions, we shall give a constructive proof.

Also, we prove that the problem $P(\lambda)$ has a bifurcation point in $(0,\lambda_1)$ when g has some rather weak restrictions which are not the ones imposed by other authors like in Proposition 2.41 of Rabinowitz [4] or Struwe [5].

Preliminaries and notations

It is well known, that $-\Delta: D(-\Delta) \subset L^2(\Omega) \to L^2(\Omega)$ is a linear, selfadjoint and injective operator with closed range $R(-\Delta)$, eigenvalues $\{\lambda_i\}, i = 1, 2, \cdots; 0 \leq \lambda_1 < \lambda_2 \leq \lambda_3 \leq \cdots$ and $\varphi_1 > 0$ is the first eigenfunction asociated to λ_1. $(-\Delta)^{-1}$ is a well defined operator from $R(-\Delta)$ onto $D(-\Delta) \cap R(-\Delta)$ and $(-\Delta)^{-1}$ is completely continuous such that

a) $(-\Delta)^{-1}(R(-\Delta) \cap L^p(\Omega)) \subset H^{2,p}(\Omega) \cap H_0^{1,2}(\Omega), p \geq 2$

b) $\|(-\Delta)^{-1}f\|_{H^{2,p}(\Omega)} \leq Cp \| f \|_{L^p}$, for all $f \in R(-\Delta) \cap L^p(\Omega)$

c) $\|(-\Delta)^{-1}\|_{L^2(\Omega)} = \frac{1}{\lambda_1}$

d) The operator $-\Delta - \lambda_1$ is injective on $\langle\varphi_1\rangle^\perp$ and satisfies properties a) and b) above.

Solutions of $P(\lambda)$

It is well known that $u \in H_0^1(\Omega)$ is a solution of $P(\lambda)$ if and only if

(1.1) $$u = (-\Delta)^{-1}[\lambda u + g(u) - h].$$

Since $(-\Delta)^{-1}$ is a completely continuous operator, then $F : L^2(\Omega) \to L^2(\Omega)$ defined as $F(v) = (-\Delta)^{-1}[\lambda v + g(v) - h]$ is completely continuous as well.

The Lyapunov–Schmidt Method

Denote by $X = \langle\varphi_1\rangle = \{t\varphi_1, t \in \mathbb{R}\}$ and $Y = X^\perp$ its $L^2(\Omega)$-orthogonal complement. $H_0^1(\Omega) = X \oplus Y$ and thus, every $u \in H_0^{1,2}(\Omega)$ can be put in the form $u = x + y, x \in X, y \in Y$. Let P and Q be the projections on X and Y respectively. Applying P and Q to (1.1) we obtain

(1.2) $$x = P(-\Delta)^{-1}\ [\lambda(x+y) + g(x+y) - h]$$
(1.3) $$y = Q(-\Delta)^{-1}\ [\lambda(x+y) + g(x+y) - h].$$

Main results

Theorem 2.1 Suppose that

a) $g \in C(\mathbb{R}, \mathbb{R}), |g(u) - g(v)| \leq k|u - v|$ for all $u, v \in \mathbb{R}$ such that $\lambda_1 + k < \lambda_2$,

b) g is bounded,

c) $h(x) \in L^\infty(\Omega)$,

then there exist $(a, b) \subset \mathbb{R}$ such that if $\int_\Omega h\varphi_1 \in (a, b), P(\lambda_1)$ has a solution, and if $\int_\Omega h\varphi_1 \notin [a, b], P(\lambda_1)$ has no solutions.

Proof. It is clear that $P(\lambda_1)$ has a solution iff the system (1.2) and (1.3), in the case $\lambda = \lambda_1$, has a solution. Let $x \in X$ be fixed and consider (1.3). By a) it is easy to see that the operator

(2.1) $$Q(-\Delta)^{-1}[\lambda_1(x + \cdot) + g(x + \cdot) - h]$$

is a contraction of Y on Y, then by contraction Banach's Theorem, (2.1) has only one fixed point $y \in Y$, then (1.3) with $\lambda = \lambda_1$ has only one solution, $y \in Y$. Let $T : X \to Y$ be the function such that $T(x)$ is the only fixed point of (2.1).

We claim that T is continuous. In fact, by (1.3)

(2.2) $$T(x) = Q(-\Delta)^{-1}[\lambda_1(x + T(x)) + g(x + T(x)) - h].$$

By hypothesis a) we have $\|T(x_n) - T(x)\|_{L^2(\Omega)} \leq \dfrac{\frac{\lambda_1+k}{\lambda_2}}{1-\frac{\lambda_1+k}{\lambda_2}} \|x_n - x\|_{L^2(\Omega)}.$

This last inequality implies that T is continuous.

Now we have the following

Lemma. There exists $c > 0$, depending on the operator $-(\Delta + \lambda_1), \|h\|_\infty$ and the bound of g such that $\|T(x)\|_{L^2} \leq c$. Whenever $\partial\Omega$ is sufficiently smooth.

Proof of the Lemma. We know that $g(u) \in L^\infty(\Omega)$ for all $u \in L^2(\Omega)$.

From property d) of $-\Delta$, for all $f \in R(-(\Delta + \lambda_1)) \cap L^P(\Omega), P \geq 2$

$$\|-(\Delta + \lambda_1)^{-1}(f)\|_{H^{2,P}(\Omega)} \leq C_P \|f\|_{L^P(\Omega)}.$$

Then, since $T(x) \in Y$ is a solution of (1.3) ($\lambda = \lambda_1$), $T(x) = [-(\Delta + \lambda_1)^{-1}]Q(g(x + T(x)) - h)$. Therefore $\|T(x)\|_{H^{2,P}} \leq C_P \|Q(g(x + T(x)) - h)\|_P, P \geq 2$. Since $g(x + T(x)) - h$ is bounded we have $T(x) \in H^{2,P}(\Omega)$ for all $P \geq 2$.

By the Sobolev embedding Theorem $T(x) \in C^{1,\alpha}(\overline{\Omega})$ and $\|T(x)\|_{C^1} \leq$ constant. Thus $\|T(x)\|_{L^2} \leq c$.

Now we consider (1.2) ($\lambda = \lambda_1$) in the following form

(2.3) $$x = P(-\Delta)^{-1}[\lambda_1(x + T(x)) + g(x + T(x)) - h].$$

Since $P(-\Delta)^{-1}(T(x)) = 0$ and $P(-\Delta)^{-1}(\lambda_1 x) = x$, (2.3) turn out to be

(2.4) $$0 = P(-\Delta)^{-1}[g(x + T(x)) - h],$$

therefore (2.4) has a solution iff there exists $x \in X$, solution of

(2.5) $$\int_\Omega h\varphi_1 = \int_\Omega g(x + T(x))\varphi_1$$

Let $x = t\varphi_1$, then the right member of (2.5) defines the function

(2.6) $$\Gamma(t) = \int_\Omega g(t\varphi_1 + T(t\varphi_1))\varphi_1.$$

Since g and T are bounded and continuous, Γ is bounded and continuous as well, let

$$a = \inf_{t \in \mathbb{R}} \Gamma(t) \quad \text{and} \quad b = \sup_{t \in \mathbb{R}} \Gamma(t).$$

From (2.5) we have, if $\int_\Omega h\varphi_1 \in (a,b)$ then (2.3) has a solution $x = t\varphi_1$ and then $t\varphi_1 + T(t\varphi_1)$ is a solution of $P(\lambda_1)$. And if $\int_\Omega h\varphi_1 \notin [a,b]$, $P(\lambda_1)$ has no solution. □

Multiplicity Results

Theorem 2.2 Suppose that $g \in C^1(\mathbb{R},\mathbb{R})$ satisfies properties a) and b) of Theorem (2.1). Additionaly suppose that

d) $g(0) = 0$

e) $g'(0) < 0$ and $\lim_{s\to+\infty} g(s) = a > 0$ and $\lim_{s\to-\infty} g(s) = -b < 0$.

Then there exist, at least, three solutions for the problem

(2.7) $\qquad \Delta u + \lambda_1 u + g(u) = 0$ on Ω; $\qquad u = 0$ on $\partial\Omega$.

Proof. It is clear that $u = 0$ is a solution of (2.7) It can be proved that $T \in C^1(X,Y)$, this is done by the use of a well known argument based upon the implicit function theorem, see [3].

Since $T(0) = 0$, $\Gamma(0) = 0$. $\Gamma \in C^1(\mathbb{R},\mathbb{R})$ and since g and T are bounded, Γ is bounded as well and $\Gamma'(0) = g'(0)$. If we suppose the hypothesis e) we have that $\Gamma'(0) < 0$, $\lim_{s\to+\infty} \Gamma(s) = +a \int_\Omega \varphi_1$, and $\lim_{s\to-\infty} \Gamma(s) = -b \int_\Omega \varphi_1$, then there exist $t^+ > 0$ and $t^- < 0$ such that $\Gamma(t^+) = \Gamma(t^-) = 0$ thus, $t^\pm \varphi_1 + T(t^\pm \varphi_1)$ are non zero solutions of (2.7). The same result is obtained if we suppose

$$g'(0) > 0, \lim_{s\to+\infty} g(s) = -a < 0, \lim_{s\to-\infty} g(s) = b > 0 \quad \square$$

Bifurcation Result

Theorem 2.3 Suppose that $g \in C^1(\mathbb{R},\mathbb{R})$ satisfies hypothesis a), b), d), e) of Theorems (2.1) and (2.2) and suppose g) $sg(s) < 0$ if $s \neq 0$. Then

(2.8) $\qquad \Delta u + \lambda u + g(u) = 0$ on Ω; $\qquad u = 0$ on $\partial\Omega$

has a bifurcation point at $(\lambda_1, 0)$.

Proof. For $\lambda < \lambda_1$, equation (2.8) has only the zero solution, otherwise we obtain a contradiction.

Since $\lambda_1 + k < \lambda_2$ then there exists $\varepsilon > 0$ such that for all $\lambda \in [\lambda_1, \lambda_1 + \varepsilon]$, $\lambda + k < \lambda_2$. As in (2.2), we have $T(\lambda,x) = Q(-\Delta)^{-1}(\lambda T(\lambda,x) + g(x + T(\lambda,x)))$, for all $\lambda \in [\lambda_1, \lambda_1 + \varepsilon]$.

We saw, in Theorem (2.1), that $T(\lambda,x)$ is continuous in x. With the same kind of argument, $T(\lambda,x)$ is continuous in λ. And its rather easy to prove the continuity of T on both variables.

We shall consider the equation $x = P(-\Delta)^{-1}[\lambda(x + T(\lambda,x)) + g(x + T(\lambda,x))]$; with $\Gamma(\lambda,t) = \int_\Omega g(t\varphi_1 + T(\lambda,t\varphi_1))\varphi_1$, it becomes equivalent to

(2.9) $\qquad 0 = \Gamma(\lambda,t) + t(\lambda - \lambda_1)$.

As in Theorem (2.2) $T(\lambda,0) = 0$ for all $\lambda \in [\lambda_1, \lambda_1 + \varepsilon]$ and $T(\lambda, \cdot) \in C^1(x, Y)$, also $\Gamma'(\lambda, 0) = g'(0)$.

By hypothesis $g'(0) < 0$, then

(2.10) $$\Gamma'(\lambda, 0) + (\lambda - \lambda_1) < 0$$

for all $\lambda \in [\lambda_1, \lambda_1 + \varepsilon] \cap [\lambda_1, \lambda_1 - g'(0)]$. By (2.9) and (2.10) and recalling the boundedness of $\Gamma(\lambda, t)$, we have that there exist $t^+ > 0$ and $t^- < 0$ such that

(2.11) $$0 = \Gamma(\lambda, t^\pm) + t^\pm(\lambda - \lambda_1)$$

for all $\lambda \in [\lambda_1, \lambda_1 + \varepsilon] \cap [\lambda_1, \lambda_1 - g'(0)]$. We conclude that $t^\pm \varphi_1 + T(\lambda, t^\pm \varphi_1)$ are two non zero solutions of (2.8). It is clear that $t^\pm$ depend on λ.

Take $t^\pm$ as the minimum and maximum real numbers satisfying (2.11), we claim that $t^\pm \to 0$ if $\lambda \to \lambda_1$. Otherwise $t^+ \to a > 0$ and $t^- \to b < 0$ and by (2.11) and continuity of $\Gamma(\lambda, t)$, $0 = \Gamma(\lambda_1, a) = \Gamma(\lambda_1, b)$. This implies that $a\varphi_1 + T(\lambda_1, a\varphi_1)$ and $b\varphi_1 + T(\lambda_1, b\varphi_1)$ are two non zero solution of (2.8). Imposible!

By the continuity of $T(\lambda, x)$ and since $T(\lambda, t^\pm \varphi_1) \to 0$ if $t^\pm \to 0$ we obtain that $(\lambda_1, 0)$ is a bifurcation point with at least two branches $t^\pm \varphi_1 + T(\lambda, t^\pm \varphi_1)$. □

References

[1] Ambrosetti, A. and Mancini, G. *Existence and multiplicity results for nonlinear elliptic problems with linear part at resonance. The case of the simple eigenvalue*, J. of Diff. Equations 28 (1978) 220–245.

[2] Kransoselśkii, M. *Topological methods in the theory of nonlinear integral equations*. Mac Millan Company, New York (1964).

[3] Landesman, E. and Lazer, A. *Nonlinear perturbation of linear elliptic boundary value problems at resonance.* J. Math. Mech 19(1970) 609–623.

[4] Rabinowitz, P. *Minimax methods in critical point theory with applications to differential equations.* CBMS 65 Regional Conference Series in Mathematics, A.M.S. (1986).

[5] Struwe, M. *A note on a result of Ambrosetti and Mancini.* Ann. Mat. Pura et Aplicada, 131 (1982) 107–115.

C. Vargas J., CINVESTAV–IPN, A.P. 14–740, 07000 México D.F.

M. Zuluaga U., Departamento de Matemáticas, Universidad Nacional, Bogotá D.E., Colombia

J VOSMANSKÝ

Zeros of solutions of linear differential equations as continuous functions of the parameter k

1. Introduction

The special functions like Bessel functions, Airy functions, classical ortogonal polynomials and others can be considered as solutions of relatively simple linear differential equations of the type $y'' + a(t)y' + b(t)y = 0$ with very strong regularity conditions on coefficients like analyticity, higher monotonicity and other. As an example consider the very well known Bessel equation. Here the function $a(t) = 1/t$ is completely monotonic on $(0, \infty)$ - we write $a \in M_\infty$ - that means $(-1)^i (1/t)^{(i)} \geq 0$ for $i = 0, 1, \ldots$, $t \in (0, \infty)$, the function $b(t) = 1 - \nu^2/t^2$ has a completely monotonic derivative.

For simplicity we shall consider here the oscillatoric equation

$$y'' + q(t)y = 0 \quad t \in (0, \infty) \tag{q}$$

and the case of complete monotonicity instead of more general regular monotonicity only.

About 30 years ago L. Lorch and P. Szego [6] initiated the investigation of relations between certain higher monotonicity properties of the coefficient $q(t)$ in (q) and the corresponding ones of its solutions. Namely the sign-regularity of the higher differences with respect to the rank k of the sequences $\{t_k\}$ of positive zeros of solutions y of (q) were studied.

Let $c_{\nu k} = c_{\nu k}(\alpha)$ denote the k-th positive zero of the Bessel function

$$C_\nu(t) = C_\nu(t, \alpha) = \cos \alpha J_\nu(t) - \sin \alpha Y_\nu(t), \ 0 \leq \alpha < \pi. \tag{1.1}$$

In particular they showed that for $|\nu| > 1/2$, we have $(-1) \Delta^{n+1} c_{\nu k} > 0$, $n = 0, 1, \ldots$, $k = 1, 2, \ldots$. In other words the sequence $\{\Delta c_{\nu k}\}$ is completely monotonic.

This result was generalized and extended by many authors (e.g. Muldoon, Lewis, Steinig, Došlá, Háčik, Laforgia, Elbert) and also by the author (see e.g.[5], [7], [8], [9]).

One can discuss, in fact, the variation of the positive zeros of $C_\nu(t, \alpha)$ with respect to any of the three variables ν, α or k (the rank of zero). However, α and k are not really independent. They may be subsumed in a single variable e.g. $\kappa = k - \alpha/\pi$. It was recently done in [4]. There extend the notion of the k-th positive zero ($k = 1, 2, \ldots$) of the above mentioned function $C_\nu(t, \alpha)$ for any suitable real $\kappa \geq 0$. This approach, however, depends heavily on the couple $J_\nu(t), Y_\nu(t)$ and cannot be used in general.

Our main purpose here is:

1° To define κ-th positive zero t_κ of any but fixed solution of the equation (q) independently on the choice of the fundamental system of solutions of (q). The important role is played here by the principal pairs of solutions.

2° To extend our earlier results and other results on sign-regularity of the higher differences (with respect to the rank k) of the sequence $\{t_k\}$ of positive zeros of a solution y of (q) to the higher derivatives with respect to κ of the zero t_κ now considered as a function of the continuous variable κ.

2. The zeros as continuous functions of the parameter κ

Consider the oscillatory differential equation (q), where $q(t)$ is continuous (or has continuous derivative if necessary) and suppose the existence of principal pairs of solutions (y_1, y_2) (see [2]), i.e. solutions, for which there exists the finite

$$\lim_{t\to\infty} v^*(t) = v(\infty) > 0, \text{ where } v^*(t) := y_1^2 + y_2^2 \text{ and } w = |y_1 y'_2 - y'_1 y_2| = 1.$$

Remember, that the existence of principal pairs is ensured if e.g. $q(t)$ is of bounded variation. The function $v(t)$ complies with the Mammana identity

$$M(v) \equiv 2v''v - v'^2 + 4q(t)v^2 = c\ (=4w^2) \tag{2.1}$$

which is in certain sense (if q' exists) equivalent with the Appel equation

$$v''' + 4q(t)v'' + 2q'(t)v = 0\ .$$

Let t_0 denote a zero of y_1. Then any solution of (q) can be expressed in the form

$$y(t) = A\sqrt{v(t)} \sin\left(\int_{t_0}^{t} [1/v(\tau)]\, d\tau - \kappa\pi\right) \tag{2.2}$$

where $v(t)$ is any solution of (2.1) with $w = 1$ and κ, A are parameters. (For more detail see [2], [8].)

Let us choose now the principal solution $v^*(t)$ of Appels equation in (2.1). In such a case we can express the principal pair y_1, y_2, up to the sign, in the form

$$y_1 = \sqrt{v^*(t)} \sin \int_{t_0}^{t} \frac{d\tau}{v^*(\tau)}, \quad y_2 = \sqrt{v^*(t)} \cos \int_{t_0}^{t} \frac{d\tau}{v^*(\tau)}.$$

Moreover if $\kappa = k = 0, 1, 2, 3 \ldots$ then the zeros t_{1k} or t_{2k} of the equations

$$\int_{t_0}^{t} [1/v^*(\tau)]\, d\tau = k\pi \quad \text{or} \quad \int_{t_0}^{t} [1/v^*(\tau)]\, d\tau = (k+1/2)\pi$$

are the k-th zeros ($k = 0, 1, 2 \ldots$) of y_1 or y_2, respectively. It is clear that for the distribution of zeros of $y(t)$ only the function

$$s(t) := \int_{t_0}^{t} [1/v^*(\tau)]\, d\tau,\ t > 0 \tag{2.3}$$

is important and the following zero occurs if $s(t)$ increases by π. This situation enable us to define the κ-th zero of $y_1(t)$ for any suitable κ in the following way.

Definition 2.1 Let $t_0 > 0$ and $y_1(t)$ denote the solution of (q) such that $y_1(t_0) = 0$ and the function $s(t)$ is defined by (2.3). We define the κ-th zero t_κ of $y_1(t)$ with respect to t_0 as the zero of

$$s(t) = \kappa\pi \quad \text{or equivalently} \quad \int_{t_0}^t [1/v^*(\tau)]\,d\tau - \kappa\pi = 0 \quad \text{or} \quad t_\kappa := s^{-1}(\kappa\pi)$$

for $\kappa \in R$ such that $t_\kappa > 0$. Here $s^{-1} \equiv t(s)$ denotes the inverse function to $s = s(t)$.

Remark 2.1 We can of course define t_κ on the base of any $\bar{v}(t)$ such that $M(\bar{v}) > 0$. It means that for $\bar{v}(t) = \bar{y}_1^2 + \bar{y}_2^2$, $\bar{y}_1, \bar{y}_2$ being suitable solutions of (q) the above mentioned formulas remain true. But in such a case t_κ may depend on the choice of the couple $(\bar{y}_1, \bar{y}_2)$ and the behaviour of $\bar{v}(t)$ is in general not "good". Our definition is independent of the choice of the principal pair, since all principal pairs can be received from any by the transformation $\bar{y}_1 = \alpha y_1 + \beta y_2$, $\bar{y}_2 = \gamma y_1 + \delta y_2$ with orthogonal matrix $\begin{pmatrix}\alpha & \beta \\ \gamma & \delta\end{pmatrix}$. This implies the uniqueness of $v^*(t)$ and t_κ.

3. Higher monotonicity properties of t_κ as continuous function of κ

Due to the oscillatory character of (q) $s(t)$ is increasing function and $s(\infty) = \infty$. So, there exist the inverse function $t(s)$ on $(0, \infty)$, increasing from $t_0 = t(0)$ to ∞. It is clear, that $s(t_\kappa) = \kappa\pi$, which implies

$$t_\kappa = t(\kappa\pi). \tag{3.1}$$

In some cases knowing certain properties of $s = s(t)$ we can derived corresponding properties of (3.1). On the first place regularity in signs of derivatives of $v(t)$ is useful. Obviously

$$ds/dt = 1/v(t) \Longrightarrow D_s t = dt/ds = v(t),$$

$$D_s^2 t(s) = \frac{d}{dt} D_s t \cdot \frac{dt}{ds} = v'(t)v(t), \quad D_s^3 t(s) = v''v^2 + v'^2 v. \tag{3.2}$$

Similar formulas hold for $D_s^i t(s)$ ($i = 4, 5, \ldots$) and it is possible to prove by induction [6] that $v \in M_\infty(t_0, \infty)$, $v(t) \not\equiv 0$ implies $D_s t(s) \in M_\infty(0, \infty)$. Here $M_\infty(t_0, \infty)$ denote the set of function f completely monotonic on (t_0, ∞), i.e. satisfying $(-1)^i f^{(i)}(t) \geq 0$, $t \in (t_0, \infty)$, for $i = 0, 1, \ldots$ and D_s denotes derivative with respect to s. Ph. Hartman proved in [3], that if $q(\infty) > 0$ and $q'(t) \in M_\infty(0, \infty)$, then (q) has (in our notation) principal pair of solutions such that

$$v^*(t) \in M_\infty(0, \infty), \quad v^*(\infty) = 1, v^{*\prime}(\infty) = v^{*\prime\prime}(\infty) = \cdots = 0 \quad \text{if} \quad q(\infty) < \infty$$

and

$$v^*(t) \in M_\infty, (0, \infty) \quad v^*(\infty) = v^{*\prime}(\infty) = \cdots = 0 \quad \text{if} \quad q(\infty) = \infty.$$

So we have the following

Theorem 3.1. Let $q(t)$ possess a derivative $q'(t)$ of class $M_\infty(0, \infty)$. Let $t_0 > 0$ be any number and $t_\kappa > 0$ denote the κ-th zero of a solution $y(t)$ of oscillatory (q) such that $y(t_0) = 0$. Then

$$(-1)^i D_\kappa^{i+1}(t_\kappa) \geq 0, \quad i = 0, 1, \ldots \tag{3.3}$$

where κ is such that $t_\kappa \geq 0$. The equality here holds only if (q) has solution such that 0 is its zero.

Proof. The theorem is direct conclusion of the above mentioned considerations. It suffices to replace t by t_κ and s by $\kappa\pi$ in (3.2).

Remark 3.1 Due to the fact that $D_\kappa t_\kappa = \pi v(t_\kappa)$ plays an important role in the proof of Theorem 3.1, we can replace $v(t_\kappa)$ by any completely monotonic function. $0 \neq W(t) \in M_\infty(0,\infty)$, $v(t) \in M_\infty(0,\infty)$, $\sigma > 0$ imply $[W(t)\{v(t)\}^\sigma] \in M_\infty(0,\infty)$[7]. So (3.3) can be replaced by $(-1)^i D_\kappa^i [W(t_\kappa)(D_\kappa t_\kappa)^\sigma] \geq 0$.

Let us consider now the case of $q(t)$ decreasing to some positive constant. (To this scope belongs e.g. the Bessel equation for $|\nu| < 1/2$.)

We understand by the symbol $f = \mathcal{O}(t^{-\alpha})$ for $t \to \infty$ the order property of the best estimation, i.e. it means that

$$\text{i)} \lim_{t\to\infty}\sup |f(t)| \, t^\alpha < \infty, \quad \text{ii)} \lim_{t\to\infty} |f(t)| \, t^{\alpha+\varepsilon} = \infty \quad \text{for any} \quad \varepsilon > 0.$$

Using the Došlá's results [1] concerning ultimate monotonicity we can prove the following

Theorem 3.2. *Let $t_0 > 0$ be any number and t_κ denote the κ-th zero of a solution $y(t)$ of (q) such that $y(t_0) = 0$. Let the function $q(t) \not\equiv const.$ satisfy $q(\infty) = 1$ and for $i = 0, 1, 2 \ldots$*

$$(-1)^i q^{(i)}(t) \geq 0 \text{ for } t > 0 \,; \quad q^{(i)}(t) = \mathcal{O}(t^{-(i+\varepsilon)}) \text{ for } t \to \infty, \varepsilon > 0 \,.$$

Then

$$0 < D_\kappa t_\kappa < \pi \quad \text{for} \quad \kappa \quad \text{such that} \quad t_\kappa > 0 \,,$$
$$(-1)^i D_\kappa^{i+2} t_\kappa \geq 0, \quad \text{for} \quad i = 0, 1, 2, \ldots \quad \text{and} \quad \kappa \quad \text{such that} \quad t_\kappa \in (\mu_i, \infty) \,.$$

Here $\{\mu_i\}_{i=1}^\infty$ is a suitable nondecreasing sequence with $\mu_{i+1} = \mu_i$ only if $\mu_i = 0$.

In some cases we are able to express the function $v^*(t)$ explicitely or to find its asymptotic expansion [5], [8]. In such a situation the following theorem is usefull. The proof follows from (3.2).

Theorem 3.3. *Let (y_1, y_2) be a principal pair of solutions of (q) and $v^*(t) = y_1^2 + y_2^2$. Then*

$$D_\kappa t_\kappa = \pi v^*(t_\kappa), D_\kappa^2 t_\kappa = \pi^2 v^{*\prime}(t_\kappa) v^*(t_\kappa), D_\kappa^3 t_\kappa = \pi^3 [v^{*\prime\prime} v^{*2} + v^{*\prime 2} v^*]_{t=t_\kappa}.$$

Remark 3.2 Similar formulas hold for $D_\kappa^n t_\kappa (n = 4, 5, \ldots)$.

4. Applications to the Bessel equation.

The normalized solution (1.1) of the Bessel equation

$$Y'' + t^{-1} Y' + (1 - \nu^2 t^{-2}) Y = 0, \quad t > 0 \tag{4.1}$$

is usually called as cylinder or Bessel function. Here $J_\nu(t), Y_\nu(t)$ are the usual Bessel function of the first or second kind, respectively.

The equation (4.1) can be transformed by $Y(t) = t^{-1/2} y(t)$ into

$$y'' + (1 - (\nu^2 - 1/4)/t^2) y(t) = 0 \qquad t > 0 \tag{4.2}$$

The couple $y_1(t) = \sqrt{\pi t/2} J_\nu(t), y_2(t) = \sqrt{\pi t/2} Y_\nu(t)$ form a principal pair of solutions of (4.2) [2]. $C_\nu(t)$ defined by (1.1) can be therefore expressed in the form

$$C_\nu(t) = C_\nu(t, \alpha) = \sqrt{2 v_\nu^*(t)/\pi t} \sin(\int_0^t [1/v_\nu^*(\tau)] dt + \alpha) \tag{4.3}$$

Let us denote by $j_{\nu\kappa}$ the κ-th zero of $C_\nu(t)$ with respect to 0 (it is possible due to the fact, that $J_\nu(0) = 0$ for $\nu > 0$). Our definition of $j_{\nu\kappa}$ coinsides with the analogous one in [3].

As the direct corollaries of our theorems we can receive the great part of results derived in [3] and some other, too. E.g.

Theorem 4.1. Let $0 < \nu < 1/2$. Then

$$0 < D_\kappa j_{\nu\kappa} < \pi, \quad D_\kappa^2 j_{\nu\kappa} > 0, \quad \text{for} \quad \kappa > 0,$$

$$(-1)^i D_\kappa^{i+2} j_{\nu\kappa} > 0 \quad \text{for} \quad i = 1, 2, \ldots \quad \text{and} \quad \kappa \quad \text{such that} \quad j_{\nu\kappa} \in (\mu_i, \infty) \, .$$

Here $\{\mu_i\}_1^\infty$ is suitable nondecreasing sequence.

Theorem 4.2. In the case $\nu > \frac{1}{2}$ there holds

$$(-1)^i D_\kappa^i (\lg[D_\kappa j_{\nu\kappa}]) > 0, \ (-1)^i D_\kappa^i (\lg[D_\kappa j_{\nu\kappa}/\pi]) > 0, \quad \kappa > 0, \quad i = 0, 1, \ldots$$

In the case $\nu < 1/2$ we have for $\kappa > 0$

$$D_\kappa[\lg(D_\kappa j_{\nu\kappa})] > 0, \quad D_\kappa^2[\lg D_\kappa j_{\nu\kappa}] < 0,$$

$$D_\kappa[\lg(D_\kappa j_{\nu\kappa}/\pi)] > 0, \quad D_\kappa^2[\lg D_\kappa j_{\nu\kappa}/\pi] < 0 \, .$$

References

[1] Došlá Z., *Higher monotonicity properties of special functions: Application on Bessel case $|\nu| < 1/2$*, Coment. Math. Univ. Carolinae **31** (1990), 232 – 241.

[2] Elbert A., Neuman F., Vosmanský J., *Principal pairs of solutions of linear second order oscillatory differential equations*, Diff. and Int. Equations (to appear).

[3] Hartman Ph., *On differential equations and the function $J_\mu^2 + Y_\mu^2$*, Amer. J. Math. **83** (1961), 154–188.

[4] Gori L. N., Laforgia A., Muldoom M. E., *Higher monotonicity properties and inequalities for zeros of Bessel functions*, Preprint, York University, North York, Report No 89 – 47.

[5] Laforgia A., Vosmanský J., *Higher monotonicity properties of generalized Airy functions*, Rendicoti di Matematica (2), Serie VII **4** (1984), 241–256.

[6] Lorch L., Szego P., *Higher monotonicity properties of certain Sturm-Liouville functions*, Acta Math. **109** (1963), 55–73.

[7] Muldoon M. E., *Higher monotonicity properties of certain Sturm-Liouville functions V.*, Proc. Royal Soc. Edinburg **77A** (1977), 23–37.

[8] Vosmanský J., *Monotonicity properties of zeros of the differential equation $y'' + q(x)y = 0$*, Arch. Math. **6** (1970), 37–74.

[9] Vosmanský J., *Distribution of zeros of solutions of linear second-order differential equations*, Diff. Equations **15** (1979), 1511–1519.

H WARCHALL
Induced lacunas in multiple-time initial value problems and unbounded domain simulation

This note points out features of hyperbolic partial differential equations whose fundamental solutions have lacunar regions of real analyticity, as is the case for the wave and Klein-Gordon equations. For linear constant-coefficient homogeneous equations, it is shown that the data at a particular time inside a ball suffices to uniquely determine the solution in a region that is much larger than the a priori domain of influence, and that includes the ball at succeeding times, provided that the support of the data at some (perhaps ancient) initial time was contained in the ball. Analogous results hold for suitable nonhomogeneous equations.

Such multiple-time initial value problems model a situation common in numerical simulation of solutions to equations formulated on unbounded spatial domains, in which it is desired to compute the restriction of a solution to a bounded computational domain, based only on knowledge of the solution inside the domain at an earlier time. The results show that, for suitable initial value problems, the Cauchy data at a single time inside a computational domain alone is sufficient to determine the solution inside the domain at all later times. This raises the question of whether there is an explicit method for propagation of waves through an artificial domain boundary, using only data at a single time inside the domain, in such a way as to duplicate exactly the propagation in unbounded space. A numerical method based on such a technique could be superior to those based on "absorbing" boundary conditions.

Introduction

Methods for numerical simulation of solutions to hyperbolic partial differential equations formulated on unbounded spatial domains must take into account the fact that computational domains are necessarily bounded. Either the unbounded domain is mapped into a bounded domain (and the problem reformulated on the image), or only values of the solution within the computational domain are calculated.

In simulations of the latter type, it is desired to compute the restriction of a solution to the computational domain, based only on knowledge of the solution inside the domain at an earlier time. This goal has been achieved to various degrees of approximation by the use of "radiation," or "absorbing," boundary conditions ([1-5, 8-10]). The approximate nature of these conditions is due to the fact that the radiation condition – that waves be outgoing at an artificial domain boundary – is generally a pseudo-differential condition, nonlocal in both space and time. Approximations to this condition give rise to boundary conditions that are local differential conditions at the computational domain boundary at a single time, and that approximately simulate unbounded space outside.

Implicit in such simulations is the assumption that the data at a single time inside the computational domain alone is in principle sufficient to determine the solution inside the domain at later times. Whether this assumption is satisfied depends on the solution under study, since, at points near the computational domain boundary, the solution's a priori domain of dependence on earlier data extends outside the computational domain.

The circumstances under which this central assumption is true, and the mechanism by which the solution is then determined by "incomplete" data, merit examination. It is conceivable that there is a way to propagate waves through an artificial domain boundary, using only data at a single time inside the domain, in such a way as to duplicate (in principle exactly) the propagation in unbounded space. A numerical method based on such a technique could be superior to those based on differential boundary conditions.

In this paper we establish uniqueness results for a class of initial value problems that model the situation described. We consider linear constant-coefficient partial differential operators whose fundamental solutions have lacunar regions of real analyticity, as is the case for the wave and Klein-Gordon equations. For homogeneous equations we show that knowledge of the data at a particular time inside a ball suffices to uniquely determine the solution in the ball at succeeding times, provided that the support of the data at some previous time was contained in the ball. For suitable nonhomogeneous equations, we show that similar conclusions hold.

These results show that, for suitable initial value problems, the Cauchy data at a single time inside a computational domain alone is sufficient to determine the solution inside the domain at all later times. While these uniqueness theorems do not directly provide a computational method for the advancement of the solution in time, they indicate mechanisms responsible for the validity of the central assumption.

These theorems are consequences of elementary facts about fundamental solutions of hyperbolic partial differential equations formulated on unbounded spatial domains. For stronger, local, results that make use of the structure of the wave front set of the fundamental solution, see [11], in which it is shown, for appropriate initial value problems, that the solution at spacetime point (x,t) is determined uniquely by data at an earlier time in the spatial region consisting of the intersection of the computational domain with the a priori domain of dependence for the solution at (x,t). The problem of determining whether for nontrivial equations there is an explicit method by which boundary values may be advanced in time, based only on data inside the computational domain, deserves further study.

Theorem 1 below shows that Cauchy data determines a solution in regions substantially larger than the a priori domain of influence, in the circumstance when the fundamental solution has an analytic lacuna, and the support of (earlier) initial data is known. The proof is essentially the observation that zero values propagate throughout regions of real analyticity of the solution, inducing lacunar regions of zero values.

Theorem 2 extends the results to suitable nonhomogeneous equations, in which the inhomogeneity has compact spatial support. This allows the observations to be applied to equations with nonlinear terms that are supported inside a computational domain.

The results of Theorem 1 are applied in Corollary 1 to show that if the initial (time t_0) data for a solution to the Klein-Gordon equation has support in the open ball B, and if $t_1 \geq t_0$, then the solution is determined uniquely in $B \times [t_1, \infty)$ by the restriction to B of the data at time t_1. Thus the solution in B is determined completely by the data at a preceding time in B alone, despite the fact that at points near the boundary of B the a priori domain of dependence for the solution falls outside of B.

Multiple-Time Initial Value Problems

For $\alpha > 0$, we define the open forward cone $V_\alpha \equiv \{(x,t) \in \mathbf{R}^{n+1} \mid |x| < \alpha t\}$. For a cone Γ with apex at 0, we define the dual cone $\Gamma^\circ \equiv \{\xi \in \mathbf{R}^{n+1} \mid \xi \cdot \theta \geq 0 \; \forall \, \theta \in \Gamma\}$.

Theorem 1. Let P be a polynomial of degree m that is hyperbolic with respect to $(0,1) \in \mathbf{R}^{n+1}$. Suppose that the corresponding differential operator P(D) has the form $P(D) = \partial_t^m + Q$ where Q is a constant-coefficient partial differential operator, of degree less than m in the variable $t \equiv x_{n+1}$. Suppose that all the real zeros of the principal part of P are contained in $V_\gamma^\circ \cup V_{-\gamma}^\circ$, for some $\gamma > 0$.

Let $E \in \mathscr{D}'(\mathbf{R}^{n+1})$ be the fundamental solution of P(D) with support in $\{(x,t) \in \mathbf{R}^{n+1} \mid t \geq 0\}$. Suppose that E has support in $\overline{V_\beta}$ for some $\beta > 0$. Suppose further that E is real analytic in V_α for some α with $0 < \alpha \leq \beta$.

Let $v \in \mathscr{D}'(\mathbf{R}^{n+1})$ be a solution of $P(D)v = 0$. Suppose that the support of the data $\partial_t^k v(t_0)$ ($k = 0, 1, \ldots, m-1$) at time $t = t_0$ is contained in the open ball $B \subset \mathbf{R}^n$ of radius R centered at 0. Let $t_1 > t_0 + \frac{R}{\alpha} - \frac{R}{\beta}$.

Then the restriction to B of the data $\partial_t^k v(t_1)$ ($k = 0, 1, \ldots, m-1$) at time $t = t_1$ uniquely determines v throughout $W \cup Y \cup Z$, where:

$W \equiv \{(x,t) \in \mathbf{R}^{n+1} \mid |x| + \beta|t - t_1| < R\}$ is the double cone based on B at $t = t_1$,

$Y \equiv \{(0, t_0 + \frac{R}{\alpha})\} + V_\alpha$ is contained in the analytic lacuna of the solution, and

$Z \equiv \{(0, t_1 - \frac{R}{\beta})\} + V_\gamma$ is the speed γ forward cone with apex at the lower apex of W.

The regions W, Y, and Z are illustrated in Figure 1.

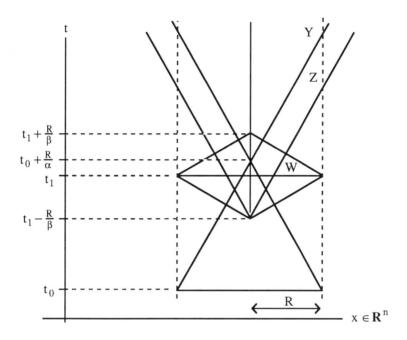

Figure 1. The spacetime regions of Theorem 1.

Proof: Because P(D) has the form $\partial_t^m + Q$, if $u \in \mathscr{D}'(\mathbf{R}^{n+1})$ solves P(D)u = 0 for t in some open interval I, then u is in fact a C^m function of $t \in I$ with values in $\mathscr{D}'(\mathbf{R}^{n+1})$. ([6], Sec. 4.4) Thus the references in the statement of the theorem to $\partial_t^k v$ at specific times make sense. Furthermore, for $t > 0$, the fundamental solution E is a C^m function of t with values in $\mathscr{D}'(\mathbf{R}^n)$. If the distributions $G_i^k \in \mathscr{D}'(\mathbf{R}^{n+1})$ are defined in terms of the data at $t = t_i$ by $G_i^k \equiv \delta(t - t_i) \partial_t^k u$ for k = 0, 1, ..., m–1, that is, $G_i^k[\phi] \equiv \partial_t^k u(t_i)[\phi(\cdot, t_i)]$ for $\phi \in \mathscr{D}(\mathbf{R}^{n+1})$, then

$$u = E*G_i^{m-1} + \partial_t E*G_i^{m-2} + \ldots + \partial_t^{m-1} E*G_i^0 \quad \text{in } \{(x,t) \mid t > t_i\}. \tag{1}$$

To prove the theorem, we first make use of the linearity of P(D). Suppose that v and w are two solutions in $\mathscr{D}'(\mathbf{R}^{n+1})$ of the homogeneous equation, both of whose data has support in B at time t_0, and whose data agrees in B at time t_1. Set $u \equiv v - w$. To establish the assertions of the theorem, we will show that u vanishes in W∪Y∪Z.

We have P(D)u = 0, with $\partial_t^k u(t_0) = 0$ in $\mathbf{R}^n \setminus B$ and $\partial_t^k u(t_1) = 0$ in B, for k = 0, 1, ..., m–1. From the hypothesized support of E in $\overline{V_\beta}$, the vanishing of $\partial_t^k u(t_1)$ in B, and (1) applied at the "initial" time t_1, it follows immediately that u vanishes in the forward half $\{(x,t) \in W \mid t \geq t_1\}$ of the double cone W. We can establish that u also vanishes in the other half of W by considering the fundamental solution of P(D) with support in $\{(x,t) \in \mathbf{R}^{n+1} \mid t \le 0\}$; its general properties ([6], Secs. 12.4-12.6) imply that P(D) is supported in $-\overline{V_\beta}$, and analogous reasoning applies for $t \le t_1$. Thus u vanishes in W.

The region $\text{sing supp}_A E$ in $\mathbf{R}^{n+1}$ where E fails to be real analytic satisfies by hypothesis $\text{sing supp}_A E \subset \overline{V_\beta} \setminus V_\alpha$. Also by hypothesis, $\text{sing supp}_A G_0^k \subset B \times \{t_0\}$. Because the analytic singular support of a convolution of distributions obeys ([6], Secs. 8.2, 8.4) $\text{sing supp}_A (E*F) \subset \text{sing supp}_A E + \text{sing supp}_A F$, it follows from (1) applied at initial time t_0 that, in particular, u is real analytic in the cone Y, none of whose points is a sum of a point in $\text{sing supp}_A \partial_t^j E$ and a point in $\text{sing supp}_A G_0^k$.

Since $t_0 + \frac{R}{\alpha} < t_1 + \frac{R}{\beta}$, the regions W and Y have open nonempty intersection. Since u is real analytic in Y and u vanishes in W, u must also vanish in Y.

Finally, to establish that u vanishes in the cone Z, note that the region $W \cup Y$ forms an open neighborhood of the open half-line $\{(0,t) \mid t > t_1 - R/\beta\}$ with endpoint at the lower apex of W. Thus Z contains an open conical neighborhood $V_\delta + \{(0, t_1 - R/\beta)\}$ of the half-line on which u vanishes. Since $\partial(V_\varepsilon) \setminus \{0\}$ is noncharacteristic for $0 < \varepsilon < \gamma$ by the hypothesis on the zeros of the principal part of P, we may apply Holmgren's uniqueness theorem [6,7] (or Theorem 8.6.8 in [6]) to conclude that u vanishes in the entire cone $Z \equiv V_\gamma + \{(0, t_1 - R/\beta)\}$. ///

Remark: Of course, u vanishes (and v is uniquely determined) throughout the whole analytic lacuna that contains the cone Y, and throughout the maximal region containing Z for which Holmgren's uniqueness theorem applies.

A modification of this proof establishes the following result for nonhomogeneous equations.

Theorem 2. Let the hyperbolic polynomial P, the differential operator P(D), and the forward fundamental solution E be as in Theorem 1.

Let $v \in \mathscr{D}'(\mathbf{R}^{n+1})$ be a solution of $P(D)v = F$, where F is a continuous function of t with values in $\mathscr{D}'(\mathbf{R}^n)$. Suppose that the support of the data $\partial_t^k v(t_0)$ ($k = 0, 1, \ldots, m-1$) at time $t = t_0$ is contained in the open ball $B \subset \mathbf{R}^n$ of radius R centered at 0. Suppose that F is supported in $B \times (-\infty, \infty)$. Let $t_1 > t_0 + \frac{R}{\alpha} - \frac{R}{\beta}$.

Then the restriction to B of the data $\partial_t^k v(t_1)$, and the restriction to $B \times \left[\min\left\{t_0, t_1 - \frac{R}{\beta}\right\}, \infty\right)$ of F, determine v throughout $W \cup Y \cup Z$, where W, Y, and Z are as in Theorem 1.

For the proof it again suffices to consider the difference between two solutions of the nonhomogeneous equation that satisfy the hypotheses and whose data agrees in B at time t_1. Details are omitted.

Klein-Gordon Equation

In the case when $P(D)v = 0$ is equivalent to the Klein-Gordon equation, Theorem 1 implies the following fact.

Corollary 1. Let $v \in \mathscr{D}'(\mathbf{R}^{n+1})$ be a solution of $P(D)v = 0$, where $P(D)$ is a linear constant-coefficient differential operator with principal part equal to the D'Alembertian $\partial_t^2 - \Delta$. Suppose that the supports of v and v_t at time $t = t_0$ are contained in the open ball $B \subset \mathbf{R}^n$. Let $t_1 \geq t_0$. Then the solution v is determined uniquely in $B \times [t_1, \infty)$ by the restrictions to B of the distributions $v(t_1)$ and $v_t(t_1)$.

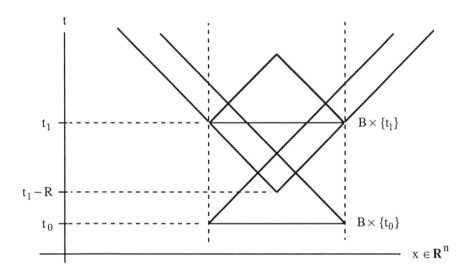

Figure 2. The spacetime regions of Corollary 1

Proof: The principal part $P_m(\xi) = -(\xi_{n+1})^2 + \sum_{k=1}^{n} (\xi_k)^2$ has all its real zeros in $V_1^\circ \cup V_{-1}^\circ$. It is well-known [6] that the forward fundamental solution E is supported in $\overline{V_1}$ and is analytic in V_1. Thus, in the notation of Theorem 1, $\alpha = \beta = \gamma = 1$.

If $t_1 = t_0$ then by hypothesis the data $\{v, v_t\}$ is known at t_0 on all of $\mathbf{R}^n$, and the conclusion follows immediately.

If $t_1 > t_0$ then trivially $t_1 > t_0 + \frac{R}{\alpha} - \frac{R}{\beta}$, where R is the radius of B, and Theorem 1 applies to the situation in which, without loss of generality, we have translated the center of B to the origin. In particular, $v(t_1)$ and $v_t(t_1)$ restricted to B uniquely determine v throughout the cone $V_1 + \{(0, t_1 - R)\}$, which contains the set $B \times [t_1, \infty)$. ///

Acknowledgements

These observations were made in connection with numerical studies of nonlinear wave equations I conducted as a visitor to the Theoretical Division and the Center for Nonlinear Studies at Los Alamos National Laboratory. It is a pleasure to acknowledge their support and hospitality. I am also pleased to acknowledge discussions with A. Castro, P. Deift, J.M. Hyman, B. Keyfitz, and P. Lax. This work was partially supported by a grant from the Texas Advanced Research Program, and by a University of North Texas Faculty Research Grant.

References

1. J.H. Chen, *Numerical boundary conditions and computational modes*, J. Comp. Phys. **13** (1973) 522-535.

2. B. Engquist and A. Majda, *Absorbing boundary conditions for the numerical simulation of waves*, Math. Comp. **31** (1977) 629-651.

3. B. Engquist and A. Majda, *Radiation boundary conditions for acoustic and elastic wave calculations*, Comm. Pure Appl. Math. **32** (1979) 313-357.

4. L. Halpern and L.N. Trefethen, *Wide-angle one-way wave equations*, MIT preprint, July 1986.

5. R.L. Higdon, *Absorbing boundary conditions for difference approximations to the multi-dimensional wave equation*, Math. Comp. **47** (1986) 437-459.

6. L. Hörmander, *The analysis of linear partial differential operators*, Vols. I & II. Grundlehren der mathematischen Wissenschaften 256 & 257, Springer-Verlag (1983).

7. F. John, *Partial Differential Equations*, 4th Edition. Springer-Verlag (1982).

8. E.L. Lindman, *Boundary conditions for the time dependent wave equation*, J. Comp. Phys. **18** (1975) 66-78.

9. Warwick D. Smith, *A nonreflecting plane boundary for wave propagation problems*, J. Comp. Phys. **15** (1974) 492-503.

10. L.N. Trefethen and L. Halpern, *Well-posedness of one-way wave equations and absorbing boundary conditions*, Math. Comp. **47** (1986) 421-435.

11. H. Warchall, *Wave propagation at computational domain boundaries*, Commun. Partial Diff. Eq. **16** (1990) 31-41.

Henry A. Warchall
Department of Mathematics
University of North Texas
Denton TX 76203-5116

G HARRIS, W HUDSON AND B ZINNER
Existence of wavefronts for the discrete Fishers equation

In the following discussion continuation and comparison methods are used to obtain existence results of traveling wavefronts for the discrete Fisher's equation

$$\dot{u}_n = d(u_{n-1} - 2u_n + u_{n+1}) + f(u_n), \quad n \in \mathbb{Z}, \tag{1}$$

where d is a positive number and f denotes a Lipschitz continuous function satisfying

$$f(0) = f(1) = 0 \text{ and } f(x) > 0 \text{ for } 0 < x < 1.$$

Note that (1) is the discrete version of the general Fisher's equation

$$\frac{\partial u}{\partial t} = D\frac{\partial^2 u}{\partial x^2} + f(u), \quad D > 0.$$

Both the discrete and continuous equations have been used as models for population dynamics and the propagation of a favoured gene in a population and have been studied in recent years by a number of mathematicians, e.g., Bell and Cosner [1], Fife [2], and Keener [4], and the above authors [7]. The discrete model might be more appropriate in certain situations, for example, population dispersal in a patchy environment.

A solution $\{u_n(t)\}_{n=-\infty}^{\infty}$ of (1) is called a *traveling wavefront* with speed $c > 0$ if there exists a function $U : \mathbb{R} \to [0, 1]$ such that $U(-\infty) = 0$, $U(\infty) = 1$ and

$$u_n(t) = U(n + ct) \text{ for all } n \in \mathbb{N} \text{ and all } t \in \mathbb{R}.$$

If $U'(z) > 0$ for all $z \in \mathbb{R}$ then the wavefront will be called strictly increasing. The problem of finding traveling wavefronts $\{u_n\}$ is equivalent to finding solutions of the functional differential equation

$$cU'(z) = d[U(z - 1) - 2U(z) + U(z + 1)] + f(U(z)) \text{ for all } z \in \mathbb{R}, \tag{2}$$

where $U(-\infty) = 0$, $U(\infty) = 1$ and $0 \leq U(z) \leq 1$ for all $z \in \mathbb{R}$. These kinds of equations are known in the literature as functional differential equations of mixed type. Further information can be found in the book by Hale [3] and recent articles by Rustichini [5, 6].

1

To prove existence of traveling wavefronts one first observes that traveling waves of speed $c = 1/\tau$ must satisfy $x_n - u_{n-1}(d, x; \tau) = 0$, $n \in \mathbb{Z}$, for some $x = \{x_n\}_{n \in \mathbb{Z}}$, where $\{u_n(d, x; t)\}_{n \in \mathbb{Z}}$ denotes the solution of (1) with initial value $x = \{x_n\}_{n \in \mathbb{Z}}$. When $d = 0$ traveling waves of any positive speed are easily found and are unique if one imposes an additional condition, for instance $x_0 = 1/2$. The idea is to continue the solution for $d = 0$, with respect to d, as far as possible.

For technical reasons approximate wavefronts will be constructed by considering, for each $N \in \mathbb{N}$, the finite dimensional initial value problem

$$\dot{u}_n = d(u_{n-1} - 2u_n + u_{n+1}) + f(u_n) \quad (3)$$
$$u_n(0) = x_n, \quad n = -N, \ldots, N, \quad u_{-N-1}(t) \equiv x_{-N}, \quad u_{N+1} \equiv 1.$$

Denote by $\{u_n(x, d; t)\}_{n=-N}^{N}$ where $x = \{x_n\}_{n=-N}^{N}$ the unique solution to (3). Now let $\tau > 0$ and define the function $F : \mathbb{R} \times \mathbb{R}^{2N+1} \to \mathbb{R}^{2N+1}$ by

$$(F(d,x))_n = \begin{cases} x_0 - \frac{1}{2} & \text{for } n = 0 \\ x_n - u_{n-1}(d, x, \tau) & \text{for } n = -N, \ldots, N. \end{cases}$$

When $d = 0$ the unique solution of $F(d, x) = 0$ is $x^0 = (x_{-N}^0, \ldots, x_N^0)$, where $x_n^0 = u(n\tau)$ and u is the solution of $\dot{u}(t) = f(u(t))$, $u(0) = 1/2$.

To use degree theory we define

$$\mathcal{O}_d = \{x = (x_{-N}, \ldots, x_N) \in \mathbb{R}^{2N+1} : d(x_{n-1} - 2x_n + x_{n+1}) + f(x_n) > 0,$$
$$0 < x_n < x_{n+1} \text{ for } n = -N, \ldots, N \text{ where } x_{-N-1} = x_{-N} \text{ and } x_{N+1} = 1\}$$

and

$$\mathcal{O} = \{(d, x) \in \mathbb{R} \times \mathbb{R}^{2N+1} : d \geq 0 \text{ and } x \in \mathcal{O}_d\}.$$

We note that $\mathcal{O}_d$ is an open bounded set in $\mathbb{R}^{2N+1}$ and that $\mathcal{O}$ is open in $[0, \infty) \times \mathbb{R}^{2N+1}$. The Brouwer degree, $\deg(F(d, \cdot), \mathcal{O}_d, 0)$, is defined as long as $F(d, x) = 0$ has no solution on the boundary of $\mathcal{O}_d$. If d^* is such that $F(d, x) = 0$ has no solution on the boundary of $\mathcal{O}_d$ for $0 \leq d \leq d^*$ then by homotopy invariance, $\deg(F(0, \cdot), \mathcal{O}_0, 0) = \deg(F(d, \cdot), \mathcal{O}_d, 0)$, $0 \leq d \leq d^*$. If f is of class C^1 then one can calculate that $D_x F(0, x^0)$ is nonsingular and hence that $\deg(F(0, \cdot), \mathcal{O}_0, 0) \neq 0$. Therefore $\deg(F(0, \cdot), \mathcal{O}_0, 0) \neq 0$ if f is Lipschitz.

The solution can escape through the boundary of $\mathcal{O}$ only if one of the following holds

(i) $x_n = x_{n+1}$ for some $n \in \{-N, \ldots, N\}$,

(ii) $d(x_{n-1} - 2x_n + x_{n+1}) + f(x_n) = 0$ for some $n \in \{-N, \ldots, N\}$, or

(iii) $x_{-N} = 0$.

Using comparison methods one can show that the first two cases, (i) and (ii), cannot occur for any $d > 0$. The third case, (iii), depends on d.

Suppose that $m > 0$ is such that $mx > f(x)$ for $x > 0$ and that $0 \leq d \leq \sup_{r>0} \dfrac{rc - m}{4 \sinh^2(r/2)}$.

Let r^* be the point at which the supremum of $\dfrac{rc - m}{4 \sinh^2(r/2)}$, $r > 0$, is assumed. Then the function $V(z) = e^{r^* z}$ satisfies

$$cV'(z) > d[V(z-1) - 2V(z) + V(z+1)] + f(V(z)), \quad z \in \mathbb{R}.$$

To reach a contradiction suppose that $x_{-N} = 0$ and let

$$U(z) = \begin{cases} 0 & \text{for } z < -N \\ u_n(\tau(z-n)) & \text{for } n \leq z \leq n+1, -N \leq n \leq N \\ 1 & \text{for } N+1 < z. \end{cases}$$

Then
$$cU'(z) = d[U(z-1) - 2U(z) + U(z+1)] + f(U(z)), \quad -N < z < N+1.$$
By a suitable shift s of V we may assume that $V(z-s) \geq U(z)$, $z \in \mathbb{R}$, and that $V(z_0 - s) = U(z_0)$ for some $z_0 \in (-N, N)$. This leads to the contradiction
$$0 = V'(z_0 - s) - U'(z_0) > d[V(z_0 - s - 1) - U(z_0 - 1)] + d[V(z_0 - s + 1) - U(z_0 + 1)] \geq 0.$$
The above discussion gives a sequence of approximate monotone wavefronts indexed by $N \in \mathbb{N}$. Using standard arguments one can argue that this sequence converges to a traveling wavefront of (1). This proves the following theorem.

Theorem 1.1 *Suppose that f is a Lipschitz continuous function on $[0, 1]$ such that $f(0) = f(1) = 0$ and for all $0 < x < 1$, $f(x) > 0$. Let $\overline{m} = \overline{m}(f) := \sup\{f(x)/x : 0 < x \leq 1\}$ and let c be any positive number. Set*
$$\overline{d} := \sup_{r > 0} \frac{rc - \overline{m}}{4 \sinh^2 \left(\frac{r}{2}\right)}.$$
If $0 \leq d \leq \overline{d}$, then there is a strictly increasing traveling wavefront of (1) with wave speed c.

2

To show that, for d large enough, there do not exist traveling wavefronts with speed c a strict subsolution U of (2) can be exhibited, i.e., a function satisfying
$$cU'(x) < d[U(x-1) - 2U(x) + U(x+1)] + f(U(x)), \quad x \in \mathbb{R},$$
$U(-\infty) < 0$, and $0 < U(\infty) < 1$, where $f(s)$ has been defined to be zero when $s \notin [0, 1]$. Such a subsolution can be obtained, provided that f is such that $f(x) \geq mx$, $0 \leq x \leq x_0$ for some $m > 0$ and some x_0 with $0 < x_0 < 1$, and
$$d > \sup_{r > 0} \frac{rc - m}{4 \sinh^2(r/2)},$$
by scaling the map
$$U(x) := \begin{cases} e^{kx} - \epsilon & \text{if } x < 0 \\ e^{kx - \epsilon x^2} - \epsilon & \text{if } 0 \leq x < k/2\epsilon \\ e^{\frac{k^2}{4\epsilon}} - \epsilon & \text{if } k/2\epsilon \leq x. \end{cases}$$
This subsolution can be used in the same manner that the supersolution was used in the previous section to give the following nonexistence result.

Theorem 2.1 *Let $D_+ f(0) = \liminf_{x \to 0} f(x)/x$ and assume that $0 < D_+ f(0) < \infty$ and that*
$$d > \sup_{r > 0} \frac{rc - D_+ f(0)}{4 \sinh^2 \left(\frac{r}{2}\right)}.$$
Then system (1) does not have a traveling wavefront with speed $c > 0$. Moreover, if $D_+ f(0) = \infty$ then system (1) does not have a traveling wavefront for any speed $c > 0$.

References

[1] J. Bell and C. Cosner, Threshold behaviour and propagation for nonlinear differential-difference systems motivated by modeling myelinated axons, Quart. Appl. Math., 42 (1984), pp. 1-14.

[2] P. C. Fife, Mathematical Aspects of Reacting and Diffusing Systems, Springer-Verlag, 1979.

[3] J. Hale, Theory of Functional Differential Equations, Springer-Verlag, New York, (1977).

[4] J. P. Keener, Propagation and its failure in coupled systems of discrete excitable cells, SIAM J. Appl. Math., 47 (1987), pp. 556-572.

[5] Aldo Rustichini, Functional differential equations of mixed type: The linear autonomous case, J. Dynamics and Differential Equations, 1 (1989), pp. 121-143.

[6] Aldo Rustichini, Hopf bifurcation for functional differential equations of mixed type, J. Dynamics and Differential Equations, 1 (1989), pp. 145-177.

[7] B. Zinner, G. Harris, and W. Hudson, Traveling wavefronts for the discrete Fisher's equation, *submitted*.